DOCUMENTA ARCHAEOBIOLOGIAE 1

Decyphering ancient bones
The research potential
of bioarchaeological collections

DOCUMENTA ARCHAEOBIOLOGIAE

Jahrbuch der Staatssammlung für
Anthropologie und Paläoanatomie München

Band 1

Herausgegeben von
Gisela Grupe und Joris Peters

DOCUMENTA ARCHAEOBIOLOGIAE

DECYPHERING ANCIENT BONES

THE RESEARCH POTENTIAL OF BIOARCHAEOLOGICAL COLLECTIONS

edited by
Gisela Grupe and Joris Peters

Verlag Marie Leidorf GmbH · Rahden/Westf.
2003

286 Seiten mit 118 Abbildungen und 26 Tabellen

Gedruckt mit Unterstützung des
**BAYERISCHEN STAATSMINISTERIUMS
FÜR WISSENSCHAFT, FORSCHUNG UND KUNST**

Die Deutsche Bibliothek - CIP-Einheitsaufnahme

Grupe, Gisela / Peters, Joris (Hrsg.):
Decyphering ancient bones ; the research potential of bioarchaeological collections.
Rahden/Westf. : Leidorf, 2003
  (Documenta Archaeobiologiae ; Bd. 1)
ISBN 3-89646-616-X

Gedruckt auf alterungsbeständigem Papier

Alle Rechte vorbehalten
© 2003

Verlag Marie Leidorf GmbH
Geschäftsführer: Dr. Bert Wiegel
Stelleroh 65 . D-32369 Rahden/Westf.

Tel: +49/(0)5771/ 9510-74
Fax: +49/(0)5771/ 9510-75
E-Mail: vml-verlag@t-online.de
Internet: http://www.leidorf.de
Internet: http://www.vml.de

ISBN 3-89646-616-X
ISSN 1611-7484

Kein Teil des Buches darf in irgendeiner Form (Druck, Fotokopie, CD-ROM, DVD, Internet oder einem anderen Verfahren) ohne schriftliche Genehmigung des Verlages Marie Leidorf GmbH reproduziert werden oder unter Verwendung elektronischer Systeme verarbeitet, vervielfältigt oder verbreitet werden.

Umschlagentwurf: Dirk Bevermann, Georgsmarienhütte und Bert Wiegel, Rahden/Westf.
Logo: Christine Kaufmann, Bamberg
Redaktion: Gisela Grupe, Joris Peters und Siew Eiselt, München
Scans: Michaela Svihla, München
Satz, Layout und Bildnachbearbeitung: Enns Schrift & Bild GmbH, Bielefeld
Druck und Produktion: DSC-Heinz J. Bevermann KG, Raiffeisenstraße 20, D-49124 Georgsmarienhütte

Table of Contents

Contributors 7

Preface 9

The Bavarian State Collection of Anthropology and Palaeoanatomy. A brief history
Gisela Grupe, Joris Peters 11

PART I: RESEARCH COLLECTIONS AND THE PUBLIC ECHO

Body of evidence: Museum Collections, why they were brought together,
their value today and public future
Theya Molleson 17

In search of the ancient Peruvians: The Pacasmayo Museum Project
Andrew Nelson, Christine Nelson 29

PART II: THE RESEARCH POTENTIAL OF BIOARCHAEOLOGICAL COLLECTIONS

Human skeletal remains from the central Balkans: A survey of the development
of human populations
Živko Mikić 51

Evaluating human fossil finds
Winfried Henke 59

Contributions of Primatological Collections to modern biodiversity research
Bernhard Wiesemüller, Hartmut Rothe 77

PART III: MAN AND HIS ANIMAL WORLD

Bone artefacts and man – an attempt at a cultural synthesis
Cornelia Becker 83

Hiding in Plain sight: The value of Museum Collections in the study of the origins
of animal domestication
Melinda Zeder 125

PART IV: DECYPHERING ANCIENT BONE

Ancient bones and teeth on the microstructural level
Simon Hillson, Daniel Antoine 141

Interpreting the trace-element components of bone – a current perspective from
the Laboratory for Archaeological Chemistry
James H. Burton, T. Douglas Price 159

Bone Collections are DNA data banks
Carles Lalueza-Fox 165

PART V: THE SOCIO-CULTURAL ASPECT AND MODERN IMPLICATIONS

Bioarchaeological Collections and the cultural heritage
Helmut Bender 177

Diversity conservation: Rare domestic farm animal breeds
Hans Hinrich Sambraus 183

PART VI: CURRENT PROJECTS OF THE COLLECTION OF ANTHROPOLOGY AND PALAEOANATOMY

Vertebrate food webs and subsistence strategies of Meso- and Neolithic populations of central Europe
Gisela Grupe, Živko Mikić, Joris Peters, Henriette Manhart 193

Histomorphometric analysis of primate and domesticated animal long bone microstructure
Karola Dittmann 215

Variations in dental microwear and abrasion in ancient human groups of southern Germany: 7500 BP to the Early Middle Ages
Irene Luise Gügel 227

Detection of *Yersinia pestis* in early and late Medieval Bavarian burials
Christina Garrelt, Ingrid Wiechmann 247

Palaeoenvironmental interpretation of fish remains from the Wadi Howar region, Northwest Sudan
Nadja Pöllath, Joris Peters 255

Holocene faunas from the Eastern Sahara: Past and future zoogeographical implications
Joris Peters, Angela von den Driesch 265

DOCUMENTA ARCHAEOBIOLOGIAE: Instructions for Authors 285

Contributors

Daniel Antoine, Institute of Archaeology, University College London, 31-34 Gordon Square, London WC1H OPY, UK.

Prof. Dr. Helmut Bender, Archäologie der Römischen Provinzen, Innstr. 40, Universität Passau, 94030 Passau, BRD. email: Helmut.Bender@uni-passau.de

Dr. Cornelia Becker, Freie Universität Berlin, Seminar für Ur- und Frühgeschichte, Altensteinstr. 15, 14195 Berlin, BRD. email: cobecker@zedat.fu-berlin.de

Dr. James H. Burton, Department of Anthropology, University of Wisconsin-Madison, 5230 Social Science Building, 1180 Observatory Drive, Madison, Wisconsin 53706-1393, USA. email: jhburton@facstaff.wisc.edu

Dipl.-Biol. Karola Dittmann, Department Biologie I, Bereich Biodiversitätsforschung/Anthropologie, Richard-Wagner-Str. 10, 80333 München, BRD. email: karola.dittmann@gmx.de

Prof. Dr. Angela von den Driesch, Staatssammlung für Anthropologie und Paläoanatomie, Karolinenplatz 2a, 80333 München, BRD.

Dipl.-Biol. Christina Garrelt, Department Biologie I, Bereich Biodiversitätsforschung/Anthropologie, Richard-Wagner-Str. 10, 80333 München. BRD.

Prof. Dr. Gisela Grupe, Staatssammlung für Anthropologie und Paläoanatomie, Karolinenplatz 2a, 80333 München, BRD. email: ASM.Grupe@extern.lrz-muenchen.de

Dipl.-Biol. Irene Luise Gügel, Department Biologie I, Bereich Biodiversitätsforschung/Anthropologie, Richard-Wagner-Str. 10, 80333 München, BRD. email: i.guegel@lrz.uni-muenchen.de

Prof. Dr. Winfried Henke, Institut für Anthropologie, Johannes-Gutenberg-Universität, Saarstr. 21, 55099 Mainz, BRD. email: erasmus@mail.uni-mainz.de

Prof. Dr. Simon Hillson, Institute of Archaeology, University College London, 31-34 Gordon Square, London WC1H OPY, UK. email: simon.hillson@ucl.ac.uk

Dr. Carles Lalueza-Fox, Secció Antropologia, Dept. Biologia Animal, Facultat de Biologia, Universitat de Barcelona, Avda. Diagonal 645, 08028 Barcelona, E. email: lalueza@bio.ub.es

Dr. Henriette Manhart, Staatssammlung für Anthropologie und Paläoanatomie, Karolinenplatz 2a, 80333 München, BRD. email: henriette.manhart@palaeo.vetmed.uni-muenchen.de

Prof. Dr. Zivko Mikic, Philosophische Fakultät, Anthropologische Sammlung, Cika Ljubina 18-20, 1100 Belgrad, YU.

Prof. Dr. Theya Molleson, Department of Palaeontology, The Natural History Museum, Cromwell Road, London SW7 5BD, UK. email: T.Molleson@nhm.ac.uk

Prof. Dr. Andrew Nelson, Department of Anthropology, Social Science Center, University of Western Ontario, London/Ontario, Canada N6A 5C2, CA. email: anelson@uwo.ca

Dr. Christine Nelson, Planning and Development, City of London, London/Ontario, Canada N6A 5C2, CA. email: cnelson@city.london.on.ca

Prof. Dr. Joris Peters, Staatssammlung für Anthropologie und Paläoanatomie, Karolinenplatz 2a, 80333 München, BRD. email: joris.peters@palaeo.vetmed.uni-muenchen.de

Nadja Pöllath, MA, SFB 389, Heinrich-Barth-Institut, Jennerstr. 8, 50823 Köln; und Staatssammlung für Anthropologie und Paläoanatomie, Karolinenplatz 2a, 80333 München, BRD. email: nadja.poellath@palaeo.vetmed.uni-muenchen.de

Prof. Dr. T. Douglas Price, Department of Anthropology, University of Wisconsin-Madison, 5230 Social Science Building, 1180 Observatory Drive, Madison, Wisconsin 53706-1393, USA. email: tdprice@facstaff.wisc.edu

Prof. Dr. Hartmut Rothe, Institut für Zoologie und Anthropologie der Universität Göttingen, Ethologische Station Sennickerode, Sennickerode 11, 37130 Gleichen, BRD. email: hrothe@gwdg.de

Prof. Dr. H. H. Sambraus, Lehrgebiet für Tierhaltung und Verhaltenskunde, Technische Universität München, 85350 Freising-Weihenstephan, BRD. email: Hans.H.Sambraus@agrar.tu-muenchen.de

Dr. Ingrid Wiechmann, Department Biologie I, Bereich Biodiversitätsforschung/Anthropologie, Richard-Wagner-Str. 10, 80333 München, BRD. email: i.wiechmann@lrz.uni-muenchen.de

Dr. Bernhard Wiesemüller, Institut für Zoologie und Anthropologie der Universität Göttingen, Ethologische Station Sennickerode, Sennickerode 11, 37130 Gleichen, BRD. email: bwiesem@gwdg.de

Dr. Melinda A. Zeder, Department of Anthropology, National Museum of Natural History, Smithsonian Institution, Washington D.C. 20560-1221, USA. email: zeder.melinda@nmnh.si.edu

Preface

The year 2002 will probably go down as a memorable year in the history of the palaeoanthropological and archaeological sciences in Bavaria. Firstly, it marks the centennial of the Anthropological Collection in München, and secondly, the establishment of the Bavarian State Collection of Anthropology and Palaeoanatomy through the merger of the Anthropological and the Palaeoanatomy Collections of München two years ago. The first results of its cooperative scientific work were made public in an international symposium. It is, therefore, right and timely that this, the first volume of *Documenta Archaeobiologiae*, as the yearbook of the newly-formed State Collection, goes to press to mark the two events.

Today, the bioarchaeological sciences stand at the threshold of a new, exciting and dynamic era. Technology, such as digitalization and microscopy, computerization, as well as new biomolecular tools are facilitating and powering the transformation of what has been narrative, speculative approaches into well-documented, data- and research-based analytical sciences. Field and laboratory data, together with new methodologies, are providing hard evidence that are forcing us to rethink and review cherished models, reject stereotypes and postulate radical new hypotheses. Besides providing information on pure anthropological issues, such as on the personality and lifestyle of the Neanderthal, the dispersal of early Homo and the numbers of speciation processes that occurred, the use of archaeological data has expanded and today constitute very valuable empirical sources for questions concerning population development, palaeoecosystem analyses, epidemiological problems and many other scientific issues. The study and analyses of animal bone finds hold the promise of being major contributors to modern biodiversity research and the understanding of many issues, including the ability to reconstruct changes in the geographical distribution of free-range vertebrates, the effects caused by the introduction of domestic animals and the development of the various domestic breeds. DNA analyses of human remains provide insights, for instance, into the relationship between ancient diseases, like plague, and their modern strains, while data on fish fossils of northwest Sudan have been used to provide insight into the palaeogeography in that area. It has provided a lexicon to the biome shift patterns of the past and may be useful in predicting the shifts in the future. Archaeological research in this case has been used as the corner-stone of natural habitat planning, certainly a futuristic role for ancient material. Such multifaceted use of archaeological research have created, at the same time, new roles for bioarchaeological collections, changing museums from the show-and-tell collections that the world is familiar with into new centres of research concerning how we, as mankind and as peoples, developed, functioned and interacted, both with our animals and our environment. While it has created in its wake a new demand for and appreciation of bioarchaeological collections, their values and unique roles in our history, museum collections still face tremendous challenges and an uncertain future. More importantly, it has created a demand for material that is in very limited, irreplaceable supply. Thus, it is with the aim of enhancing dialogue between the natural sciences and archaeology and placing particular emphasis upon research carried out using bioarchaeological collections that *Documenta Archaeobiologiae* has been founded.

This first volume is dedicated to an international workshop held in July 2002 to mark the centennial of the Anthropological Collection in München. Its publication has been made possible by generous funding from the Bavarian Ministry for Science, Research and Art, to whom we are most indebted. Held on July 19th and 20th, 2002, and entitled "Deciphering Ancient Bones: The Research Potential of Bioarchaeological Collections", the Anthropological Collection Centennial Workshop drew scientists of international reputation together, not only to discuss current research perspectives, but also the particularities inherent to such collections, the work of which often remains hidden behind the well-recognized scientific publications. A special note of thanks must be made here to all the speakers who readily accepted our invitation, generously gave of their time and presented those valuable contributions. Their presence were proof of the huge research potential that archaeological collections hold as well as the spectrum of research that are dependent on them. While many who attended the meeting were from Germany, there were many too from Austria, Switzerland, Norway and France, and as far away as Australia. What was particularly encouraging was the considerable number of young scientists present, since it testifies to a current and possibly growing interest in such interdisciplinary, but nevertheless, small academic subjects.

The workshop was opened by Dr. U. Kirste, representative of the Bavarian Ministry of Science, Research and Art and by Prof. Dr. J. Grau, Vice-General Director of the Bavarian State Collections of Natural Sciences. The scientific part comprised six plenary sessions, which in itself demonstrates how well physical anthropology and palaeoanatomy fit together beyond their own subject-specific projects. In the first session, "Research Collections and the Public Echo", the focus was on the necessity for museum collections, their values, and the possible ways of saving this cultural heritage, as well as the difficulties in raising funds and possible ways of successfully achieving this goal. These topics were addressed by Prof. Dr. Theya Molleson, London/UK, and Prof. Dr. Andrew Nelson/Dr. Christine Nelson, London/CA. Again, while museum collections are also centres of modern biochemical research, their futures are unfortunately not necessarily guaranteed, and many have to fight for existence.

The second session was entitled "The Research Potential of Bioarchaeological Collections", and here, the subjects of physical anthropology, palaeoanthropology and primatology were referred to by Prof. Dr. Zivko Mikic, Belgrade, Prof. Dr. Winfried Henke, Mainz, and Bernhard Wiesemüller/Prof. Dr. Hartmut Rothe, Göttingen. Due to unforeseen events, both Prof. Mikic and Prof. Rothe were unfortunately unable to attend the workshop, but Prof. Mikic's paper arrived in time for it to be read by one of the organisers. Therefore, the meeting did not suffer from any loss of scientific content.

In the third session, "Man and his Animal World", Dr. Cornelia Becker, Berlin, and Dr. Melinda Zeder, Washington D.C., led discussions on the coexistence of man and the higher vertebrates, a relationship that has existed long before the beginnings of livestock husbandry and animal domestication. It was followed by short contributions on current projects of the Collection of Anthropology and Palaeoanatomy, the majority of which were or are currently financially supported by the German Science Foundation.

The second day's session, "Deciphering Ancient Bone", focussed on the information hidden in the microstructure (Prof. Dr. Simon Hillson/Daniel Antoine, London/UK) and the crystalline composition of bone (Dr. James Burton/Prof. Dr. T. Douglas Price, Madison/Wisconsin), and on the research potential of preserved DNA relics (Dr. Carles Lalueza-Fox, Barcelona). While gross morphological inspections and diagnoses of bone finds give clues to many important aspects of the population biology of man and animals, they also constitute the necessary bases for any further, small-scale reconstructions of the past. It was thrilling to learn how the microstructure of vertebrate teeth provide clues to life on a day-to-day level, and how the mineral composition of bone on the trace element and stable isotope levels permit the reconstruction of migration and origins of individuals as well as populations. The research on ancient DNA preserved in corporeal relics is rapidly progressing, but the limitations and technical problems associated with it should not be neglected.

The archaeologist's perspective was presented by Prof. Dr. Helmut Bender, Passau, in the final session on "The socio-cultural aspect and modern implications". Prof. Bender impressively demonstrated how changing methods and theory can lead to revisions on results previously obtained on specific skeletal series over time and pointed out the absolute need to curate bone finds for further research, even if they have been previously thoroughly worked on. Unlike a variety of other finds, bones, generally, are of no commercial value. While nobody would ever think of discarding artefacts or treasures, bones are frequently treated less carefully, which can often result in considerable scientific loss, not only for the biosciences, but for the social sciences, like archaeology, as well. The last paper of this session was presented by Prof. Dr. Hans Hinrich Sambraus, München, who gave an unusual perspective to rare domestic farm animal breeds, their histories and specific properties, and the current danger of a loss of biodiversity through extinction.

All in all, the meeting was a success, not only because of the number of highly scientific contributions from a variety of academic disciplines presented, but also because of the congeniality that was felt throughout the workshop and which supported the establishment of new scientific contacts, communication and exchange, as well as new joint projects.

This workshop could not have been realized without the financial support of the Bavarian Ministry of Science, Research and Art, the German Science Foundation, the Bavarian State Collections of Natural Sciences, and the Zeiss Company. Special thanks are due to the staff members of the Collection and the respective university institutes for their engagement and help in organising and operating the meeting and to Siew Eiselt who assisted with the manuscripts from contributors whose native language is not English.

Gisela Grupe, Joris Peters
Editors

The Bavarian State Collection of Anthropology and Palaeoanatomy.
A brief history

Gisela Grupe, Joris Peters (Directors)

The Bavarian State Collection of Anthropology and Palaeoanatomy is an extra-university research institution and archive for archaeological human and animal bone remains. In many countries, at least in Europe, both physical anthropology and palaeoanatomy belong to the so-called "small academic subjects" which are, however, interdisciplinary by nature. Bioarchaeological collections serve as an interface between the physical and social sciences, and as the two disciplines fit very well together beyond their respective subject-specific research, the merger of the Anthropological and Palaeoanatomical Collections in München into the current single, twin-discipline institution in the year 2000 is a major step forward for both academic subjects.

Both physical anthropology and palaeoanatomy have their own specific histories in München, and physical anthropology in particular has a very long tradition, having already enjoyed the support of King Ludwig I in the 19th century. The Munich Society for Physical Anthropology, Ethnology and Prehistory dates its founding to as far back as 1870. In 1886, Prof. Johannes Ranke became the first professor elected for physical anthropology in Germany, and in August 1902, the anthropological collection of which he was the founder came into independent existence (Glowatzki 1977). The collection grew constantly, but not much of it was left after April 25, 1944, when the "Old Academy" in München was destroyed. Many finds were lost forever, together with a good portion of the original documentation which was equally disastrous. Fortunately, several valuable finds were rescued in time, among them the Mesolithic finds from the Ofnet Cave (5 500 BC; Figure 1), the Epipalaeolithic find from Neu-Essing (16 200 BC; Figure 2), and the first human find from the Olduvai Gorge, Olduvai Hominid I (OH 1, 15 000 BC; Figure 3).

In 1972, the collection acquired its own premises in a charming 19th-century villa in Karolinenplatz 2a in downtown München. Built and designed originally as a stable for horses and carriages (Yblagger 1977), the building has been remodelled several times and was completely renovated in the year 2000. The present premises provide a stimulating atmosphere and a very comfortable place to work in, but is by far incapable of housing all the human and animal archaeological finds the collection has curatorial charge of. These finds cover the periods from the Epipalaeolithic to Early Modern times, and thanks to the excavation activities of the cultural heritage institutions, have constantly grown in number. Till today, no adequate, permanent storage facility has been found and the research objects remain distributed over many parts of the city of München. Special thanks are due to the Bavarian Ministry for Science, Research and Art and the Bavarian State Collections of Natural Sciences for their constant efforts and help in finding homes for all these prehistoric people and their animals.

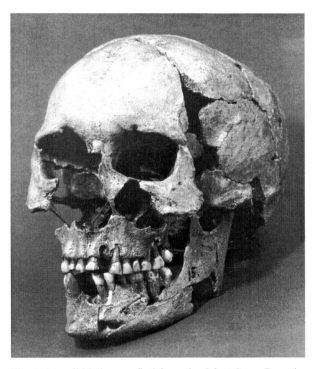

Fig. 1: Mesolithic human find from the Ofnet Cave, Bavaria, 5500 BC.

In the course of a hundred years, the aims and scopes of physical anthropology have changed, just as they have changed in other scientific disciplines. Modern physical anthropology today largely concentrates on population biology, that is, beyond single finds that are outstanding because of their age, status etc., skeletal series that have been properly excavated and documented are in the focus of modern research. These are materials that constitute very valuable empirical sources for questions concerning population development, palaeoecosystem analyses, epidemiological problems and many other scientific issues. The study and analyses of animal bone finds hold the promise of being major contributors to

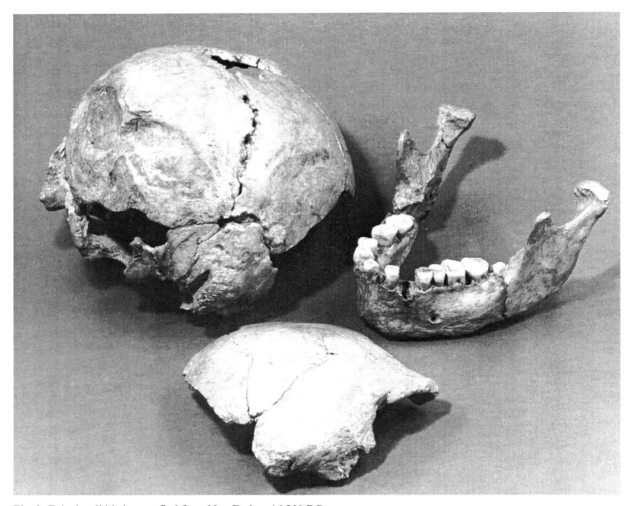

Fig. 2: Epipalaeolithic human find from Neu-Essing, 16 200 BC.

modern biodiversity research and the understanding of many issues, including the ability to reconstruct changes in the geographical distribution of free ranging vertebrates, the effects caused by the introduction of domestic animals, and the development of the various domestic breeds. Of special interest are man's closest relatives, the non-human primates, species that are highly endangered today. In addition, both methods and theory have switched from being descriptive to being analytical. Even if the bodily relics of ancient populations are dead tissues, they nevertheless constitute "tissue banks" that record several thousand years of human population development. Fortunately, reburial is not a big issue in Germany yet, but it is growing and one will have to find ways to solve or at least cope with it. For other countries, this aspect has been of major importance for several years, which is one of the major reasons why the collection in München is sought after by many colleagues worldwide.

Due to the rapid technical and methodological progress, especially in the field of archaeometry, one is increasingly capable of deciphering detailed information of past individual and collective daily life recorded in the crystalline and biomolecular composition of bones and teeth. Therefore, questions that have long been asked can now be answered, e.g., concerning palaeodiet or genetic relationships. As a consequence, one can never claim that a certain skeletal series has been investigated to its very end: it is impossible to predict whether additional samplings will be necessary, especially for finds which have already been extensively worked on, morphologically diagnosed and used for some archaeometric approaches, i.e., where important data have already been accumulated. The scientific value of a skeletal series is less dependent on the number of individuals in the series than on the archaeological documentation and the diagnoses previously established on it. Hence, the more data a series accumulates, the higher its value is. Reburial as a solution to the storage problems is therefore in no way justified but rather results in a destruction of valuable empirical sources.

At this point, another particularity that is sometimes overlooked by those scientists who do not have curatorial responsibilities for research collections should be

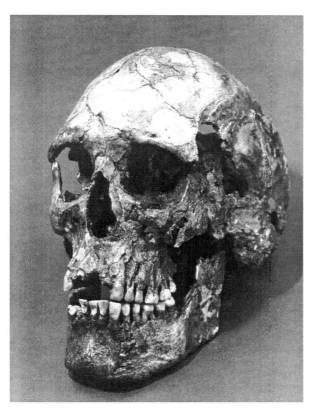

Fig. 3: Olduvai Hominid I, the first human find recovered at Olduvai Gorge.

addressed. The situation in München is very favourable since all the directors of Bavarian State Collections are concurrently university professors for the respective subjects. Scientifically, this situation offers two advantages: students enjoy easy and direct access to a wide variety of research objects that are regularly used in academic teaching and research projects benefit since the dependency on the availability of adequate research objects is minimized. At the same time, the existence of collections closely tied to the university enables scientists to start their research with a hypothesis, and immediately follow it through by direct access to the appropriate material for testing the hypothesis. In other words, a synergy effect is realised. Therefore, the collection, restoration, diagnoses and curation of the finds is as important as further scientific research.

Certainly, research objects housed in the collections are also open to scientists from other scientific institutions. Indeed, the finds are sought after by many colleagues and also by physicians, anatomists, dentists, forensic scientists and veterinarians. Since modern research methods are often invasive, not every request for sampling can be accommodated, although it would solve the storage problem within a rather short time because the finds would be pulverised in various laboratories. However, this is only one reason why every request has to be carefully checked.

Another reason is that material has to be preserved and collections must not be used as quarries. A close cooperation of all scientists involved is absolutely necessary. A piece of bone or a tooth is easily sampled, but someone has to first identify the species a specific specimen has been taken from, as well as whether it is from an adult male, female or a child, and whether this particular individual exhibited any features of pathology including metabolic disorders. One should keep in mind that someone has to provide all these informations, and not just anyone, but only a scientist highly experienced in osteology who has performed these important and equally time-consuming diagnoses will suffice. Although current research is dominated by molecular biology, there is no reason why morphology should be less valued – archaeometry without these basic information may be compared to diagnoses by physicians without the appropriate anamnestic queries. In addition, while gross morphology is thoroughly investigated on the phenomenological level, major deficiencies still exist in terms of functional morphology, especially on the micro- and ultrastructural level.

The hundred years of history of the Anthropological Collection of München, like the history of any scientific institution are marked by progress as well as by regression. Special progress took place in August 2000, when the anthropological collection merged with the palaeoanatomical collection to form a bigger, again explicitly interdisciplinary bioarchaeological institution, the Bavarian State Collection for Anthropology and Palaeoanatomy. The latter research discipline, named Archaeozoology by the international scientific community, was institutionalised decidedly later than its anthropological counterpart, namely in 1965, when the *Institut für Palaeoanatomie, Domestikationsforschung und Geschichte der Tiermedizin* (Institute of Palaeoanatomy, Domestication Research and History of Veterinary Medicine) at the Veterinary Faculty was founded. Four decades of research on animal bones in Munich resulted in a huge data base concerning the Late Quaternary animal world in Late Pleistocene and Holocene Europe, Asia and Africa. Due to the lack of adequate facilities and an inability to store the majority of the materials analysed, we could not but return most of the materials studied to the excavators until the end of the 1980s. In view of the multitude of new methods and techniques which can now be applied to these materials (see this volume), this is very unfortunate, particularly since in some instances these samples appear definitely lost.

Besides the analysis of archaeofaunal assemblages, it should be mentioned that from the 1960s onwards, immense efforts were made to prepare comparative specimens of all vertebrate groups. These activities resulted in one of the finest archaeozoological collections worldwide of recent vertebrates, especially of fish and birds.

From a historical perspective, however, it is noteworthy that the idea of combining research on human and animal bones in a single institution was already formulated by Ferdinand Birkner at the turn of the 19th century. Birkner, then assistant at the Anthropologic-Prehistoric Collection of State in Munich under the directorship of Johannes Ranke, once concluded an oral presentation about dog bones from Roman sites in Germany at a 1900 meeting of the *Deutsche Gesellschaft für Anthropologie, Ethnologie und Urgeschichte* (*German Association of Anthropology, Ethnology and Prehistory*) as follows: "The until now almost completely neglected study of dogs and other domestic animals of the Romans in Germany constitutes a valuable contribution to the history of modern dog breeds and of other domestic breeds. The study collection available, however, is still small; I therefore would like to ask anyone involved in excavations of Roman forts and settlements to collect every piece of bone carefully and to send these to the Anthropologic-Prehistoric Collection of State in Munich" (Birkner 1901). Almost a century elapsed before Birkner's vision became reality in the year 2000. Looking through the files, one realises that the merger alone took five full years of intensive and hard discussions, many letters and memos before it finally came to pass. Now, two years later, we are convinced that nobody will regret this step of combining animal and human bone research. The conference held on the occasion of the 100th anniversary of the Anthropological Collection in München, together with the papers presented during this meeting and published in this volume, demonstrate that it is indeed possible to bring various scientific disciplines with their diverse research subjects, with all focussing on bioarchaeology, into an intense dialogue on the unravelling of population histories of man and his animal world, thereby contributing to the understanding of the present – a task that is impossible without a profound understanding of the past.

References

BIRKNER F., 1901.
Über die Hunde der Römer in Deutschland. *Korrespondenzblatt der Deutschen Gesellschaft für Anthropologie, Ethnologie und Urgeschichte* **33**: 165-162.

GLOWATZKI G., 1977.
Zur Geschichte der Anthropologischen Staatssammlung München. In: Schröter, P. (ed.). *75 Jahre Anthropologische Staatssammlung München 1902-1977*: 15-18. München: Selbstverlag der Anthropologischen Staatssammlung.

YBLAGGER W., 1977.
Chronik des neuen Dienstgebäudes der Anthropologischen Staatssammlung (München, Karolinenplatz 2 a). In: Schröter, P. (ed.). *75 Jahre Anthropologische Staatssammlung München 1902-1977*: 19-28. München: Selbstverlag der Anthropologischen Staatssammlung.

For further detailed informations on the Collection for Anthropology and Palaeoanatomy please visit the homepage of the Bavarian State Collections (www.naturwissenschaftlichesammlungenbayerns.de and proceed by clicking on our logo).

PART I:

RESEARCH COLLECTIONS AND THE PUBLIC ECHO

Body of evidence:
Museum Collections, why they were brought together, their value today and public future

Theya Molleson, The Natural History Museum, London

Abstract / Zusammenfassung

The variety and history of museum collections is reviewed and attention drawn to the educational, research and aesthetic value of human skeletal collections in the modern world. Presentation of material extending from the simple static museum display through interactive and virtual displays puts greater emphasis on the material itself and inevitably exposes it to greater handling. New approaches are constantly being developed, funding restrictions notwithstanding. Curatorial and political responsibility for our collections, many of which cannot be replaced, has never been greater.

Die unterschiedlichen musealen Sammlungen und deren Geschichte werden referiert, wobei insbesondere auf den Wert menschlicher Skelettsammlungen in der heutigen Zeit in bezug auf den Bildungssektor, die wissenschaftliche Erschließung und auch ästhetische Aspekte hingewiesen wird. Eine Präsentation des Materials, welche über die einfache statische Ausstellung im Museum hinausgeht und auch interaktive und virtuelle Möglichkeiten nutzt, steigert den Stellenwert des Materiales selbst, setzt es aber unausweichlich stärkeren Manipulationen aus. Es werden ständig neue wissenschaftliche Zugänge zu dem Material entwickelt, trotz zunehmender Mittelkürzungen. Zu keinem Zeitpunkt war die Verantwortlichkeit für unsere Sammlungen, von denen viele unersetzlich sind, sowohl in konservatorischer als auch in politischer Hinsicht größer.

Key words: Collections History, Research Techniques Value
Sammlungsgeschichte, Bedeutung von Untersuchungsmethoden

Introduction

The secure residence of ten thousand human remains is SW7 5BD, one of the most expensive of London addresses, The Natural History Museum at South Kensington. Some of the bones are sleeping but most are very busy, so busy that there is a booking system for anyone wanting to see them. They are scrutinized, measured, photographed, radiographed and tested. They appear in publications but rarely in public. Skeletons and mummified bodies are, however, on display at the British Museum, Bloomsbury, and draw intense interest; while an exhibition of excavated skeletons at the Museum of London was hugely successful. Our need to know about ourselves and others is perpetual, undiminished and ever changing and comes in part through the study of the once living. We have moved from categorizing heads into racial types to looking at the impact of lifestyle on the bones and teeth. Our collections are part of the National Heritage and undoubtedly have a future.

Collections are the offspring of individuals. The first collectors amassed artifacts of cultures but they also collected the remains of the people. They spent their personal fortunes on the task, and yours too given the opportunity. 'Collecting is an hereditary disease, and I fear incurable' claimed A. W. Franks, first Keeper of British and Medieval Antiquities at the British Museum (Caygill and Cherry 1997, 318).

Types of collections

1. The National Collections

The British Museum: arguably the world's greatest Museum.

The British Museum was long in recognizing the need to represent British history in its galleries, preferring manuscripts, exotica, even natural history, which was outstandingly popular with its nineteenth century public. The Museum was open to the studious and the curious; its brief to educate and entertain. The history of the nation was not initially a priority and the Museum was reproached for neglecting the antiquities of its own country while accumulating those of other lands (Caygill 1997, 163ff). After a long battle Angustus Wollaston Franks entered the employ of the British

Museum in 1851 and eventually, in 1881, was appointed the first Keeper of the Department of British and Medieval Antiquities, which he had a remit to illustrate. The growth in the study of Anglo-Saxon archaeology was fueled by nineteenth century interest in national identity – many discoveries being made during urban development and the building of the railways (Caygill 1997, 165). Space to accommodate the ever growing collections became critical. While the Trustees were loath to split the collections, particularly to separate them from the library, the Government refused adequate funding.

Eventually, in order to gain space for the British collections of artifacts, the natural history specimens were relocated to South Kensington with Richard Owen as Superintendent. The move was expedient and gave Owen a position and opportunity to influence the design of the museum that was to be built to house the collections (Stearn 1981). His aim was to display an index collection of natural history specimens, fixed and immutable. The Natural History Museum at South Kensington, designed by Waterhouse, was opened to the public in 1881. The building was and is a Romanesque Temple of Nature. Fortunately for science, Richard Owen retired and William Flower, anatomist and osteologist, was appointed the first Director in 1884. Thus, the concept of display and collection was to be within an evolutionary framework. His collections of human anatomy formed at the Hunterian Museum of the Royal College of Surgeons, however, only came to the Natural History Museum in 1953.

New collections are added as they become available and by deliberate initiative. Recently collections of human skeletal material have come to us through destruction, development or restoration of a cemetery. Newark Bay, Poundbury, Spitalfields, Abingdon are examples. These were Christian places of rest and the Christian belief that the soul leaves the mortal remains at death enables such remains to be recovered, indeed treated with care, but open to study.

2. Regional Museums

Many of these are a joy to visit. They put on enterprising, innovative exhibitions based on their collections, that would enliven any dull afternoon and might be the enduring memory a visitor takes home.

Increasingly Regional Museums are the repository for local material and have to find accessible suitable storage space – a garage subject to flooding is not suitable storage for cardboard boxes. Yet they have to fight for every penny from their local Councils, who may have little comprehension of the responsibility due to such collections. The museums are not short of collectors' items – granny's china dolls will often end up in the local Museum but who is to conserve, restore and catalogue them? Can a Museum in the Outer Hebrides or the Orkney Islands be expected to have experts in horse brasses, trilobites or Viking brooches. At present, the National Museums will usually provide help and there is a network of experts that can be called upon.

These Museums will be part of the Museums Association and will generally prefer to employ staff having a Museums Diploma. So standards are high. The National Museums are less likely to be so closely linked to the Museums Association. They are financed by Direct-Grant-in-Aid from the Government, enhanced by sponsorship and donations. It is Government policy that Museums should be free. And increase in attendance since this policy was implemented at the beginning of 2002 has been dramatic, both in terms of numbers and the range of people visiting.

3. University Museums

Many of these were created to present student teaching collections, but are now open, in part at least, to the public, albeit often on a restricted basis. Many have collections that rank with the National Collections, and they may have teaching displays for the use of students that are not normally accessible to the general public, although this varies greatly.

4. Specialist and Private Museums

Often the product of serious collecting by enthusiasts who had personal fortunes to spend both on collecting and on housing their collections. Their museums may have restricted access but usually have a Gallery open to the public. They are important sources of specialist collections for teaching and research. They often have a private endowment which enables them to keep going at a 'holding level' at least.

5. Repositories

They are not open to the public and are largely unexploited by researchers.

6. Temporary Collections

Collections on loan, perhaps prior to reburial.

The requirement to rebury excavated material can provoke a coordinated research program drawing together researchers from a wide range of disciplines, as is happening with the Spitalfields Collection of known age individuals.

Functions of collections

Collections for the people: public education and entertainment

Any mummy on display in the British Museum serves a real need as children and adults alike confront the death but not the relative. Unfortunately in some museums whether to display or not has become overtaken by 'political correctness' at the expense of educational value. Yet human fossils can be displayed. Does having no name for the remains mean that we are justified in displaying them? Can we exhibit somebody's hat but not his skull; his portrait but not his effigy? Some cultures or belief systems would say not. Yet there is on display in Karaman Museum, a traditional Islamic town in Turkey, a mummified body that is just as startling as any in the British Museum. And it draws the same fascinated interest from visitors.

Exhibition galleries of Museums have been transformed in recent years to be enjoyed by the general public, who now participate to much a greater degree. We have attempted to show them that Museums are also places of research by taking groups on tours of the work areas, developing privileges and events for friends and taking on volunteers of all ages, whether school leavers, the graduate unemployed, or the retired. Despite a tendency to collect those in need of 'a day care center for the intellectually alert', there is in fact a reservoir of people, who can and want to work on collections, but whom we have yet to appreciate and provide for.

A 'Dead Zoo' is how one child described the Natural History Museum. Museums display objects, categorized according to type but without value judgement; a big lion is not better than a small mouse, a flea than the plague bacterium in carries. And this surely is how it should be – at first level. But as we handle more information, become interdisciplinary, we also need to show how climate and food and health interact. Why the species diversity in Africa, not northern Europe, why does rickets exist in Britain and hardly in Spain, why is cystic fibrosis the commonest genetic disease in northern Europe but virtually unknown in Japan if we all came out of Africa? The curators of the collections know the answers to some of these questions, and there is a need to tell the public about the living parts of our dead zoos, especially if we want our collections to survive. We can broadcast our findings from a neutral environment and so perhaps avoid some of the tragedies of modern fads – the vegan who feeds his baby on soya milk and peanuts – the baby died. I do not know of any public displays that address these sort of issues. Our Human Biology Gallery is twenty years old and does make some attempt to describe physiological processes but is far too anodyne to tackle behaviours that are incompetent through ignorance.

Collections for Research

The anthropology collections at the Natural History Museum are increasingly used for research projects and increasingly sampling for analytical and technical procedures is deemed necessary to understand the biology or cultural history of our specimens.

The Department is reviewing its policies and procedures for the loan of specimens where these are being subject to any physical process. We are reminded that if any of the processes listed below, or other physical treatments, are to be carried out on any accessioned specimens from our collections, permission must be obtained in writing from the Keeper of Palaeontology. Such requests for permission should be accompanied by a recommendation from the appropriate Curator and Collections Manager.

Our existing policy reads: consent to carry out physical treatment of a specimen must first be obtained, in writing, from the Keeper of Palaeontology. All requests must describe the proposed treatment and the materials to be used. This applies to any destructive techniques and to treatment which may alter the state of the specimen including (section 2):

2.1 Development of specimens from their surrounding matrix, or removal of any part of the specimen.
2.2 Sectioning of a specimen in the form of a polished surface or in the form of a thin section. All sections remain the property of The Natural History Museum and must be returned with the remainder of the specimen.
2.3 The removal or application of any adhesives and consolidants.
2.4 Staining or coating of a specimen for photography, scanning electron-microscopy, or other purposes.
2.5 Preparation of moulds or casts in any material.
2.6 Sampling for chemical or DNA analysis.
2.7 The removal or application of labels.

In addition this 'requesting of permission from the Keeper' should also apply to non-staff wishing to carry out physical procedures on accessioned specimens while studying them during a visit to the Department, i.e. specimens which are not on loan.

Collection based research: towards an understanding of ourselves

I. The ancient human occupation of Britain (AHOB)
A major research initiative which will examine the total context of Pleistocene hominids, the associated fauna, and environmental background for the evolution of human societies in Britain.

In October 2001, The Leverhulme Trust awarded the Museum and partners a grant of 1.2 million pounds for a five-year study of the Ancient Human Occupation of Britain (AHOB). Some of the key research questions they hope to address include study of the general factors controlling human occupation in Britain, such as climate and geography, as well as specific phases of human presence and absence during the Middle and Upper Pleistocene.

This project will draw on collections of human and animal bones throughout the British Isles and into Europe. The range and variation in species will be documented and evaluated by our most eminent palaeontologists; the fossils will be tested isotopically and their migrations assessed; all in the hope of understanding why the islands were largely uninhabited during most of the Pleistocene.

II. Deserted Britain: declining populations in the British late Middle Pleistocene

One of the major questions of the AHOB project is the apparent absence of people in Britain from 186,000 to possibly 57,000 years ago. One way to test whether this pattern is merely a bias in preservation of the data is to look at artifact accumulation within a single flight of river terraces, thus documenting changes through time. The Middle Thames valley suggests that the pattern is real, and points to a population decline prior to the absence. The reasons for the population decline and absence are also investigated, concentrating on two potential factors. The first is the adaptation at this time of human populations to more open conditions, particularly the mammoth-steppes of eastern Europe and Asia, while the second is the effect of the isolation of Britain during warm stages after the catastrophic breach by the sea of the Straits of Dover.

New collections, new research

New collections are added as they become available and by deliberate initiative. The Spitalfields collection of named individuals of known age, sex, and often occupation, is one such. The crypt of Christ Church, Spitalfields had to be cleared as part of a program of restoration. It was estimated that the parochial and private vaults beneath the nave of the church held about a thousand individuals and funding was sought for a two year program of excavation and data collection. Just over a thousand individuals were recovered of which nearly four hundred had legible coffin plates. One third were juveniles and females and males were equally represented – a common finding of settlement cemeteries from the Christian era. The average age at death was 56 for both sexes, considerably older than is generally given for archaeological samples and a challenge to accepted teaching.

The age related studies of the Spitalfields remains emphatically showed that human beings in the past were not all dead at thirty and that aging is not linear (Molleson et al. 1993). Some had old bones at thirty, others were still young nearing 80. Rates of aging evidently vary according to the individual so we shall never be able to determine age at death of adults precisely.

In fact the driving motive behind the initiative to study the Christ Church skeletons had been concern that standard methods for estimating skeletal age at death were inadequate. This was the opportunity to follow skeletal aging against a background of evidence for ethnicity, occupation, diet, climate and health. Children, however, appeared to be under-represented in the sample, but it has not been possible to establish whether this reflects higher survival rates among this predominantly middle class Huguenot population or burial practice.

Skeletal age was examined using standard methods of cranial suture closure, pubic symphysis involution, trabecular thinning of the humerus and femur proximal end, rib end involution and cortical thinning of the femur. The first four methods were also assessed by application of the Complex Method of Acsádi and Nemeskéri (1970).

When compared to known chronological age at death the Complex Method tended to over-age the young and under-age the old. The tendency was the same whichever of the four methods was considered (Molleson et al. 1993). It would seem that, in this cemetery sample, the young individuals already had old bones while those who died at a great age retained young bone – they were survivors. The same pattern was obtained when amino acid racemisation, a chemical method, was used on dentine from teeth (Gillard et al.1990). Evidently we were observing rate of aging, not age, and in a cemetery sample at that; our findings probably would not apply to the living. Worse, the fact that the methods did not work on this known age sample means that they cannot be applied to samples of unknown age since they might not work there but there would be no means of checking.

That rate of aging might have differed in the past from that found in modern urban peoples was considered. DEXA, a radiographic technique, can be used to evaluate trabecular and cortical thinning of bone. It is used particularly to estimate the risk of osteoporosis in the femoral neck and in vertebrae. Our studies of the femora from Spitalfields indicated that females retained bone for longer than do modern women. This applied to femoral neck and to Wards triangle (Lees et al. 1993).

Age related studies of cortical thickness of the femoral mid-shaft revealed a very important pattern: that, in females, the cortical thickness reached a peak at about thirty years (as it does in modern samples) and remained at this level through maturity, only beginning to decline from about age fifty (Fig. 1). This contrasts with the

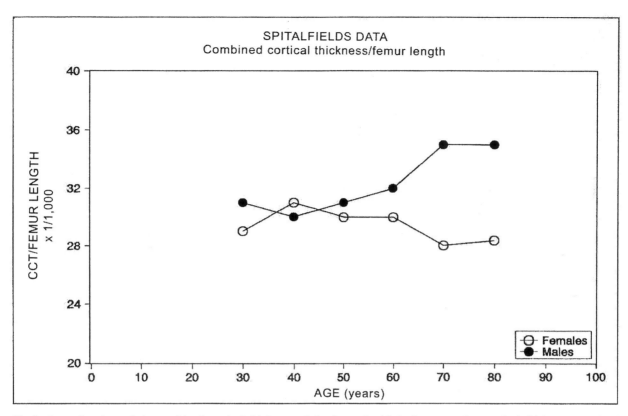

Fig.1: Age related trends in combined cortical thickness of the femoral mid-shaft suggest that cortical thickness remained relatively stable after reaching peak thickness in early adulthood until old age.

modern pattern of steady decline from peak thickness at around thirty years. Some of the reasons for this decline towards osteoporosis risk have been identified in modern populations (Stephenson et al. 1989). They include activity, smoking, alcohol, and, during the reproductive years, pregnancy and lactation.

We knew the occupations of sufficient of the named sample to be able to place them into one of four occupational groups. There were some significant differences in cortical thickness among the women, but these appeared to be random. For the males any differences in cortical thickness related to occupation were not significant. Presumably they had all experienced similar forces during growth.

Some occupational habits were influential, perhaps because more sustained. The traits of horse riding were identified among master weavers, who would have regularly ridden between one of their journeymen weavers to another; their habit of mounting a horse from the left pre-disposed to osteo-arthritis of the medial condyle of the left knee, in contrast to the general pattern of right knee involvement recorded for other groups. Journeymen weavers, from ten years of age had spent twelve hours a day, six days a week sitting at the weaving loom. Since they started work when they were ten years old and still growing, they developed what is described as a static sacro-iliac articulation (Kapandji 1988). This is a locked L-shaped articulation, in contrast to the dynamic C-shaped one that allows rotation around the joint.

Relationships demonstrated between bone morphology and activity have enabled us to set up criteria by which we can infer patterns of behaviour in past populations. We can use the present as a model for the past. Craftsmen have emerged to solve problems of equipment, baskets to carry produce, preparation of hard cereals, by pounding and grinding. Many of their methods have persisted through the ages and can be seen even today.

Basket makers from Orkney to New Zealand and North America have been recorded pulling the cane across their teeth (Inoue 1993, Larsen 1985, Robertson 1991, Pl. 30, 113). The action both moistens the fibre and uses the teeth as a third hand. It leaves distinctive grooves across the front teeth, that were easily recognized in Neolithic jaws from Abu Hureyra, Syria (Molleson 1994a). The evidence for basket making was found in several individuals from one area on the site. Moreover the jaws were distinctive in shape, suggesting a family of specialist craftsmen that quite possibly originated outside the region, even from as far away as Africa (Rosas & Molleson in preparation). Are these the first signs of a caste system (Molleson 1996, 348)? Evidence for activity patterns can bring us knowledge of the emergence of division of roles and role specialization in early societies; and we can follow the economic impetus that can transfer a division of roles based on sex into a caste system based on role specialization within the family.

It is all too rare that an individual is excavated together with the tools of its trade. An exception is the female excavated at al-'Ubaid near Ur, Iraq (Hall & Woolley 1927). She was found with a large stone pounder and a ceramic spindle whorl. Her enormously expanded clavicle and sternum, indicating development of the Pectoralis major muscles, suggest a lifetime spent pounding – perhaps ochre to make dyes. Of course, she might have been a cripple and used a crutch since this activity also results in development of the P. major muscles, as seen in the woman from Bronze Age Jericho on display with her crutch in the British Museum (Tubb 2000). Alternative causes for a given morphology must always be borne in mind, as must alternative ways of making a given product.

New research from existing collections

The re-examination of old collections can be rewarding. The British Museum which once gave us their bones now has a policy of not splitting collections. So it was particularly rewarding recently to work on the collections from Ur in the British Museum and to have immediate access to the excavator's notebooks and photographs as well as to the library. This proved to be most efficient; it not only facilitated the work but enhanced the quality of the study since facts could be checked on the spot. The only difficulty that I can see is that conditions of storage for artifacts are not always ideal for bones.

Sir Leonard Woolley, Sir Max Mallowan, Agatha Christie, A.T. Hall excavated at Ur and nearby Al-'Ubaid in Iraq from 1919. The collections were divided between the Museum in Baghdad, the Museum of the University of Pennsylvania and The British Museum. It is arguably the most important collection held by the British Museum after the Oxus Treasure (that was acquired by A.W. Franks). "Our object was to get history, not to fill Museum cases with miscellaneous curios" (Woolley & Moorey 1982, 52). To this end Woolley describes how they decided to remove an entire skull with all its ornaments in position on it, commenting that while the wreaths and chains and necklaces re-strung and arranged in a glass case might look very well, but it would be more interesting to see them as they were actually found, and therefore a few heads on which the original order of the beads and gold work was best preserved were laboriously cleaned with small knives and brushes, the dirt being removed without disturbing any of the ornaments and then boiling wax was poured over them solidifying them in one mass. The lump of wax could be lifted from the ground by undercutting. Mounted in plaster, with the superfluous wax cleaned off, the beads form an exhibit which is not only of interest in itself but proves the accuracy of the restorations which we have made of others (Woolley & Moorey 1982, 80). Two of these crushed waxed skulls are still on display to the public in Gallery 56, the Gallery of the Ancient Near East at the British Museum. These heads are included in a new study and new interpretation of the human bones that were brought back from Iraq to be described originally by Sir Arthur Keith at the Royal College of Surgeons and which are now housed in the Natural History Museum at South Kensington (Molleson & Hodgson in press). A preview can be seen on the web page for the British Museum. (So demonstrating how modern technology can make collections accessible). The British Museum web page: http://www.thebritishmuseum.ac.uk/science/whatsnew/ur.htm

PG789. Body 46 (121414). Skull of a soldier in helmet facing right. In addition to the flattening the whole has been crushed sideways so that the lower jaw is displaced to the back and the wisdom tooth lies behind the wisdom tooth of the upper jaw. The teeth are worn flat, a type of wear associated with masticating cereal based bread. The wear, however, is moderate, which may indicate a wheat based bread rather than barley bread which is more abrasive. The radiograph reveals the teeth of the left jaws, not otherwise visible. The teeth are all fully developed, with closed apices even of the wisdom teeth, indicating an adult and, from the size of the pulp cavities, a young adult. There are no dental caries.

"The best example of the death-pit was that of our royal grave RT 1237. The death-pit was intact. Six men-servants lay near the entrance, in front of them stood a great copper basin, and by it were the bodies of four women harpists. Over the rest of the pit's area there lay in ordered rows the bodies of sixty-four ladies of the court. All of them wore some sort of ceremonial dress. Around the neck was worn a 'dog-collar' of lapis lazuli and gold together with other looser necklaces of gold, silver lapis lazuli and carnelian beads; in the ears were very large crescent-shaped earrings of gold and silver. Twenty-eight of these court ladies wore golden hair-ribbons, the rest silver."

The head of Body 53 (122294) superimposed on the legs (tibias) of body 67 is displayed on a plaster plaque. The skull is much more difficult to interpret than that of the soldier. The head faces to the left. The silver comb part of the headdress has fallen to the back of the skull; the choker beads, two gold, one paste are at the throat. The ear-ring is more to the front: but there is only one. The skull bone, top left, is rounded and has the contour of a frontal of a girl. If this orientation is correct the teeth have been pushed well to the back of the skull. Only two teeth are visible in the specimen and it is not clear which they are; they are very little worn. In order to get a better idea of the age of the individual the whole specimen was radiographed. This clarified that, to the right of the teeth, bone of the jaw, rather than tooth was visible, and, most important, that hidden under a gold leaf of the headdress was an upper incisor. Although only two teeth are visible on the specimen, the radiograph reveals that at least six upper and six lower teeth are present. These include the visible teeth. Identification of the teeth and developmental stage was not easy, especially as sections of the teeth which have an open pulp cavity can look like the perforated stone beads; both give a darker radiographic image where the density of tooth/stone is less. The teeth are all permanent.

The radiograph reveals that in order to lift the skull it must have been bisected leaving behind the right side. There is only one ear-ring, the choker consists of only three beads, and there is only a partial dentition suggests that the skull was cut in half to lift it. This attendant, in lying down to die cradled her head on the legs of the attendant in the row above her. The shafts of the shins of this attendant can be seen across the top of the skull and in front of the face. After consultation with a number of dental researchers, Professor Christopher Dean, Dr Helen Liversidge and her MSc student Boonwadee Poolsanguan, we agreed that the tooth had a large pulp cavity and recently closed root apex, indicating an age at death of about 20 years – a young adult. The lack of wear on the teeth suggests a soft, perhaps privileged, diet. The bone of the upper and lower left jaws is poorly preserved and disorganized. No caries were noted in the teeth. The rest of the skull is also badly crushed and displaced.

New techniques, new knowledge

In the past, changes in diet, as from hunting and foraging to agriculture must have led to changes in activity patterns. Modern studies, even of old collections, that integrate analytical data with morphological, can provide insights into the social organization of past cultures. Stable isotopes of carbon and nitrogen recovered from ribs of a collection can identify not only what the main protein source was, meat fish or plant, but also when the infants were weaned and at what age the children were integrated into the adult work force (Richards & Molleson in preparation).

Children, in the past, were often required to work as long and as hard as were adults. In addition, they were often under-nourished given their energy output. This shows as a major growth deficit when compared to growth achievement of modern children. An inadequate weaning diet, once pottery had been developed, further exacerbated the growth retardation – a retardation from which there was often no recovery. The result was an adult population that was of small stature, below the potential for human beings.

But, it is the difference between male and female stature – the dimorphism index (stature females/stature males x 100) that can reveal the 'environmental health' of a community. Studies of mammal social structures have shown that where the resources are rich, males grow to their full potential size, while females are physiologically ready to reproduce early and remain small. This results in a high dimorphism between the sexes (Wilner & Martin 1985). Such societies are male dominated, have a high fertility and there is high juvenile mortality. When environmental resources are restricted there is delayed fertility, so females grow for longer, while males may not achieve their full growth potential. The result is a low dimorphism between the sexes, an egalitarian society and low fertility. These principals have been applied to human samples with some success (Harvey & Bennett 1985, Molleson 1994b).

There is an additional aspect to this response to environmental pressures and it is that populations under very severe stress from predation, hunting and perhaps disease start to reproduce at an ever smaller body size and younger age. This phenomenon is seen today in the cod, in island faunas and quite possibly in the Level VI sample from Neolithic Çatalhöyük. After a thousand years of occupation of this backswamp site the large population must have severely depleted the local resources (Molleson in prep.). It is frustrating that stable isotope analyses still cannot tell us how much protein is being consumed, only what kind; so that, although the indications are that animals were the main source of protein the quantity is hidden. It would seem that most calories came from such carbohydrate rich plants as tubers which are not revealed in the nitrogen isotopes. It is notable that there are indications of a C4 plant input (Richards et al. 2002).

These complicated clues to Neolithic society, that derive from newly excavated material from Çatalhöyük, Turkey, have been brought about by the collaboration of specialists working on the different collections – faunal, plant, human and analytical. Such an interdisciplinary approach is very much the signature of the present use of collections. The results are made available to all through the Internet web pages for the sites. Visitor Centres are constructed on site for tour groups and local people to see what the material finds are and how the interpretation has come about. Collections may no longer leave the site, or at least the country of origin, which may lead to a certain insularity and a lack of comparative appreciation.

Collections for the future

The consequence of this tendency to isolation is to render the old collections even more precious than ever. Yet old Museums struggle to survive; to such an extent that the need for funding and the ease by which money can be obtained for some projects and not others is in danger of driving some institutions 'off-course'.

There are still major issues to be addressed as far as our understanding of the working of human biology in societies of the past and the present are concerned. We still cannot determine satisfactorily the sex of juvenile skeletons, estimate the number of pregnancies that a woman had had or the relative amounts of meat, fish or beans in the diet of different elements of society. Ways of understanding migrations and inter-population movement are opening up while displays that demonstrate early technologies have proved an invaluable way to reach a new generation of archaeological explorers.

This is a very exciting time for anthropology. New questions are being asked and there are new techniques ready to help answer them.

There is a need for people's museums – new buildings for the new millennium – that present who and what we are. Biologically there are opportunities for education and for entertainment. The exhibition "London Bodies" at the Museum of London was a splendid array.

Issues of conservation, preservation, restoration, recording, archive, publication and storage are all being revolutionized as new technologies become available. Details of many of our collections are now widely available on our web sites, which means that more people want to see and handle them; while funding to support such intense interest is increasingly difficult.

The National Collections are, or at least have been, less vulnerable to dispersal than are those in private hands which may be sold or simply buried or destroyed. But they are now vulnerable to interference from Government; and as the search for funds takes priority over research may be in danger of dispersal. The British Museum is now to be subject to 15% cuts just as other Museums less motivated and less innovative have been. It seems that quality is no protection and that curtailment has a momentum from the international political echelons. Why? The money saved is negligible. It is possibly futile to fight – de Lumley's battle for the Musée de L'Homme in Paris may only have been effective in the short term though worthwhile. Tourists will visit regardless of what is displayed.

Museums seem to be viewed by politicians as left-overs from the 19th century, monumental repositories of the past, that should be self-funding. They are not seen as having any function other than as items on a tourist itinerary. Places of fascination and wonder, yes; that they might be vehicles of understanding and peace is difficult to get across.

Conclusions

We need collections that are accessible if we are to achieve:
1. Knowledge of Peoples, of kingdoms, invasions and migrations. Nature or nurture. Many of the traits that we used to look at as being genetically determined are in fact influenced by the environment.
2. Knowledge of society: Food and life-style. Work, craftsmanship, caste and hierarchy.
3. Biological knowledge: Growing up and growing old.
4. Epidemiology of diseases: famine, epidemics and war. Selection: cystic fibrosis, tuberculosis, malaria.
5. Knowledge of the dead: exposure or burial? Disposal of the dead versus exhibition; the need for the Turner prize.

Summary / Zusammenfassung

Collections are the offspring of individuals. The first collectors, like A.W. Franks of the British Museum, amassed artifacts of cultures but they also collected the remains of the people. Space to accommodate the ever growing collections became critical. Eventually, in order to gain space for the British collections of artifacts, the natural history specimens were relocated to the Natural History Museum at South Kensington. The Museum, designed by Waterhouse, was opened to the public in 1881. William Flower, anatomist and osteologist, was appointed the first Director.

Exhibition galleries of Museums have been transformed in recent years to be enjoyed by the general public, who now participate to a much greater degree. We have attempted to show them that Museums are also places of research by taking groups on tours of the work areas, developing privileges and events for Friends and taking on volunteers of all ages, whether school leavers, the graduate unemployed, or the retired. There is a reservoir of people, who can and want to work on collections, but whom we have yet to fully appreciate and provide for.

The Anthropology Collections at the Natural History Museum are increasingly used for research and increasingly sampling for analytical and technical procedures is deemed necessary to understand the biology or cultural history of our specimens. New collections are added as they become available and by deliberate initiative. The Spitalfields collection of named individuals of known age, sex, and often occupation is one such. The crypt of Christ Church, Spitalfields had to be cleared as part of a program of restoration.

The age related studies of the Spitalfields remains emphatically showed that human beings in the past were not all dead at thirty and that aging is not linear (Molleson et al. 1993). Some individuals had old bones at thirty, others were still young nearing 80. Evidently we were observing rate of aging, not age, and in a cemetery sample at that; our findings probably would not apply to the living. Worse, the fact that the methods did not work on this known age sample means that they cannot be applied to samples of unknown age since they might not work there but there would be no means of checking.

Age related studies of cortical thickness of the femoral mid-shaft revealed a very important pattern: that, in females, the cortical thickness reached a peak at about thirty years (as it does in modern samples) and remained at this level during adulthood, only beginning to decline from about age fifty. This contrasts with the modern pattern of steady decline from peak thickness at around thirty years.

We knew the occupations of sufficient of the named sample to be able to place them into one of four occupational groups. There were some significant differences in cortical thickness among the women, but these appeared to be random. For the males any differences in cortical thickness related to occupation were not significant. Presumably they had all experienced similar forces during growth.

Some occupational habits were influential, perhaps because more sustained. The traits of horse riding were identified among master weavers, who would have regularly ridden from one of their journeymen weavers to another; their habit of mounting a horse from the left pre-disposed to osteo-arthritis of the medial condyle of the left knee, in contrast to the general pattern of right knee involvement recorded for other groups.

Relationships demonstrated between bone morphology and activity have enabled us to set up criteria by which we can infer patterns of behaviour in past populations. We can use the present as a model for the past. Craftsmen have emerged to solve problems of equipment: baskets to carry produce, preparation of hard cereals, by pounding and grinding. Many of their methods have persisted through the ages.

New research from existing collections adds to our knowledge of the structure of society in the past. Who died in the Great Death Pits at Ur excavated by Leonard Woolley and the construction of the golden headdresses are revealed by radiography and displayed for all to see on the British Museum web-page.

Complicated clues to Neolithic society emerging from newly excavated material from Çatalhöyük, Turkey, have been brought about by the collaboration of specialists working on the different collections – faunal, plant, human and analytical. Such an interdisciplinary approach is very much the signature of the present use of collections. The results are made available to all through the Internet web pages for the sites. Visitor Centres are constructed on site for tour groups and local people to see what the material finds are and how the interpretation has come about. Collections may no longer leave the site or at least the country of origin, which may lead to a certain insularity and a lack of comparative appreciation.

The consequence of this tendency to isolation is to render the old collections even more precious than ever. Yet old Museums struggle to survive; to such an extent that the need for funding and the ease by which money can be obtained for some projects and not others is in danger of driving some institutions 'off-course'.

There are still major issues to be addressed as far as our understanding of the working of human biology in societies of the past and the present are concerned. We still cannot determine satisfactorily the sex of juvenile skeletons, estimate the number of pregnancies that a woman had had or the relative amounts of meat, fish or beans in the diet of different elements of society. Ways of understanding migrations and inter-population movement are opening up while displays that demonstrate early technologies have proved an invaluable way to reach a new generation of archaeological explorers.

This is a very exciting time for anthropology. New questions are being asked and there are new techniques ready to help answer them.

There is a need for people's museums – new buildings for the new millennium – that present who and what we are. Biologically there are opportunities for education and for entertainment.

Sammlungen beruhen auf individuellen Leistungen. Die ersten Sammler wie A.W. Franks vom Britischen Museum, haben Artefakte unterschiedlicher Kulturen angehäuft, aber ebenso bereits die körperlichen Relikte von Menschen. Das Raumproblem zur Unterbringung der ständig anwachsenden Sammlungen wurde kritisch. Um Platz für die Britische Artefaktensammlungen zu schaffen, wurden die naturhistorischen Objekte in das Natural History Museum in South Kensington überführt. Dieses von Waterhouse entworfene Museum wurde 1881 der Öffentlichkeit zugänglich gemacht, und der Anatom und Osteologe William Flower zu seinem ersten Direktor ernannt.

Die Ausstellungsräume der Museen sind in den letzten Jahren umgestaltet worden, um einem breiten Publikum gerecht zu werden, welches heute ein deutlich höheres Interesse zeigt. Wir haben versucht, den Besuchern zu zeigen, dass Museen auch Forschungsinstitutionen sind und veranstalten hierzu Gruppenführungen durch die Arbeitsräume, bieten Sonderkonditionen und Veranstaltungen für die Freunde der Museen an, und laden Freiwillige aller Altersstufen zur Mitarbeit ein, ob Schulabgänger, graduierte Arbeitslose, oder Rentner. Es gibt durchaus ein großes Reservoir von Personen, die mit Sammlungsmaterial arbeiten können und wollen, und die wir noch viel mehr würdigen und in größerer Zahl heranziehen sollten.

Die anthropologischen Sammlungen des Natural History Museum werden zunehmend für wissenschaftliche Forschung herangezogen, und eine zunehmende Beprobung für analytische und methodische Zwecke ist unabweisbar, um die Biologie oder die Kulturgeschichte des Sammlungsgutes zu verstehen. Sobald verfügbar, oder auch aufgrund gezielter Initiativen, werden neue Sammlungen hinzugefügt. Eine davon ist die Spitalfields-Sammlung von namentlich bekannten Individuen, von denen das Sterbealter, das Geschlecht, und häufig auch die berufliche Tätigkeit bekannt sind. Sie wurden aus der Krypta der Christ Church in Spitalfields geborgen, welche im Zuge von Restaurierungsarbeiten geräumt werden musste.

Die Untersuchungen zur Sterbealtersbestimmung anhand der Skelettfunde von Spitalfields haben nachdrücklich gezeigt, dass Menschen vergangener Epochen nicht alle bereits im Alter von 30 Jahren verstarben, und dass der Alterungsprozeß als solcher nicht linear verläuft (Molleson et al. 1993). Das Skelett einiger 30jähriger Individuen erschien biologisch alt, jenes einiger fast 80jähriger dagegen noch jung. Was wir offensichtlich beobachten ist die Alterungsrate, nicht das tatsächliche Alter, so dass die Befunde für eine Friedhofsserie wahrscheinlich nicht auf die Lebendpopulation übertragbar sein werden. Noch bedenklicher ist angesichts der Tatsache, dass die Methoden zur Sterbealtersbestimmung an dieser altersbekannten Serie versagten, dass diese nicht auf altersunbekannte Serien angewendet werden können. Sie könnten dort ebenfalls versagen, wobei es keine Möglichkeit zur Überprüfung gibt.

Untersuchungen zum Alternsgang der Kompaktadicke des mittleren Femurschaftes haben ein signifikantes Muster erkennen lassen: Bei Frauen wurde die maximale Schaftdicke mit etwa 30 Jahren erreicht (vergleichbar mit modernen Populationen) und auf diesem Niveau gehalten, bis von etwa 50 Jahren an die Wandstärke wieder abnimmt. Dies steht im Gegensatz zu dem modernen Alternsgang der stetigen Abnahme der Schaftdicke nach Erreichen des Maximums mit etwa 30 Jahren.

Von einer hinreichenden Anzahl der namensbekannten Individuen war auch deren Beruf bekannt, so dass sie jeweils einer von vier Berufsgruppen zugeordnet werden konnten. Es zeigten sich einige signifikante Unterschiede in der Femurschaftdicke bei den Frauen, welche jedoch eher zufällig sein dürften. Bei den Männern ergaben sich keinerlei Unterschiede in der Kompaktadicke in bezug zum Beruf. Vermutlich waren sie alle während der Wachstumsperiode ähnlichen mechanischen Beanspruchungen ausgesetzt.

Einige Aktivitätsmuster haben sich am Skelett ausgeprägt, vermutlich aufgrund deren ständiger Wiederholung. So etwa häufiges Reiten bei den Webermeistern, welche die Wegstrecken von einem Gesellen zum anderen zu Pferd zurücklegen mussten. Das übliche Aufsitzen auf das Pferd von links führt zu einer Prädisposition für Arthrose des medialen Condylus des linken Knies, im Gegensatz zur häufigen Affektion des rechten Knies bei anderen Berufsgruppen.

Body of evidence   27

Derartige Beziehungen zwischen Knochenmorphologie und Aktivitätsmustern haben es uns ermöglicht, die Kriterien für das Rückschließen auf Verhaltensmuster früherer Populationen formulieren zu können. So kann die Gegenwart als Modell für die Vergangenheit herangezogen werden. Demnach entwickelte sich das spezialisierte Handwerk im Zuge der Problemlösung für Ausrüstungsgegenstände: Herstellung von Körben zum Transport von Waren, das Zerstampfen und Zermahlen harten Getreides. Viele dieser Methoden haben sich bis in die heutige Zeit gehalten.

Neue Untersuchungen an bestehenden Sammlungen erweitern unser Wissen über frühe Sozialstrukturen. Durch Röntgenuntersuchungen konnte geklärt werden, wer in den Great Death Pits in Ur, ergraben durch Leonard Woolley, verstorben war, und wie der goldene Haarschmuck gefertigt wurde. Diese Befunde sind auf der Webseite des British Museum veröffentlicht.

Komplizierte Anhaltspunkte zur Struktur einer neolithischen Gesellschaft wurden anhand von neu ergrabenem Material aus Catalhöyük/Türkei durch Kooperation jener Spezialisten gewonnen, welche mit den verschiedenen Sammlungen arbeiten – den Tierknochen, den Pflanzenresten, den menschlichen Skelettfunden, sowie Analytiker. Derartige interdisziplinäre Forschung ist bezeichnend für die zeitgemäße Erschließung von Sammlungsmaterial. Die erzielten Ergebnisse werden für alle über das Internet zugänglich gemacht. An den Ausgrabungsstätten werden Besucherzentren für Reisegruppen und Einheimische eingerichtet, damit diese sich darüber informieren können, welche Funde gemacht wurden, und wie diese interpretiert werden.

Sammlungen brauchen nicht länger vom Ausgrabungsort oder sogar außer Landes verbracht werden, was zu einer gewissen Isolierung mit verminderter vergleichender Würdigung führen könnte.

Als Folge dieser Tendenz sind bestehende Sammlungen heute wertvoller denn je. Dennoch kämpfen traditionsreiche Museen um ihr Überleben und zwar in einem solchen Maße, dass die Notwendigkeit der Finanzierung, gemeinsam mit der Tatsache, dass Geld für manche Projekte leichter, für andere dagegen schwerer zu beschaffen ist, die Gefahr mit sich bringt, dass manche Institutionen in eine bestimmte Forschungsrichtung gedrängt werden.

Es gibt noch eine Reihe bedeutender Forschungslücken in bezug auf unser Verständnis der Auswirkung biologischer Mechanismen auf frühe und heutige Gesellschaften. Wir können noch immer nicht das Geschlecht von Kinderskeletten in zufriedenstellender Weise bestimmen, ebenso wie die Anzahl von Schwangerschaften einer erwachsenen Frau, oder die relativen Anteile von Fleisch, Fisch oder Bohnen in der Ernährung der verschiedenen Gesellschaftsschichten. Gegenwärtig eröffnen sich Möglichkeiten zur Rekonstruktion von Migration oder Wanderungsbewegungen zwischen den Bevölkerungen, während Ausstellungen über frühe Technologien sich als unschätzbarer Weg erwiesen haben, eine neue Generation archäologischer Forscher zu motivieren.

In bezug auf die Anthropologie leben wir in sehr spannenden Zeiten. Neue Fragen werden gestellt, und neue Techniken für deren Beantwortung stehen zur Verfügung.

Es besteht Bedarf an Museen für die breite Öffentlichkeit – neue Gebäude für das neue Jahrtausend – welche zeigen sollen, wer wir sind und was wir sind. Entsprechend eröffnen sich für die Naturkunde Möglichkeiten sowohl im Bereich der Bildung als auch auf dem Freizeitsektor.

Bibliography

Acsádi G., & Nemeskéri J., 1970.
History of Human Life span and Mortality. Budapest.

Caygill M., 1997.
'Some recollection of me when I am gone': Franks and the Early Medieval Archaeology of Britain and Ireland. In Caygill M. & Cherry J. (eds). *A.W. Franks. Nineteenth-Century Collecting and the British Museum*: 160-183. London: British Museum Press.

Caygill M., & Cherry J., (eds). 1997.
A.W. Franks. Nineteenth Century Collecting and the British Museum. London: British Museum Press: 372pp.

Gillard R.D., Hardman S.M., Pollard A.M., Sutton & Whittaker D.K., 1990.
An improved method for age determination from the measurement of D-aspartic acid in dental collagen. *Archaeometry* **32**: 61-70.

Hall H.R. & Woolley, C.L., 1927.
Ur excavations Volume 1, Al- 'Ubaid. London: Publications of the Joint Expedition of the British Museum and of the Museum of the University of Pennsylvania to Mesopotamia.

Harvey P.H., & Bennett P.M., 1985.
Sexual dimorphism and reproductive strategies. In Ghesquiere, J, Martin, RD, Newcombe, F (eds). *Human Sexual Dimorphism*: 43-59. London: Taylor and Francis.

Inoue N. (ed.), 1993.
Culture of Food and Oral Health in Maori. Tokyo: Therapeia Publishing Co.

Kapandji I.A., 1988.
The Physiology of the Joints. Vol.3 The Trunk and Vertebral Column. London: Churchill Livingstone, 251pp.

Larson C.S., 1985.
Dental Modifications and Tool Use in the Western Great Basin. *American Journal of Physical Anthropology* **67**: 393-402.

Lees B., Molleson T.I., Arnett T.R., & Stevenson J.C.,1993.
Differences in proximal femur bone density over two centuries. *The Lancet* **341**: 673-675.

Molleson T., 1994a.
The eloquent bones of Abu Hureyra. *Scientific American* **271**: 70-75.

Molleson T., 1994b.
Can the degree of sexual dimorphism provide an insight into the position of women in past populations? *Actes des 6e Journées Anthropolgiques, Dossier de Documentation Archéologique* no.17: 51-67. Paris: CNRS Editions.

Molleson T., 1996.
Skeletal Evidence for Identity and Role in the Neolithic. In L'identité des Populations Archéologiques, *XVI Rencontres Internationales d'Archéologie et d'Histoire d'Antibes*: 345-350. Sophia Antipolis.

Molleson T., Cox M., Whittaker D.K., & Waldron A.H., 1993.
The People of Spitalfields. York: Council for British Archaeology.

Molleson T.I., & Hodgson D. in press.
The Human Remains from Woolley's Excavations at Ur. *Iraq*.

Richards M.P., Pearson J.A., Molleson T.I., Russell N., & Martin L., 2002.
Stable isotope evidence of diet at Çatalhöyük, Turkey. *Journal of Archaeological Science*.

Robertson J.D.M. (ed.), 1991
An Orkney Anthology. The Selected Works of Ernest Walker Marwick. Edinburgh, Scottish Academic Press, 491pp.

Stearn W.T. 1981.
The Natural History Museum at South Kensington. A History of the British Museum (Natural History) 1753-1980. London: Heinemann.

Stevenson J.C., Lees B., Devenport M., Cust M.P., & Ganger K.F., 1989.
Determination of bone density in normal women: risk factors for future osteoporosis? *British Medical Journal* **298**: 924-928.

Tubb J.N. 2000.
Two examples of disability in the Levant. In Hubert J. (ed.) *Madness, Disability and Anthropology of 'Difference'*: 81-86. London: Routledge.

Willner L.D., & Martin R.D., 1985.
Some basic principles of mammalian dimorphism. In Ghesquiere J. Martin R.D. Newcombe F. (eds). *Human Sexual Dimorphism*: 1-42. London: Taylor and Francis.

Woolley L., & Moorey P.R.S., 1982.
Ur of the Chaldees. London: The Herbert Press.

In search of the ancient Peruvians: The Pacasmayo Museum Project

Andrew J. Nelson, Department of Anthropology, University of Western Ontario, Canada
Christine S. Nelson, Planning and Development, City of London, Ontario, Canada

I. Introduction

Research collections serve a wide variety of important roles in the academic and public spheres. This is particularly true in the field of bioarchaeology (archaeology + biological anthropology), where much of our work involves the analysis and reanalysis of material housed in research collections. We live at a critical juncture in the development of our discipline – new technologies are enabling us to ask new questions and to gain a much richer understanding of the people of the past, at the same time as laws are being promulgated which mandate the reburial of the physical remains of those same people of the past. Thus, one perspective sees bioarchaeological research collections as bastions of rational inquiry, where researchers using methods and theory from the social and natural sciences explore such issues as: how do biology and culture interact?; what is the history of human adaptation?; how do people change over time? These are questions fundamental to our understanding of what it means to be human. At the same time, the opposite perspective sees these same collections as bastions of colonialism, where researchers inflict their paradigm of scientism on the aboriginal communities of the world. The latter perspective is not held without reason, as physical anthropologists of the late 1800s and early 1900s published work focused on the identification of whether individuals were "uncivilized", "inferior", or "backwards" on the basis of their skeletal remains (see Gould 1981). It is probably safe to say that all the contributors to this workshop and proceedings volume would fall in the former camp. We approach our work with a passion and empathy that is at its essence humanist, and we work within a paradigm that emphasizes the importance of both synchronic and diachronic cross-cultural studies that is far removed from the caricature of the unsympathetic colonialist of the last century. Furthermore, we see the enormous value of these research collections, and appreciate the efforts of those who established them and who struggle today to maintain them.

The objective of this paper is to explore the question: *how does one go about establishing a bioarchaeological research collection today*? The two authors have been working for the past three years to attempt to do just that: to establish a cultural history museum and a bioarchaeological research institute in the town of Pacasmayo on the North Coast of Peru. This is the *Pacasmayo Museum Project*. In the process, two "dirt archaeologists" have been transformed into international development workers necessitating the examination of such basic questions as: what is the mission of a bioarchaeological research collection?; how does one organize such an institution "from scratch"?; and, perhaps most fundamental of all, how does one find the financial resources to start and then sustain such a project? It is the consideration of these questions that leads us to examine **Research and the Public Echo**.

II. Existing Bioarchaeological Research Collections – The Context of the Discipline

Museums and their study collections originated as material manifestations of the fascination with the ancient world that characterized the Neoclassicist and Romanticist ages of the late 18th and 19th centuries. Antiquarians traveled the world gathering animals, artifacts and human remains (often mummies), which were displayed at home to demonstrate the collectors' worldliness and sophistication. However, by the late 1700s, learned societies were starting to arise and by the time of publication of Darwin's "*Origin of Species*" in 1859 and Lyell's "*The Geological Evidences for the Antiquity of Man*" in 1863, research into the history of humankind was recognized as a legitimate area of scientific inquiry. Human skeletal remains, such as the Upper Paleolithic skeleton known as the "Red Lady of Paviland" and the Gibraltar Neandertal, played prominent roles in the development of this research.

It was in this intellectual context that the British Museum was founded (in the late 1750s) as the repository of the collection of the English physician Sir Hans Sloane. The natural history collections were transferred to the South Kensington site in the 1880s to found the Natural History Museum. These collections were greatly expanded under the direction of Robert Owen as he sought to amass collections that were "worthy of this great Empire" (NHM 2002). In 1889, The Anthropological State Collection (*Anthropologische Staatssammlung*) was formed in Munich to function as an archive and research center for human remains recovered from archaeological contexts in the state of Bavaria and from other countries where Bavarian scholars worked (e.g. Tanzania, Peru) (SAPM 2002). The Anthropology department at the US National Museum of Natural History (*The Smithsonian*) was established in 1897, with Physical Anthropology added in 1904. Its mandate was "to procure collections... illustrating the natural history of the country and more especially the physical history, manners and customs of the various tribes of aborigines of the North American continent", which has been "broadened to include a global study of all aspects of human beings, from the earliest origins of the species up to the modern day" (SI 2002). In our own country of Canada, The Canadian Museum of Civilization's Anthropology division was established in 1910 in order to "to conduct research on the archaeology of the Native peoples or First Nations of Canada – from the earliest beginnings to the period of European settlement." (CMC 1996).

These institutions carefully curate invaluable collections of skeletal remains that span hundreds of thousands of years and touch on all corners of the world. They include many thousands of human remains, allowing populationally oriented research that simply cannot be done on isolated skeletons or on small samples. The specific focus of these collections varies from state to national to global. The institutions which house the collections have employed and welcomed thousands of researchers over the past 100+ years, leading to substantial contributions to our understanding of human-kind.

These institutions were all founded following state or national acts of government and have historically received the majority of their funding from government sources. However, as times have changed from the intellectual climate of the late 1800s to today, so too have government attitudes about funding. More and more, museums must cultivate "strategic partnerships" with the private sector, rely on grass-roots level fund raising campaigns, and turn to "block buster" exhibitions to bring in funds from paying patrons. This shift is starkly illustrated by the Canadian Museum of Civilization's web site, where the museum is described as a *corporation* (the CMC was established as a Canadian federal Crown Corporation in 1990), and "*research*" (including curatorial activities) is listed as a "*business activity*" (CMC 1996).

This shift in how such institutions are perceived has placed the institutions themselves and their invaluable collections in great danger. It is becoming increasingly difficult to obtain the funding necessary to expand storage space, to upgrade curatorial facilities, and to provide the environment necessary to support ongoing research. Furthermore, the "reburial climate", typified by the North American Graves Repatriation Act is stripping many of these institutions of their collections.

It is a terrible irony that at the same time that the collections are in danger, we are experiencing great advances in analytical methods that allow us to recover more information from these remains, to ask more sophisticated questions and have richer interpretations. The recovery and analysis of ancient DNA (see Fox, this volume), the analysis of trace elemental and isotopic composition of bones and teeth (see Burton, this volume) and the analysis of bones and teeth at the microstructural level (see Hillson, this volume) are all examples of cutting edge technical analyses that have only reached their full fruition in the last decade. At the same time, advances in bioarchaeological theory have led to studies of populations from a much more dynamic point of view than that of the typological "racial studies" of the late 1800s and early 1900s. Processual archaeology has provided the theoretical context for many studies of human skeletal remains, emphasizing the broad scale environmental context of human adaptation. In many cases, these studies require large numbers of individual "cases" in order to detect meaningful patterns over time (e.g. Steckle et al. 2002) – something not possible without large curated collections. Post-processual theory also has great promise for bioarchaeology, through an increased emphasis on the individual (emerging from the "osteobiography"). From this perspective, the individual is not merely a passive actor subsumed by culture, but an active agent, involved in the creation and modification of culture. This point of view is just beginning to be felt in our field (e.g. Spence 1998, Nelson et al. 2000b, Dolphin 2002). These new methods and theoretical perspectives allow us to collect information such as how ancient populations related to each other, what they ate and how diseases affected their skeletons, and to place these elements in an interpretive context that yields a rich tapestry that tells us who these people were, how they lived, and how they died.

It is a belief in the importance of the contributions that bioarchaeology can make, these advances in our field, and the promise that bioarchaeology holds for the future that drives the project described in this paper. However, in order to make it a success we must understand the history of our discipline and acknowledge the fact that the project must unfold within the current academic, political and economic context. The world of the early 21st century is a very different place than that of the 18th!

III.

III.a Horizons, Deserts and Graves – The Geographic Context

This project seeks to establish a cultural history museum and bioarchaeological research center in Peru. In order to understand and ultimately to present the prehistory and history of Peru one must appreciate the complexity of its geography, and the role that geography has played in cultural development (Burger 1992, von Ha-

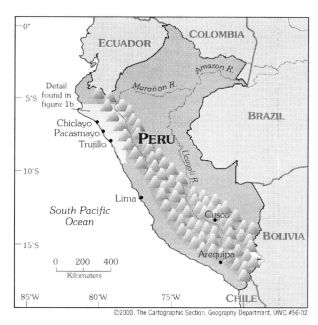

Fig. 1a: General view of Peru, showing major cities and landmarks.

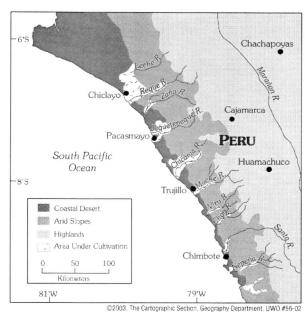

Fig. 1b: Detail view of the north coast of Peru. Note the irrigated areas around the rivers and the narrow arid littoral desert along the coast.

gen & Morris 1998). Peru's cultural geography is dominated by two major zones: the highlands and the coast. The "highlands" refers to the Andes Mountains and includes mountains, high plateaus and highland basins. The plateaus and basins of the highlands are separated by ridges, making communication between them difficult. In prehistoric time foot traffic was readily able to move north-south. Now, vehicular traffic must descend to the coast, go north or south there, then ascend again into the highlands. The "coast" is a narrow strip of relatively flat land that lies between the base of the Andes and the Pacific Ocean. This strip varies between 2 and 20 kilometers wide and is largely desert, except where it is punctuated by rivers that emerge from the Andes (see Figure 1). Civilization, both ancient and modern, clusters around these rivers and spreads as far as irrigation allows crops to grow.

This peculiar geography has shaped the development of Peruvian cultures. At times, individual cultures existed in individual basins, or individual river valleys and had little contact beyond that. Thus, they developed unique local flavors. At other times, particularly strong cultural influences were felt throughout the highlands and the coast, leading to the elaboration of common cultural themes all over the region. Peruvian prehistory is generally organized into a series of "Horizons" and "Intermediate Periods". Horizons are periods where influences are widespread, as in the Inca Period, where a single empire controlled almost a million square kilometers. Intermediate Periods are the times when wide spread influences broke down, and local cultures fluoresced (Rowe 1962).

Geography and subsistence are also inextricably intertwined. Cultures which lived along the coast would have had access to a wide variety of marine resources. Fisheries along the Peruvian coast are extremely rich, due to the upwelling of the nutrient rich Humboldt Current. Irrigated portions of the coast were (and are) productive agricultural zones, yielding maize, beans, squash and so on. The highland environments supported herds of camelids, such as llamas and alpacas, and tuber crops, such as the potato (von Hagen & Morris 1998).

Finally, geography and climate have clearly had profound negative effects on culture. The El Niño phenomenon is named for the rains which strike the Peruvian coast when the Humboldt Current is forced out of its regular flow. There is evidence for huge El Niño events which ravaged ancient settlements, irrigation systems and fields. The severity of the impact on the cultures is well illustrated by large scale human sacrifices at the Temple of the Moon (Huaca de la Luna) in modern Trujillo, where large numbers of victims were sacrificed and thrown onto the ground, made muddy by torrential rains (HDL 2002). Earthquakes, the product of the Nazca plate sliding under the South American continental plate, may also have had devastating effects on culture. Uplift accompanying a large quake in the 6th century may have made irrigation systems unworkable, devastating the local agricultural economy (Burger 1992, Moseley 1992, von Hagen & Morris 1998).

III.b Horizons, Deserts and Graves – The Archaeological Context

The initial occupation of Peru took place some 12,000 years ago. There is a great deal of debate regarding the first occupation of the Americas, but people would have been in Peru by at least 10,000BP, as they would have had to pass through Peru to get to the Chilean site of Monte Verde (dated to approximately 13,000BP). The earliest sites are located in the highlands, and include Lauricocha and Pikimachay Cave (Moseley 1992). Slightly later sites have been found along the north coast in an area that is now a desert, called the Pampa de Paijan.

The first period when cultural influences were first felt over a large area of Peru is called the Early Horizon and is characterized by the Chavin cult. The Early Horizon was followed by the First Intermediate Period, which saw the evolution of regional polities, including the well known Moche culture in the north and the Nazca culture in the south. The Middle Horizon Period was characterized by the wide-spread influence of the Huari culture, which originated in the southern highlands. The Late Intermediate Period saw the development of the Chimu empire in the north and Ica in the south. The Late Horizon was the time of the Inca, who controlled all of modern Peru, and parts of Ecuador, Bolivia, Argentina and Chile (Burger 1992, Moseley 1992) (see Figure 2).

IV.

IV.a The Jequetepeque Valley – The Local Context

The Jequetepeque (he-ke-te-PE-ke) Valley lies on the North Coast of Peru, some 620 km north of Lima. The valley occupies a strategic location, with rich fishing grounds off its Pacific coast to the west, the area of Cajamarca (which was a major center during Inca times and contains rich mineral deposits) in the highlands to the east and the important Lambayeque and Moche Valleys to the North and South respectively (see Figure 1b).

The Jequetepeque Valley is best known for the discovery of the tomb of a late Moche (ca. 750AD) "priestess" at the site of San Jose de Moro (Donnan & Castillo 1992). The importance of this discovery, along with another at the site of Sipan, is that "characters" who played roles in thematic presentations in Moche art were identified as specific individuals on the basis of paraphernalia found in the tombs. This priestess was "Figure C", an individual who played an important role in the "presentation theme", by passing a goblet filled with the blood of sacrificial victims (Donnan 1975, Donnan & McClelland 1999). This figure, now known as *La Sacerdotista* has become an icon for the community of Chepen. A large reconstruction of *La Sacerdotista* now stands by the Pan American Highway as it passes Chepen (see Figure 4).

Most recently, a site from this valley, Dos Cabezas, was featured in National Geographic magazine (Donnan 2001). Dos Cabezas is also a Moche site, but from the early part of that time period (ca. 100-450AD). The site is a large adobe huaca (pyramid), including associated architecture and many burials. Of particular interest is a series of 5 tombs, that contained the remains of young males between the ages of 15 and early 20s. These young men were buried with considerable wealth, including elaborate ceramics and gold and copper objects.

The Jequetepeque Valley has a long and spectacular archaeological history. Its archaeological record contains material relevant to major cultural research questions such as the origins of agriculture, the origins of state, and the rise and fall of several major civilizations. Furthermore, the interactions between culture and the environment are well documented here, including such events as earthquakes and floods. The valley's strategic location, and circumscribed geography make it an ideal laboratory for archaeologists, geographers and other scientists to examine these key issues of cultural evolution.

IV.b – The Jequetepeque Valley – Bioarchaeological Research

The same qualities that make the Jequetepeque Valley an ideal laboratory for archaeologists and geographers make it an ideal setting for bioarchaeological research. Many of the sites in the valley include substantial collections of skeletal remains, ranging from the earliest occupation at Paijan (ca. 8000BC) to the time of the Inca. The first large sample of burials comes from the Initial Period and Early Horizon site of Puemape. The Early Intermediate Period is well represented in the valley with good samples of Early Moche (100 to 450AD) burials coming from the site of Dos Cabezas and Middle and Late Moche (450 to 800AD) burials coming from the sites of Pacatnamu and San Jose de Moro. Transitional Period (800 to 950AD) and Lambayeque (950 to 1100AD) burials are known from Pacatnamu and San Jose de Moro. Late Intermediate Period burials are documented from Pacatnamu. Finally, recent work at the site of Farfan (Mackey 2001, in press) has produced an excellent sample of Late Horizon (1470 to 1532AD) material. In all, there are approximately 1000 skeletons from the Jequetepeque Valley.

The Paijan sample is quite small (Lacombe 2000). However, the Initial Period and Early Horizon material from Puemape is much more plentiful. This site has burials from three time periods, 1,200 to 900BC, 900 to 500BC

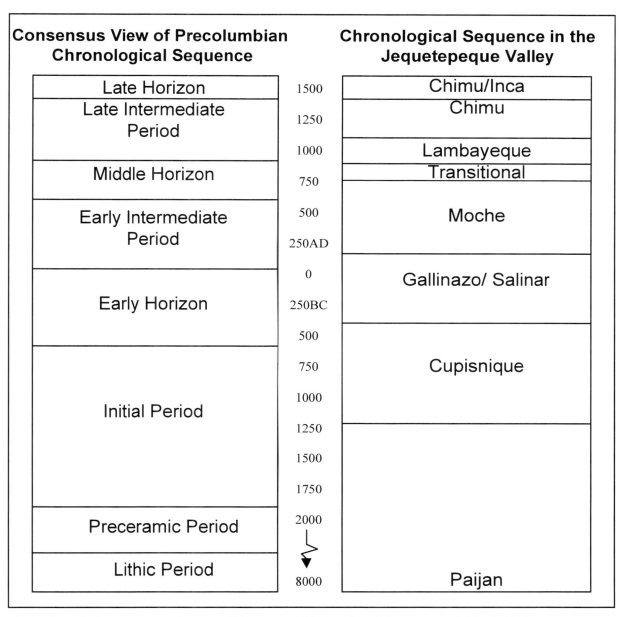

Fig. 2: Chronological sequences (after Rowe 1962, Burger 1992, Moseley 1992, von Hagen & Morris 1998)

and 400 to 100BC (Gillespie 1998, Elera 1998). Excavations yielded 24, 41 and 53 individuals from each respective time period.

There is more extensive material from the Moche and later cultures. This includes a collection that was the focus of the first systematic osteological work to be undertaken on skeletal material from this valley. H. Ubbelohde-Doering (1983) worked at the site of Pacatnamu in 2 campaigns, in the 1930s & 1950s, and G. & W. Hecker excavated there in the 1960s (Hecker & Hecker, 1985). The human skeletal material from the 1930s campaign, including material from the Moche, Lambayeque and Chimu cultures, was collected and shipped back to Munich, Germany (unfortunately, the material from the 1950s and 1960s was not kept). Many of the artifacts from Ubbelohde-Doering excavations are housed in the *Staatliches Museum für Völkerkunde*, while the skeletal remains are housed in the *Staatssammlung für Anthropologie und Paläoanatomie*, both in Munich.

The skeletal remains from Doering's excavations at Pacatnamu have formed the basis of three diploma theses (Burger 1978, Heine 1979, Zink 1993) and many research projects (e.g. Parsche et al. 1994, Wiechmann & Grupe 1997, Grupe & Turban-Just 1998, Wiechmann et al. 1999). This research has demonstrated the presence of distinctive regional patterns of cranial deformation (Burger 1978, Heine 1979), highlighted several significant pathological conditions (including *spina bifida*, immobilization osteoporosis trauma and

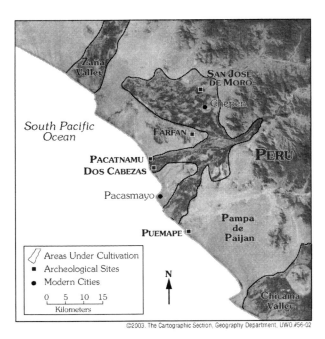

Fig. 3: The Jequetepeque Valley showing the major archaeological sites.

Fig. 4: *La Sacerdotista*.
This reconstruction of the priestess figure found in Moche art and in a tomb at San Jose de Moro stands beside the Pan American highway in Chepen (photo A. Nelson).

possibly treponemal disease [Zink 1993]), and explored the excellent preservation of organic matter in the bone matrix (Grupe & Turban-Just 1998, Wiechmann et al. 1999).

The site of Pacatnamu was excavated again from 1983 to 1987 by *Proyecto Pacatnamu*, directed by C. Donnan and G. Cock, of the University of California, Los Angeles (Donnan & Cock 1986, 1997). They determined that this large urban center was first occupied in Moche times (phase V, approximately 600AD), but the majority of architecture visible at the site dates to Lambayeque times (approximately 1100-1150AD – note: phase terminology used here follows Donnan & Cock 1997). The skeletal material recovered from this campaign formed the basis for a PhD dissertation (Verano 1987), and then provided the material for several subsequent publications (Mann & Verano 1990, Verano 1992, 1994, 1995, 1997, 1998, Verano & DeNiro 1993). There have been two major themes to Verano's research. The first is the differentiation of individual populations within the site, leading to the hypothesis of kin group specific burial organization. The second is the detailed characterization of the Moche population from Pacatnamu. These individuals demonstrate little evidence of nutritional stress, but considerable evidence for an active lifestyle, particularly for the males.

Our work in the Jequetepeque Valley began in 1995, when we joined with L.J. Castillo (Pontifica Universidad Catolica del Peru) and C. Mackey (California State University Northridge) to excavate at the site of San Jose de Moro. San Jose de Moro was mentioned earlier as the site where the Moche priestess was found in the early 1990s. We excavated in the cemetery component at Moro, which covers much the same time period as Pacatnamu, with the earliest dates coming in at 450AD, and the latest around 950AD. There is also a large Chimu administrative compound at the site that dates to a later Chimu occupation (ca. 1200AD) (Nelson et al. 1996a, b).

Over 3 consecutive archaeological seasons (1995 to 1997) we excavated 52 burials, which were analyzed over the subsequent 2 years. Results of this work have been presented in some 15 conference papers, 5 journal articles and 1 masters thesis (Castillo et al. 1997, Conlogue & Nelson 1998, 1999, Lichtenfeld 2001, Martin et al. 2002, all Nelson references). Our results echo those from both sets of Pacatnamu material in the documentation of distinctive forms of cranial modification for the Moche and Lambayeque periods, but have added additional information regarding shifting patterns of health (particularly dental health) and diet (as documented by isotopic analysis) between these two time periods. Preliminary analysis of ancient DNA recovered from the San Jose de Moro material (Nelson et al. 2000c) suggests that there was biological continuity between these two periods, despite differences in health and cultural practices.

Recently we have had the opportunity to visit the Munich collections to gather information from the Doering collection that is comparable to that gathered at San Jose de Moro. The excellent organic preservation noted by Grupe & Turban-Just (1998) and Wiechmann et al. (1999) led us to sample bone and hair from this collection for analysis of ancient DNA and trace elements. Analysis of these data is still underway, although preliminary chemical analysis of the hair has been presented in 2 conference papers (Nelson et al. 2002a and Martin et al. 2002). In this work we are exploring the effects of diagenesis on the original biogenic signal.

Important bioarchaeological work has been done at the early Moche (ca. 450AD) site of Dos Cabezas (mentioned above) by C.B. Donnan and A. Cordy-Collins (Donnan 2001). Cordy-Collins has been the osteologist on this project, but there has been a wide series of consultations, recently presented as a session at the European Paleopathology Association meetings (Cordy-Collins 2002a, b, Geyer et al. 2002, Heflin 2002, Lombardi 2002, Nelson & Nelson 2002b, Nelson et al. 2002b, Ryser 2002, Tyson 2002). Dos Cabezas has yielded a series of early Moche individuals, but of particular importance is a sample of 5 high status male individuals. These individuals all demonstrate varying degrees of a skeletal syndrome that led to increased stature and skeleton-wide osteoporosis and other abnormalities. The occurrence of 5 individuals sharing similar symptoms suggests a genetic disorder shared among closely related individuals, which could have important implications for our understanding of elite Moche mating practices. However, the combination of increased stature with the other abnormalities has, as yet, defied definitive diagnosis (Cordy-Collins et al. 2001).

In the last few years we have been collaborating with C. Mackey at the Chimu/Inca (ca 1470-1532AD) site of Farfan. This site is extremely important, as it brings the Jeqetepeque skeletal population up to the time of the Inca. Analysis of this material is still ongoing, but one finding of particular note is the presence of an individual who apparently suffered from the same malady as the individuals at Dos Cabezas. This finding is extremely valuable, as it suggests that the Dos Cabezas cases were not isolated, and either the conditions causing the disease or the disease itself were stable over the course of 1000 years (Nelson & Nelson 2002b).

In sum, there is a tremendously rich archaeological and osteological record in the Jequetepeque Valley. Osteological collections span approximately 9,500 years and have a particularly detailed coverage of the Moche, Lambayeque and Chimu/Inca periods. While many individual studies have been done on this material, much of it is unpublished and there has yet to be a major synthesis. The major reason for a lack of synthesis is the difficulty involved with getting access to all the collections. Ubbelodhde-Doering's material is housed in Munich (it is possible that some of this collection remained in Lima), and the other collections are distributed among the cities of Chepen, Jequetepeque, Pacasmayo Trujillo and Lima (all in Peru).

V. Train Stations, Research Collections and the Interested Public: The Pacasmayo Museum Project

Over the past four years, we have been attempting to examine as much of the osteological material from the Jequetepeque Valley as possible, in an effort to work toward a synthesis, but the expense and logistical difficulties have been considerable. Thus, in 1999, we initiated discussions with archaeological colleagues and members of the local community with the objective of bringing the collections together into a single locale. That is, we are seeking to create a regional osteological research collection.

Our discussions began with a colleague, CB Donnan, who recommended that we talk to a prominent member of the Pacasmayo community, LF Arbaiza. Sr. Arbaiza had been exploring the possibility of starting a community museum as a project by the *Casa de la Cultura* (House of Culture). The *Casa de la Cultura* is a local nonprofit association established by the late, highly respected former mayor of Pacasmayo, Carlos Arbaiza Strohmeier, to nurture community cultural and historic projects. We met with Sr. Arbaiza, and it became immediately clear that we shared objectives that stood the best chance of success if we moved forward together. Thus, in the intervening years we have worked to form an entity that has two components: 1) a museum of the cultural history of the Jequetepeque Valley, and 2) a center for bioarchaeological research. The mission of the museum is to celebrate the history and prehistory of the Jequetepeque Valley by reaching out to audiences at the local, national and international levels. The mission of the research center is to collect and curate the osteological collections from the Jequetepeque Valley in order to permit populational level, multidisciplinary bioarchaeological research on this remarkable corpus of osteological material. The combined entity will operate as part of the *Casa de la Cultura* and will be housed in the historic 1870s Pacasmayo train station building (Figure 5).

V.a Reality Sets In

The project concept quickly gained support, both at the local and international levels. We started to rally support from our archaeological colleagues working in the area and from colleagues who could be involved in important related fields such as conservation (J. Figari, Insti-

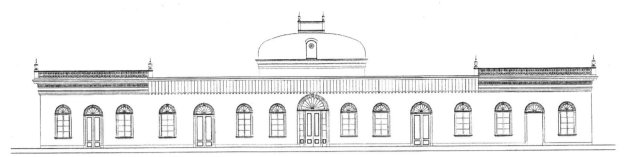

Fig. 5: Blueprint elevation of the platform view of the train station. The station was built in the 1870s from pine wood imported from Oregon.

tuto Superior Yachay Wasi, Lima) and radiography (G. Conlogue, Quinnipiac University, USA). The University of Western Ontario's Office of International Research (F. Keenan director, S. James project officer) got involved in an effort to identify possible funding sources. Between the two authors we had experience with museum exhibit design, heritage conservation projects, and fundraising as well as the key areas of archaeology and human osteology. Thus, we were very optimistic that we could bring the project to fruition in a short period of time.

The reality was that we still had a great deal to learn. The world of museums is a very different place now than it was in the late 1800s and early 1900s when countries viewed museums as items that are high on the national funding priority list. Today, while Peru clearly wants to be able to fund more museums, it is a country that must use its funds in other ways. Peru currently holds an estimated $33.1 billion debt (2001 estimate), and 55% of Peruvians live in poverty, 24% in extreme poverty (CIA 2002). Thus, Peru's immediate priority is the alleviation of poverty. The strategy of the Toledo government is to combat poverty by means of the creation of jobs, as outlined in the program *a trabajar* (to work) (PICIC 2001). The need to address issues of poverty and the need for fundamental infrastructure also shapes the priorities of international development agencies such as the World Bank, USAID and the Canadian International Development Agency.

V.b The Rules of the Game

In addition to funding issues, while we had a great concept and a building that would provide an excellent venue for the museum/research center, there were important design and bureaucratic issues that were yet to be discovered. In particular, the bureaucratic details of how we must proceed with this project have taken some time and effort to determine. However, it has always been clear that the government representative in this project must be the *Instituto Nacional de Cultura* (National Institute of Culture). The INC oversees museums, archaeology and most cultural organizations in Peru. The INC is highly centralized, with its main office in Lima. The provincial office of the INC that is responsible for archaeology in the Pacasmayo area is in Trujillo, a city 100km south of Pacasmayo. The head of the INC Trujillo office, Sra. Ana Maria Hoyle, has been supportive of this project.

The most critical step in the process of moving this project forward was the signing (in July 2002) of a *convenio* (agreement) with Sra. Lenor Cisneros, the then national director of the INC in Lima. The *convenio* links the INC, the *Casa de la Cultura* (represented by Sr. Arbaiza) and the University of Western Ontario (represented by us), formalizing the concept of the project. This document outlines the responsibilities of the three parties, and gives official national legal recognition to the project and the parties involved. The road to this agreement was paved by numerous consultations with colleagues, members of the INC and by the signing of a *carta de intencion* (letter of intent) in 2001.

V.c Time to Pay the Piper – Funding a Museum Project

The preparation of a project such as this must include the preparation of budgets. While all archaeologists are familiar with preparing budgets for granting organizations, institutional operating budgets and business plans are unfamiliar territory. With the assistance of architect and museography colleagues, and particularly with that of F. Keenan (UWO, Office of International Research) we have prepared a restoration budget of $180,000 (all figures in US$), a museography budget of $220,000 and we estimate a balanced operating budget of some $60,000 per annum. It is only this year that these budget figures have crystallized, but we have been making efforts to identify potential funding sources since the earliest days of this project. The reality of the time in which we live is that it is extremely difficult, if not impossible, to find a funding agency that will simply fund a "museum". Rather, we must seek funds from a variety of sources: from research councils, from international development sources and from private corporations. It is clear that in order to be attractive to these various funders, this project must be much more than simply a repository for bones. Thus, in order to be successful, we must understand the nature of these different funding sources, in order to see how we fit their priorities and criteria.

V.c.i Research Councils

Our work in Peru has been largely funded from two sources. The first is a private foundation called the GL Bruno Foundation. The other is the Canadian federal granting agency called the Social Sciences and Humanities Research Council (SSHRC). We received funding for our initial work in Munich from the Vice President Research at the University of Western Ontario, B. Bridger. We have also been fortunate to receive material support from Victorinox, Kodak Health Sciences (Canada), Novacks of London (Ontario), and Aeroperu. While granting agencies fund ongoing research projects, their mission does not include an investment in restoration of buildings nor in museographic installations in foreign countries. Once the facility is established we will apply to SSHRC and other organizations such as the Wenner-Gren Foundation for funds to support bioarchaeological work, and those funds will be used to help defray operating costs. However, the problem is to get the facility established. We initially had high hopes for a recently established Canadian federal research infrastructure granting program (Canadian Foundation for Innovation or CFI), but this program would not fund building renovation and at that time required that all equipment be based in Canada. Thus, we have had to look for other funding opportunities.

V.c.ii International Tourist Development

Very early on, we decided that a sensible approach would be to pursue funds from international development agencies, packaging the project as a local development engine which would attract new tourist dollars into an underdeveloped area. This is consistent with the International Council of Museum's 1968 Munich resolutions, where international organizations including UNESCO, governments and museum authorities were "urgently requested" to "recognize museums as major institutions in the service of development" and to "include the creation of the development of museums in their long- or short-term plans for national development" (ICOM 1968).

With ICOM's words giving us encouragement, and with considerable support from UWO's Office of International Research we approached several international development organizations, including the Canadian International Development Agency (CIDA), the International Development Research Corporation (IDRC), the World Bank, the Inter-American Development Bank, the *Fundo Contravalor* Peru-Canada, the British International Development Agency (DFID) and the Spanish International Cooperation Agency in Lima. In Peru, tourism is the 3rd largest export, falling only after mining and fishing as a contributor to the national economy (IDRC 1999). Thus, it is clear that tourism is responsible for bringing an enormous amount of foreign currency into the country – and if nationally – why not to the north coast?

Unfortunately, we found that most international development agencies do not see tourism as a viable means of fostering sustainable economic development. This is because tourism development projects have had a decidedly mixed record in economic terms, with more failures than successes in countries as diverse as Costa Rica and Bali. Furthermore, a panoply of studies has lined up to decry the negative effects of tourism on local communities.

"Tourism" can be defined in a variety of ways. *Mass tourism* is generally thought of as people in search of the 3-s – "sun, sea and sand". Mass tourism is often characterized as wreaking havoc in host communities by (among other things) shifting patterns of labor, increasing economic stratification and freezing or irrevocably changing cultural ceremonies (e.g. Schlüter 1994, Stronza 2001). Projects that have fostered mass tourism without careful planning are the ones responsible for the attitude of the international development agencies.

The "havoc" wrought by tourism can be attributed to a variety of factors. Perhaps foremost is the fact that many major tourism ventures are foreign owned, so that profits "leak away" from the local community (Graburn 1989). Ethnic and cultural tourism, where the guests may expect the hosts to behave in what they (the guests) perceive as "authentic" ways, lead to great tensions between the guest and host communities, as the hosts are expected to become frozen in time (Graburn 1989, Stronza 2001). Management of tourism ventures has often been problematic, with either foreign "experts" being brought in with no real knowledge of the local context, or local leaders acting in their own political self interest (Schlüter 1994). Finally, there is typically a large asymmetry in power brought about by the economic power of the guests, which brings about tensions between the two groups and within the host group, as subgroups strive to achieve their own economic power (Mowforth & Hunt 1998).

As opposed to mass tourism, *alternative tourism* (or "new tourism" after Mowforth & Hunt 1998), is defined as "forms of tourism that are consistent with natural, social and community values and which allow both hosts and guests to enjoy positive and worthwhile interaction and shared experiences" (Eadington & Smith 1992: 3). The objective of this type of tourism is to avoid the negative consequences of mass tourism. Within alternative tourism, a variety of subtypes can be defined, including ethnic tourism, cultural tourism, heritage (or historic) tourism, and environmental tourism (Graburn 1989). Ecotourism combines environmental tourism with any of the other three subtypes (Scheyvens 1999).

Tourist development in Latin America dates back to the 1930s, when the Argentinean government tried to use resort development as a means of exerting political control over the contentious area of San Carlos de Bariloche (Schlüter 1994). In the 1950s the 3-s tourist resorts took hold in Mexico and the Caribbean. Peru developed a plan in collaboration with UNESCO in the 1960s that led to the development of the Machu Picchu-Cuzco area, basing the country's tourism strategy on heritage tourism (Schlüter 1994). Recently a major tourism development plan for Peru was developed by the Japanese International Cooperation Agency (JICA) and the Peruvian government to invest $6.2 billion over 17 years. That plan sought to contribute to economic development as a competitive "export-oriented" industry for Peru, to social development by creating employment and providing local opportunities for small businesses contribution, and to the conservation of natural and cultural heritage using tourism as an economic incentive and catalyst for awareness (Pacific Consultants International 1999). Unfortunately, in the past few years the political climate in Peru has not favored Japanese government projects, so the JICA plan has not been put into action.

A new alternative tourism development initiative in Peru is "The Northern Circuit Economic Development, Poverty Reduction & Tourism Program" (Nizette & Goodwin 2002), which seeks to develop a tourist circuit in Northern Peru as an alternative to the well known route from Lima-Cuzco-Machu Picchu. Their proposed circuit runs from Trujillo north up the Panamerican highway, through Pacasmayo to Chiclayo, then east to the cloud forest and the Chachapoyas area, south to the highland city of Cajamarca, then back to the coast at Pacasmayo (see Figure 6). This project is based on "pro-poor" tourism, which is defined as tourism that is sustainable from the social, environmental and economic points of view, with the specific goal of creating benefits for poor communities (WTO 2002). This kind of development program seeks to avoid the shortcomings of previous tourist development projects by explicitly addressing the issues of economic leakage and sustainability, by emphasizing local linkages and the engagement of multiple stakeholders, and by diversifying the "product", both geographically and in terms of what is actually offered. The emphasis is to provide the poor with access to the tourist market, to encourage the creation of small and medium enterprises and to provide employment and training (WTO 2002).

The NTC is firmly based on pro-poor principles, first by seeking to diversify the tourist landscape in Peru (which still emphasizes the Lima-Cuzco-Machu Picchu axis developed in the 1960s), and second, by seeking to create sustainable development in Northern Peru (Nizette & Goodwin 2002). The project includes 134 microprojects, ranging from development of food provision capacity at roadside stops, to archaeological conservation, to development of local craft production, to museums. The goals of the NTC are to "reduce poverty, counter rural depopulation and urban drift, provide increased employment and supplementary incomes and generate resources for the maintenance of natural and cultural heritage" (Nizette & Goodwin 2002: 6). Sustainability is to be ensured by emphasizing local community participation, by creating inter-sectoral linkages, by encouraging the conservation of natural and cultural heritage resources and by small and medium enterprise development, job creation and poverty reduction (Nizette & Goodwin 2002).

On July 11, 2002, A. Nelson and Sr. Arbaiza made a presentation on the Pacasmayo Museum project to the principals of the Northern Tourist Circuit. At that point we were invited to become part of the circuit project. The Pacasmayo Museum project is consistent with the goals of the NTC in the following ways:

- it is a locally driven enterprise, ultimately to be managed by the *Casa de la Cultura*, thus there will be no economic leakage
- it will help to conserve heritage resources (from the historic building to the archaeological material housed within). This conservation effort will include both direct curation of material and proactive education programs designed to reduce looting of archaeological sites
- it will create direct employment opportunities for staff to manage and maintain the facility, and indirect opportunities for docents, tour guides, and craft producers
- it will have several inter-sector linkages, including the national government (the INC), local government (civic and provincial), private associations (the Casa de la Cultura), the archaeological/conservation sector (Yachay Wasi, University of Western Ontario, Quinnipiac University among others), and (hopefully) the business sector (both locally and at the departmental level – see below).

At the time of writing, the NTC consortium has applied to the Italian international funding agency – *Fondo Italo Peruano* – to fund the first phase of the project, with a total request of $1.7million. This phase is to provide a "skeleton" for the circuit, establishing project governance, infrastructure and training. Funds allocated to the Pacasmayo Museum project are to cover the restoration of the building.

Northern Tourist Circuit

Fig. 6: The Northern Tourist Circuit (Nizette & Goodwin 2002).

V.c.iii Corporate Philanthropy

We have also worked hard to find corporate support for the project. In this endeavor we have been fortunate to be able to draw upon the support of the Peruvian Consul General based in Toronto, Sr. C. Gamarra, and the Canadian Ambassador in Peru (initially G. Clark, now H. Rousseau). We have focused on Canadian companies that are active in Peru, reasoning that a cultural project with both Peruvian and Canadian components would be beneficial from the perspective of the public perception of their company.

There are two particularly good models of successful corporate philanthropy at work in the archaeological world in Peru. The first is a collaboration between the Southern Peru Copper Corporation and an archaeological group referred to as *Programa Continsuyo* (Owen 2000). These two entities combined forces to establish *Asociación Contisuyo*, a non-profit organization formed to establish and manage an archaeological museum, *Museo Contisuyo*, in the town of Moquegua (see also http://members.aol.com/contisuyo/MuseoE.html). This successful collaboration even survived a corporate takeover of the Southern Peru Copper Corporation by *Grupo México*.

The second example is a collaborative effort to fund an archaeological and community development project, *Proyecto Arqueológico Cajamarquilla*, at the site of Cajamarquilla, in Lima (Segura et al. 2001). The corporate partner in this venture is a Canadian mining company called Teck Cominco, and the project is managed by a conservation institute in Lima, the *Instituto Superior Yachay Wasi*. Teck Cominco runs a zinc refinery near the site of Cajamarquilla and has supported this work since 1996 (Agg & Perrier 2002).

Why do corporations support these sorts of projects (and hopefully ours)? The primary responsibility of any business is to maximize the return on investment to its shareholders, so at first glance it would seem that "giving away money" would be a bad business practice. In light of this responsibility to the shareholders, purely altruistic corporate philanthropy is probably the exception rather than the rule. Thus, the key to a successful relationship between a nonprofit project and a corporation is to resolve how best to serve the interests of both (Levy 1999).

Philanthropic activity can be either *directed*, where a specific potential market segment is targeted for support, or *general*, where there is no clear potential for market benefit. This activity can either be *reactive*, where the strategy is in response to a particular (presumably negative) triggering event, or *proactive* where the philanthropy arises without a particular trigger (Ricks 2002). Decisions regarding the allocation of corporate resources are made in the context of the specific business interests of the firm, which must include external stakeholders. In this context, corporate philanthropy is sometimes referred to as "strategic giving" or "charitable investing" (Yankey 1996).

It is clear that reactive, directed corporate giving in an attempt to mitigate negative publicity is well within the interests of a business. It is less clear why companies should engage in general, proactive giving. However, a corporation can benefit through, among other things, ultimate financial gain (broadening and deepening customer base and increasing access to markets), increased image (strengthening of the corporate image or of a specific brand), facilitating or enhancing relationships with key public sector officials, and by improving employee morale (Lewin & Sabater 1996, Levy 1999, Ricks 2002). Lewin & Sabater (1996) showed that the relationships among business performance, employee morale and community involvement are systematically positively correlated and mutually reinforcing. The importance of corporate charity to the ultimate bottom line is well illustrated by the active and substantial philanthropic programs of such large companies as Dayton-Hudson, Levi-Strauss, the American International Group (AIG), IBM and AT&T (Levy 1999).

While corporations can derive great benefit from a program of philanthropy, any organization seeking funds must recognize that corporate sponsorship can be a double edged sword if it comes with "strings attached", or if the sponsor has a negative profile within the fundee's stakeholder community (e.g. Harrison 1988 and Trigger 1988). The relationship between the corporate donor and the nonprofit organization is a complex one. The onus is clearly on the nonprofit organization to do some careful homework to fully understand the goals of the target corporation and to clearly indicate how the particular project is consistent with these goals, and how this support can also advance the interests of the company. Careful research will increase the likelihood of receiving a positive hearing. In addition, any potential conflict will be reduced by careful choice of the corporate partner. Finally, any dealings with a corporation will require vigilance and hard bargaining (Levy 1999).

With all due consideration to the points outlined above, we have made approaches to several Canadian companies that are active in Peru. Our logic is that such a company would gain "increased image" in both Peru and Canada, due to the involvement of both Peruvian and Canadian partners in the project. In general, the companies we have made contacts with have been supportive of the project, but most have not felt that they had discretionary funds to spare, given the present uncertain economic climate in Peru. However, this past July, the Consul General of Peru, Sr. Gamarra, arranged a meeting for us and our Peruvian partner Sr. Arbaiza, with Sr. J. Quijandria, the Peruvian Minister of Energy and Mines. Sr. Quijandria indicated his enthusiastic support for the project and offered to intercede with Canadian mining companies on our behalf. Following his suggestion, we have now approached a Canadian mining company (that will remain unnamed) that is active in Northern Peru, in order to obtain support for the museographic component of this project.

A funding application prepared for a corporation is very different from that which one typically prepares for research councils. A corporation wants to see a business plan – an instrument with which few archaeologists have experience. Fortunately in this endeavor we have (again!) had the active involvement and support of F. Keenan, director of UWO's Office of International Research. A business plan must contain several key elements: an executive summary, a description of the business, a description of the target market, an analysis of the competition, a description of the management team and marketing strategy. The overall strategy of the applicant must be to present the project as being consistent with the corporation's goals, to outline how the corporation will benefit, and to demonstrate that the project is sustainable. In our case, responsible cultural resource management is consistent with archaeological mitigation that the mining company must undertake in its new area of operation. In the preparation of the business plan, we assembled research data to demonstrate that large numbers of tourists could reasonably be expected to visit the museum, which would both ensure the museum's financial health, and provide large exposure to the sponsor's name.

In addition to benefits likely to accrue from broad exposure of the corporate image, mining companies in Peru would be likely to benefit from supporting a cultural project such as a museum in other ways. In the course of their archaeological mitigation, the companies must work closely with the *Instituto Nacional de Cultura*. The INC will be an important component of our project, as they ultimately own the historic building, and they will be involved in decisions about design and displays in the museum and research center. Thus, in the parlance outlined above, they would be "facilitating or enhancing relationships with key public sector officials". The project could also provide an outlet for volunteerism within the corporation. Finally, employee morale could be bolstered by seeing a vital cultural project succeed, due to the support of their company.

As a project with important cultural and environmental (promoting the preservation of archaeological sites) components, we are well aware that mining companies have in the past, occasionally had to resort to reactive, directed donations in order to mitigate the environmental and social consequences of their mining operations (for some examples see Project Underground, 2002). However, the same process of globalization that is increasing access for these companies to third world countries, is encouraging an increase in both social and environmental awareness and an appreciation for sustainable community development (McNeilly 2000, Cooney 2002, Newmont 2002). In particular, these corporations are beginning to recognize the need to broaden their definition of who are their "stakeholders". Now the definition extends beyond simply the shareholders, to include communities and organizations where they work (Cooney 2002). The document "World Heritage and Mining" (WHC 2000) outlines the potential mutual benefits for both mining companies and heritage organizations that will arise from constructive dialogue and cooperation.

The company we have targeted is clearly using responsible methods in its new operation. In addition, we feel that constructive engagement may be the best way to encourage this increase in the appreciation for culture and the environment. In sum, we have high hopes that we will be able to establish a synergistic relationship with our corporate target. As of time of writing, our business plan has been received by the corporation, and is being considered by its public relations department.

VI. Where Do We Go From Here?

In parallel with our efforts to garner funding for this project, we are working on several projects that will be extremely important to the operation of the museum. The first is the preparation of a manual that will incorporate "best-practice" guidelines for the operation of a successful museum. The second is the preparation of a detailed plan for the lay-out of the museum, referred to in Spanish as the *expediente técnica*.

We are basing our operations manual on the "Standards for Community Museums in Ontario" (MOC 2000). These standards are for publicly funded museums in the province of Ontario, Canada and are based on similar documents prepared by the Canadian Museums Association and the International Council on Museums. MOC (2000) lays out a set of 10 standards written to provide museums with a guide to good practice, and specifically to provide guidelines as to what the provincial granting authority looks for in grant proposals. We are using it in the former sense. The standards include governance, finance, collections, exhibition, interpretation and education, research, conservation, physical plant, community and human resources.

The education standard is a particularly important one from the perspective of interaction with the local community. Through this standard the museum seeks to enhance opportunities for interaction between museum personnel and the community. As an example of one particular kind of interaction, we are supporting a post-doctoral application to the Social Sciences and Humanities Research Council of Canada by E. Weinstein. Dr. Weinstein plans to build on work she did in Ecuador, by working with local women to train them as museum docents, tour guides and outreach educators to the local schools. This component of our project will not only provide economic benefits, but it will empower the local women, providing important social benefits.

An important step will be the preparation of an *"expediente técnica"*, a dossier of detailed technical documents relating to the exact design of the museum, details of the building design and of museographic displays. To this end, we are working with architects J. Flores and C. Arbaiza of CITEmadera (center of technology and innovation focusing on wood products and structures – the train station is a timber frame building) in Lima to develop plans for the building restoration. In addition we have initiated discussions with the Peruvian museographers H. Fiestas and P. Carcedo. These individuals undertook the museographic component of the recently opened Sican Museum, in Fereñafe, Peru. A concept layout is presented in Figure 7. The full elaboration of this plan will require extensive consultation with archaeologists in the area and with officials from the INC in Trujillo and Lima. As part of the preparation of this document, we will need to negotiate with the various repositories of both artifacts and skeletal remains in order to have them transferred to this museum. It is extremely important to note that Prof. Dr. Grupe, the director of the anthropological collections of the *Staatssammlung für Anthropologie und Paläoanatomie München* has indicated an interest in repatriating Ubbelohde-Doering's osteological collection to our facility, provided the long term security of the collection can be assured.

The final step will be to formally constitute the project as a museum, and to bring the museum into the national system of museums. This will be done by means of a *resolución de creación*, which is in essence, another *convenio* with the museums arm of the INC.

As we proceed with this project, we must learn the lessons taught by previous tourist development projects and from corporate/cultural collaborations. Many of the concerns of tourism development in general are addressed by the pro-poor approach of the Northern

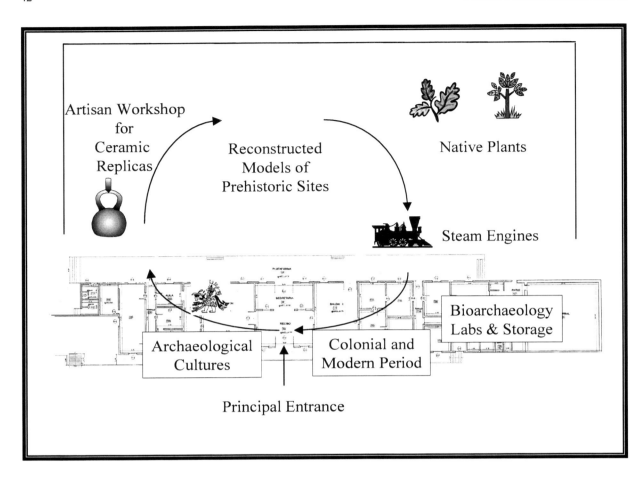

Fig. 7: Concept layout of the Pacasmayo Museum Project. The Steam Engines are two small shunt engines that are currently in the station yard.

Tourist Circuit. In particular, the circuit seeks to provide geographic diversity to tourism in Peru, reduce economic leakage and to create local inter-sectoral linkages. These elements will work together to promote the sustainability of the circuit as a whole. With a specific focus on our project, we will have no economic leakage and we are working to create local linkages. To that end, we have established close working relationships with the Pacasmayo Chamber of Commerce, the local tourism board and the *alcaldes* (mayors) of both the City and Province of Pacasmayo. C. Nelson's experience as the Heritage Planner for the City of London (Ontario, Canada) and member of the London Heritage and Museum Coordinating Committee (HMCC) will be invaluable in this process. London's HMCC was established and is functioning to celebrate and market London's heritage, to develop partnerships and to leverage funding allocations for the benefit of all regional heritage resources (see Dryden & Fleming 2002).

Finally, we hope to include local businesses as partners by requesting "in-kind" support, for example of building materials. In addition, we plan to have a gift shop in the museum, where local craft producers can sell their products. These efforts are consistent with the goals described above, that is that a project like ours should

VII. Conclusion

Our initial goal of bringing together a collection of human bones for osteological analysis has led us on an ever expanding journey involving *convenios*, international development, and business plans. It all seemed so simple back in 1999, when we had our first conversation with CB Donnan and then with our Peruvian partner, Sr. Arbaiza. However, while it has been a long and winding road, the Pacasmayo Museum Project as it now stands, is an extremely exciting mission. This is a genuinely international collaborative effort, based on a firm local foundation – ultimately this is a Peruvian project that will be owned and operated by Peruvians. The benefits of the project will be clearly felt by the people of Pacasmayo and the surrounding area. The first benefit will be to their pride in their heritage. The fact that archaeological heritage has a role to play in present day society is well illustrated by the effort that went into the construction of the statue of *La Sacerdotista* in Chepen (see Figure 4). Second, there will be numerous economic benefits, arising from tourists visiting the city, from direct employment at the museum, and from indirect opportunities for entrepreneurs. Third, there will be a clear benefit to archaeologists working in the valley, who will have an excellent research facility. Finally, there will be a definite benefit to bioarchaeologists seeking to better understand the actual people who lived in this valley hundreds and thousands of years ago. Put together, these benefits will create a synergy, where the whole will be much greater than the sum of its parts.

VIII. Acknowledgements

Many, many people have had important roles in supporting our work in Peru and with this project in particular. This project would not be possible without the support and encouragement of all of them. Sr. Luis Felipe Arbaiza M. is our Peruvian partner whose initial idea for a Pacasmayo museum gave shape to this project. Arq. Christian Arbaiza M. is has been an important collaborator throughout, especially from the architectural point of view. This project will succeed because we share a common vision.

Partners in Project Development – Arq. José Armando Alcazar F. (Architect, CITEmadera); Sr. Hugo Fiestas and Sra. Paloma Carcedo (museographers, Lima); Sra. Jenny Figari de Ruiz (Instituto Superior Yachay Wasi, Lima); Mr. Gerald Conlogue and Dr. Ron Beckett (Quinnipiac University, USA); Dr. Fred Keenan (Office of International Research, UWO); Dr. Gisela Grupe, (Staatssammlung für Anthropologie und Paläoanatomie München); Dr. Elka Weinstein (University of Toronto); Dr. Antoine Zalatan (University of Ottawa); Dr. Christine White (Anthropology, UWO).

Funding and In-kind Assistance – Social Sciences and Humanities Research Council of Canada; G.L. Bruno Foundation; VP Research, University of Western Ontario; Office of International Research, UWO; Kodak Canada, Health Sciences Division; Victorinox, Canada; Novacks, London, Canada.

Logistical Support and Encouragement in Peru – Srta. Ana Maria Hoyle M. (Director, INC, Trujillo), Sra. Lenor Sisneros (National Director, INC in 2002), Sr. Enrique Gonzales C. (National Director, INC, 2001), Sr. Jaime Quijandria S. (Minister of Mines and Energy); Ambassador Jaime León P. (Executive Director, Cultural Promotion of Ministry Foreign Affairs, Peru); Mr. Enrique Sánchez M. (Director, Industry, Tourism and International Commercial Negotiations, CTAR, La Libertad); Sr. Ontere Giura A. (Alcalde, Province of Pacasmayo & Chair of the Pacasmayo Tourism Board); Sr. Victor Alayo León (Alcalde, City of Pacasmayo); Sra. Luz A. Rivera R. de Luna (President, Chamber of Commerce, Industry and Production, Province of Pacasmayo); Mr. C. Clark and Mr. Hugues Rousseau (Canadian Ambassadors to Peru) and Mr. Douglass Challborn (Political Consul, Canadian Embassy in Lima); Dr. Alfredo Narvaez (Director, INC, Lambayeque); Mr. Mike Agg (Manager, Tech Cominco, Peru).

Logistical Support and Encouragement in Canada – Sr. Carlos Gamarra M. (Peruvian Consul General in Toronto); Ms. Susan James & Mr. Peter Ross (Office of International Research, UWO); Dr. Bill Bridger & Dr. Nils Petersen (Vice Presidents Research, UWO); Dr. Peter Neary & Dr. Brian Timney (Deans, Social Sciences, UWO); Dr. Chris Ellis & Dr. Chet Creider (Chairs, Department of Anthropology, UWO); Mr. Kevin Goldthorp (Alumni Affairs, UWO); Ms. Margaret Dryden & Mr. John Fleming (City of London).

Archaeological Colleagues in the Jequetepeque Valley – Dr. Christopher Donnan (UCLA); Mr. Guillermo Cock (Lima); Dr. Carol Mackey (CSUN); Prof. Luis Jaime Castillo B. (PUCP); Dr. Alana Cordy-Collins (U. San Diego); Dr. Bill Sapp (UCLA); Maria Sidoroff.

References

Agg J. & Perrier C., 2002.
Uncovering Secrets at Cajamarquilla. http://www.teckcominco.com/articles/operations/caj-archdig.htm

Burger E., 1978.
Vergleichende Unterschungen an menschlichem Skelett – und Mumienmaterial von der Nord – und Südküste Altperus. (Teil I: Ergebnisse). Diploma Thesis: Ludwig-Maximilians Universität, München, Institut für Anthropologie und Humangenetik.

Burger R.L., 1992.
Chavin and the Origins of Andean Civilization. Thames and Hudson, Ltd.: London.

Castillo L.J., Nelson A.J., Nelson C.S., 1997.
Maquetas Mochicas S., San Jose de Moro. *Arkinka* **22**: 121-128

CIA, 2002.
The World Factbook 2002 – Peru. http://www.cia.gov/cia/publications/factbook/geos/pe.html November 12, 2002

CMC, 1996.
Corporate Profile. Canadian Museum of Civilization Corporation. Summary of the Corporate Plan (1996-1997 to 2000-2001). http://www.civilization.ca/societe/corpsm96/cs96c01e.html

Conlogue G. & Nelson A.J., 1998.
The use of Polaroid photographic imagery systems to produce radiographic images at a field site in Peru. *Paper presented to the Paleopathology Association*.

Conlogue G. & Nelson A.J., 1999.
The use of the Polaroid photographic imaging system to produce radiographic images at a field archaeological site in Peru. *Radiologic Technology* **70(3)**: 121-128.

Cooney J.P., 2002.
The Importance Of Partnerships Pre-And Post WSSD Presentation To The Panel On Sustainable Development: The Road To Johannesburg Summit 2002. http://www.wmmf.org/2002wmmf_files/sd-joburg/sust_devel_conney_sup.pdf

Cordy-Collins A., 2002a.
Archaeological context of the Dos Cabezas giants. *Paper presented to the 14th Biennial European Meeting of the Paleopathology Association*. Coimbra, Portugal.

Cordy-Collins A., 2002b.
Posterior parietal thinning: five Moche giants from prehistoric Peru. *Paper presented to the 14th Biennial European Meeting of the Paleopathology Association*. Coimbra, Portugal.

Cordy-Collins A., Conlogue G., Garvin G., Nelson A.J., Toyne J.M. & Holdsworth D., 2001.
Radiographic and paleopathologic diagnosis of A52 T1 B1 (an ancient Peruvian giant) *Paper presented to the Paleopathology Association*.

Dolphin A., 2002.
Beautiful ideas: a biocultural approach to the body and its modification. *Paper presented to the Canadian Association for Physical Anthropology*.

Donnan C.B., 1975.
The thematic approach to Moche iconography. *Journal of Latin American Lore* **1(2)**: 147-162.

Donnan C.B., 2001.
Ancient tombs of Peru: Pre-Inca treasures. *National Geographic* March: 58-73.

Donnan C.B. & Castillo L.J., 1992.
Finding the tomb of a Moche priestess. *Archaeology* Nov/Dec 38-42.

Donnan C.B. & Cock G., 1986.
The Pacatnamu Papers: Volume 1. UCLA Fowler Museum of Cultural History: Los Angeles.

Donnan C.B. & Cock G., 1997.
The Pacatnamu Papers: Volume 2. The Moche Occupation. UCLA Fowler Museum of Cultural History: Los Angeles.

Donnan C.B. & McClelland D., 1999.
Moche Fineline Painting. Its Evolution and its Artists. UCLA Fowler Museum of Cultural History: Los Angeles.

Eadington W.R. & Smith V.L., 1992.
Introduction In: Smith V.L. & Eadington W.R. (eds.), *Tourism Alternatives: Potentials And Problems In The Development Of Tourism*: 1-12. Philadelphia, PA: University of Pennsylvania Press.

Elera C.G., 1998.
The Puemape Site and the Cupisnique Culture: a case study in the origins and development of complex society in the Central Andes, Peru. Unpublished PhD Thesis, Department of Archaeology, University of Calgary, Alberta.

Geyer P., Larson T.S. & Stroik L., 2002.
Palynological investigation of the Dos Cabezas giants. *Paper presented to the 14th Biennial European Meeting of the Paleopathology Association*, Coimbra, Portugal.

Gillespie R., 1998.
Health at the North Coast Site of Puemape During the Peruvian Formative Period. Unpublished MA Thesis, Department of Anthropology, University of Calgary, Alberta.

Gould S.J., 1981.
The Mismeasure of Man. WW Norton & Company: New York.

Graburn N.H.H., 1989.
Introduction. In: Smith V.L. (ed.), *Hosts and Guests. The Anthropology of Tourism* (second edition). University of Pennsylvania Press: Philadelphia.

Grupe G. & Turban-Just S., 1998.
Amino acid decomposition of degraded matrix collagen from archaeological human bone. *Anthropologischer Anzeiger* **56(3)**: 213-226.

Harrison J., 1988.
"The spirit sings" and the future of anthropology. *Anthropology Today* **4(6)**: 6-9.

HDL, 2002.
Human Sacrifices at the Huaca de la Luna. http://www.huacas.com/page140.htm

Hecker G. & Hecker W., 1985.
Pacatnamú y sus Construcciones. Centro Religioso Prehispánico en la Costa Norte Peruana. Verlag Klaus Dieter Vervuert: Frankfurt.

Heflin T., 2002.
Rib trauma in a Moche giant. *Paper presented to the 14th Biennial European Meeting of the Paleopathology Association*, Coimbra, Portugal.

Heine E., 1979.
Zur Variabilität und geographischen Verbreitung künstlicher Schädeldeformationen in Alt-Peru. Diploma Thesis: Ludwig-Maximilians Universität, München, Institut für Anthropologie und Humangenetik.

ICOM, 1968. Resolutions adopted by ICOM's General Assembly, 1968. http://icom.museum/resolutions/eres68.html (posted to the WWW 2001).

IDRC, 1999.
Analisis Subsectoral del Turismo en el Peru. Report published by the International Research and Development Corporation.

Lacombe J.P., 2000.
Les Hommes de Paijan (Péreu) dans le Contexte Preceramique Archaique et Paleoindian de L'Amerique Du Sud. Unpublished doctoral thesis, Université Bordeaux I. Bordeaux, France.

Levy R., 1999.
Give and Take. A Candid Account of Corporate Philanthropy. Harvard Business School Press: Boston.

Lewin D. & Sabater J.M., 1996.
Corporate philanthropy and business performance. In: Burlingame D.F. & Young D.R. (eds.), Corporate Philanthropy at the Crossroads: 105-126. Indiana University Press: Bloomington.

Lichtenfeld M.J., 2001.
Artificial Cranial Modification on the Jequetepeque Valley, Peru. MA Thesis: Department of Anthropology, University of Western Ontario.

Lombardi G., 2002.
Moche giants: a CT-scanning survey. Paper presented to the 14th Biennial European Meeting of the Paleopathology Association, Coimbra, Portugal.

Mackey C.J., 2001.
Transformacion socio-economica de Farfan durante la conquista Inka. Paper presented to The IV Simposio Internacional de Arqueologia Pontificia Universidad Catolica del Peru, 16-18 agosto 2001.

Mackey C.J., in press.
La Ocupacion de dos centros Administrativos Chimu del Valle de Jequetepeque: El Algarrobal de Moro y el Complejo VI de Farfan. Sian.

Mann R.W. & Verano J.W., 1990.
Congenital spinal anomalies in a prehistoric adult female from Peru. Case Reports in Paleopathology No 13. Paleopathology Newsletter **72**: 5-6.

Martin R.I., Nelson A.J., Beisinger M., Naftel S.J., Kempson I.M. & Skinner W.M., 2002.
Mechanisms for the accumulation of metals in human hair from an archaeological context: studies using ultra-trace analysis. Paper presented to the Canadian Association for Physical Anthropology.

McNeilly R.J., 2000.
A Mining Company for the 21st Century. http://www.globalmining.com/further_info/documents/100400.pdf

MOC, 2000.
Standards for Community Museums in Ontario. Ministry of Culture, Government of Ontario. available at: http://www.culture.gov.on.ca/culture/english/culdiv/heritage/mustand.htm

Moseley M., 1992.
The Incas and Their Ancestors: The Archaeology of Peru. Thames and Hudson: London.

Mowforth M. & Hunt I., 1998.
Tourism and Sustainability. New Tourism in the Third World. Routledge: London.

Nelson A.J., 1996.
Skeletons and society in the Moche times: bioarchaeology at San Jose de Moro, Peru. Paper presented to the Canadian Association for Physical Anthropology.

Nelson A.J., 1998.
Wandering bones: archaeology, forensic science and Moche burial practices. International Journal of Osteoarchaeology **8**: 192-212.

Nelson A.J., 2000.
Death is just a phase: The cycle of life on the ancient Peruvian North Coast. Paper presented to the Brock University Archaeological Society.

Nelson A.J., 2001.
The anthropometry/osteology interface: body size and form in Precolumbian populations on the North Coast of Peru. Paper presented to the Canadian Association for Physical Anthropology.

Nelson A.J. & Castillo L.J., 1997.
Huesos a la deriva: tafonomia y tratamiento funerario en entierros Mochica tardio de San Jose de Moro. Boletin de Pontifica Universidad Catolica del Peru **1**: 137-163.

Nelson A.J. & Conlogue G., 1997.
Field radiology in archaeology: penetrating the problems and illuminating research in osteology. Paper presented to the Canadian Association for Physical Anthropology.

Nelson A.J. & Mackey C.J., 1997.
Wandering bones – secondary burial in the Moche. Paper presented to the Institute of Andean Studies.

Nelson A.J., Castillo L.J. & Mackey C.J., 1996a.
El Complejo de San Jose de Moro – 1995 Season, Jequetepeque Valley, Peru. Paper presented to the Society for American Archaeologists.

Nelson A.J., Conlogue G., Hennessy W. & Gauld S., 1999.
A preliminary study to determine the most suitable radiographic projection to document intentional cranial deformation. Paper presented to the Paleopathology Association.

Nelson A.J., Lichtenfeld M.J., Conlogue G., Toyne J.M. & Pool S., 2000a.
Cranial modification in the Jequetepeque Valley. Paper presented to the Northeastern Andean Archaeology and Ethnohistory.

Nelson A.J., Mackey C.J. & Castillo L.J., 1996b.
Bottles, bones and buildings: cultural dynamics at the site of San Jose de Moro, Peru. Paper presented to the Northeastern Andean Archaeology and Ethnohistory.

Nelson A.J., Martin R., Biesinger M.C., Naftel S.J., Kempson I.M., Skinner W.M., Jones K.W., 2002a.
Trace Element Analysis of Human Hair from the Precolumbian Site of Pacatnamu, Peru: An Exploration of Biogenic and Diagenetic Signals. *Paper presented to the North American Paleopathology Association.*

Nelson A.J., Nelson C.S., Castillo L.J. & Mackey C., 2000b.
Osteobiografia de una hilandera precolumbina. La mujer detras de la mascara. *Íconos* **4(2)**:30-43.

Nelson A.J., Wang J. & Hegele R.A., 2000c.
Ancient DNA from the Peruvian North Coast. *Paper presented to the Canadian Association for Physical Anthropology.*

Nelson A.J. & Nelson C., 2002a.
In search of the ancient Peruvians: The Pacasmayo Museum Project. *Paper Presented to Deciphering Ancient Bones: The Research Potential of Bioarchaeological Collections; International Workshop Celebrating 100 Years of Anthropology in the Collection of Anthropology & Palaeoanatomy at the Staatssammlung für Anthropologie und Paläoanatomie München,* Munich, Germany.

Nelson A.J. & Nelson C., 2002b.
Gigantism and the individuals from Dos Cabezas, Peru. *Paper presented to the 14th Biennial European Meeting of the Paleopathology Association,* Coimbra, Portugal.

Nelson A.J., Cordy-Collins A., Conlogue G., Beckett R., Garvin G., Holdsworth D. & Ford N., 2002b.
Radiographic and endoscopic analyses of the giants from Dos Cabezas, Peru. *Paper presented to the 14th Biennial European Meeting of the Paleopathology Association,* Coimbra, Portugal.

Newmont, 2002.
Normandy mining limited – our corporate responsibility. http://www.normandymining.com/contents/social.htm

NHM, 2002.
About The Natural History Museum. http://www.nhm.ac.uk/info/index.html

Nizette P., Goodwin H., 2002.
Proposal for the Financing of Phase 1 – 18 Months of the Northern Circuit Economic Development and Poverty Reduction Tourism Programme. Proposal submitted to the Fondo Italo Peruano.

Owen B., 2000.
The Museo Contisuyo: a successful partnership across multiple boundaries. *SAA Bulletin* **18(4)**: 27-30.

Pacific Consultants International, 1999.
Master Plan Study on National Tourism Development in the Republic of Peru (Phase-1). Final Report Summary. Report prepared for the Japan International Cooperation Agency and the Peruvian Ministry of Industry, Tourism, Integration, and International Trade Negotiations.

Parsche F., Nerlich A., Zink A. & Wiest I., 1994.
Collagen immunohistology in paleopathology. Evidence for active bone remodeling in a Peruvian tibia. *Journal of Paleopathology* **6(1)**: 103-108.

PICIC, 2001.
Social Emergency, Economic Recovery and Democratic Governance in the New Peru. Report prepared by the Peruvian Interministerial Commission for International Cooperation. http://www.peruembassy-uk.com/Cooperation/atrabajar.pdf.

Project Underground, 2002.
Newmonster-The Golden Meanie: 10 problems with gold mining. http://www.moles.org/ProjectUnderground/mining/newmont/n_goldmining0502.html

Ricks J.M. jr., 2002.
The Effects of Strategic Corporate Philanthropy on Consumer Perceptions: An Experimental Assessment. Unpublished PhD Dissertation, Interdepartmental Program in Business Administration, Louisiana State University and Agricultural and Mechanical College.

Ryser G., 2002.
Paleoethnobotany at the site of Dos Cabezas. *Paper presented to the 14th Biennial European Meeting of the Paleopathology Association,* Coimbra, Portugal.

Rowe J.H., 1962.
Stages and periods in archaeological interpretation. *Southwest Journal of Anthropology* **18(1)**: 40-54.

SAPM, 2002.
Staatssammlung fuer Anthropologie und Palaeoanatomie Muenchen (State Collection of Anthropology and Palaeoanatomy) Department of Anthropology. http://www.lrz-muenchen.de/~NatSamm/Anthropologie/Anthro.html

Scheyvens R., 1999.
Ecotourism and the empowerment of local communities. *Tourism Management* **20**: 245-249.

Schlüter R.G., 1994.
Tourism development: a Latin American perspective. In: Theobald W.F (ed.), *Global Tourism. The Next Decade*: 246-260. Butterworth & Heinemann: Oxford.

Segura R., Vega MdC. & Landa P., 2001.
Recent investigations at the Site of Cajamarquilla: Advances in the Study of Precolumbian Mortuary Practices on the Peruvian Central Coast. Translated by A. Nelson. http://www.teckcominco.com/articles/operations/caj-llanos.pdf

SI, 2002.
A History of the Department of Anthropology, Smithsonian Institution. http://www.nmnh.si.edu/anthro/outreach/depthist.html

Spence M., 1998.
Congenital deformity in a Late Woodland burial from Southwestern Ontario. Northeast Anthropology **55**: 31-46.

Steckle R.H., Rose J.C., Larsen C.S. & Walker P., 2002.
Skeletal health in the western hemisphere from 4000 B.C. to the present. *Evolutionary Anthropology* **11**: 142-155.

Stronza A., 2001
Anthropology of tourism: forging new ground for ecotourism and other alternatives. *Annual Reviews of Anthropology* **30**: 261-283.

Trigger B., 1988.
Reply. *Anthropology Today* **4(6)**: 9-10.

Tyson R., 2002.
Toes of Atlas, a young Moche giant. *Paper presented to the 14th Biennial European Meeting of the Paleopathology Association*, Coimbra, Portugal.

Ubbelohde-Doering H., 1983.
Vorspanische Gräber von Pacatnamú, Nordperu. Materialien zur Allgemeinen und Vergleichenden Archäologie. Verlag C. H. Beck: Munich.

Verano J.W., 1987.
Cranial Microvariation at Pactnamu: A Study of Cemetery Population Variability. Unpublished PhD dissertation, Department of Anthropology, University of California, Los Angeles.

Verano J.W., 1992.
Prehistoric Disease and Demography of Andean South America. In: Verano J.W. & Ubelaker D.H. (eds.), *Disease and Demography in the Americas*: 15-24. Washington, DC: Smithsonian Institution Press.

Verano J.W., 1994.
Características físicas y biología osteológical de los Moche. In: Uceda S. & Mujica E. (eds.), *Moche: Propuestas y Perspectivas, Travaux de l'Institut Français d'Etudes Andines* **79**: 307-326.

Verano J.W., 1995.
Where do they rest? The treatment of human offerings and trophies in Ancient Peru. In: Dillehay T.D. (ed.), *Tombs for the Living: Andean Mortuary Practices*: 189-227: Dumbarton Oaks Research Library and Collection: Washington, DC.

Verano J.W., 1997.
Physical characteristics and skeletal biology of the Moche population at Pacatnamu. In: Donnan C.B. & Cock G.A. (eds.), *The Pacatnamu Papers, Vol. 2*: 189-214. Fowler Museum of Cultural History, University of California, Los Angeles: Los Angeles.

Verano J.W., 1998.
The Moche: profile of an ancient Peruvian people. In: Selig R.O. & London M.R. (eds.): *Anthropology Explored: The Best of Smithsonian AnthroNotes*: 167-177. Washington, D.C. and London: Smithsonian Institution Press.

Verano J.W. & DeNiro M.J., 1993.
Locals or foreigners? Morphological, biometric, and isotopic approaches to the question of group affinity in human skeletal remains recovered from unusual archaeological contexts. In: Sandford M.K. (ed.): *Investigations of Ancient Human Tissue: Chemical Analysis in Anthropology*, 361-386. New York: Gordon and Breach.

von Hagen A. & Morris C., 1998.
The Cities of the Ancient Andes. Thames and Hudson, Ltd.: London.

Wiechmann I. & Grupe G., 1997.
Serumproteine in menschlichem skelettmaterial. *Anthropologischer Anzeiger* **55(1)**: 143-146.

Wiechmann I., Brandt E. & Grupe G., 1999.
State of preservation of polymorphic plasma proteins recovered from ancient human bones. *International Journal of Osteoarchaeology* **9**: 383-394.

WHC, 2000.
World Heritage and Mining. Report of a Technical Workshop, IUCN Headquarters, Gland Switzerland. WHC-2000/CONF.203/INF.7. http://www.naturalresources.org/minerals/generalforum/docs/pdfs/World%20Heritage%20and%20Mining%20-%20Tech%20Workshop%20Report%20-%20Sept%202000.pdf

WTO, 2002.
Tourism and Poverty Alleviation. World Trade Organization: Madrid.

Yankey J.A., 1996.
Corporate support of nonprofit organizations. In: Burlingame D.F & Young D.R. (eds.), *Corporate Philanthropy at the Crossroads*: 7-22. Indiana University Press: Bloomington.

Zink A., 1993.
Paläopathologische Untersuchungen an ausgewähltem historischem Skelettmaterial aus der präkolombianischen Zeit Perus. Diploma Thesis: Ludwig-Maximilians Universität, München, Institut für Anthropologie und Humangenetik.

Part II:
The Research Potential of Bioarchaeological Collections

Human skeletal remains from the central Balkans: A survey of the development of human populations

Živko Mikić, Philosophische Fakultät, Anthropologische Sammlung,
Univerzität Belgrad

Abstract / Zusammenfassung

This essay is about the development of human populations in the central Balkans by tracing the brachycephalisation process during the Iron Age. One should emphasize that the oldest brachycephalic skulls were found in the osteological material from the Iron Gates. The brachycephalisation process was disrupted by the infiltration of other populations during the Roman and Migration period. However, while the full extent of the process became apparent in the late Middle Ages, the process itself has not stopped during the last two millennia and the outcome is the emergence of the Dinaric anthropological type as a substrate of the modern population of the region.

Dieser Essay behandelt die menschliche Bevölkerungsentwicklung auf dem Zentralbalkan, wobei der Brachycephalisationsprozeß während der Eisenzeit erkennbar wird. Es sollte betont werden, dass die ältesten brachycephalen Schädel bereits im osteologischen Material vom Eisernen Tor aufgefunden wurden. Der Prozeß der Brachycephalisation, welcher sein volles Ausmaß im ausgehenden Mittelalter erreicht, wurde durch die Einwanderung anderer menschlicher Bevölkerungen während der Römer- und Völkerwanderungszeit unterbrochen, kam jedoch während der letzten beiden Jahrtausende nicht völlig zum Erliegen. Hieraus resultiert die Entstehung des dinarischen anthropologischen Typus, das Substrat für die heutige Population in der Region.

Keywords: Central Balkans, Mesolithic population, Neolithic human remains, gracilisation, brachycephalisation, dinarisation, present-day population
Zentralbalkan, mesolithische Bevölkerung, neolithische menschliche Skelettfunde, Grazilisation, Brachycephalisation, Dinarisierung, heutige Bevölkerung

There are two issues to consider when discussing research on the complex development of human populations in the central Balkans. The first is that in this area, which is now the Republic of Serbia, anthropology on the whole, and subsequently physical anthropology, is not a completely institutionalized science. There are still neither appropriate institutes nor departments where anthropology is or can be studied as a major, or practiced as a science. At present, physical anthropology is taught only as a compulsory but supplementary course for students of archaeology and ethnography at the Faculty of Philosophy in Belgrade. The Chair of Physical Anthropology has been established at this department since 1998. In addition, there are only a few individuals in the region who more or less professionally practice physical anthropology. It seems redundant and unnecessary to explain or comment upon the consequences of the present state of affairs. The author himself was obliged for many years, to conduct his research in liaison with the Anthropological Institute of Gutenberg-University, Mainz, and to consider *that* his own institutional base. In the past couple of years, contact has mostly been with the State Anthropological Collection of Bavaria. The result of this collaboration has been short study periods both in Munich and Belgrade and the realisation of a number of significant projects, such as the complex study of the Iron Gates Mesolithic population and the study of several anthropological aspects of the Great Movement of Peoples in the Early Middle Ages (Mikić 1999).

The second issue relates to the cooperation, or lack of, between anthropologists and archaeologists. Archaeologists often excavate necropolises without the on-site involvement of anthropologists. One can hardly be satisfied with the existing cooperation. Consider that during the 20th century, archaeologists excavated 400 necropolises from all periods of human history in Serbia. Out of this astounding total, osteological remains from only around 70 necropolises were studied. Human remains from the remaining more than 300 necropolises were not curated for anthropological examination. For the science these are lost forever (Stefanović 2000). It follows that one can only approach the topic contained within the title of the essay with these short qualifications about the research and study circumstances.

To begin, a chronology of the work done to date in the field of anthropology in this region is necessary.

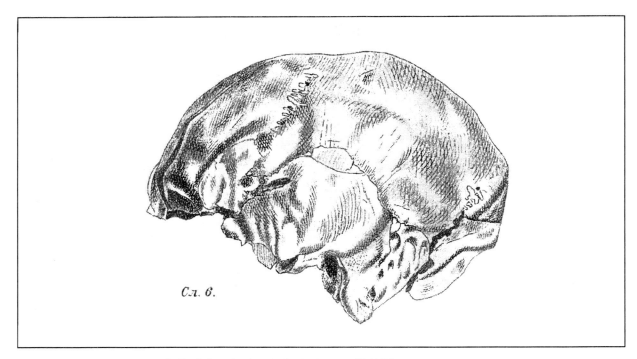

Fig. 1: Neanderthal Belgrade Skull – left projection (After Đ. Jovanović 1892).

The earliest fossil and oldest remains of humans in this region were discovered at the end of the 19th century. In 1890, a professor of the High School in Belgrade, Dr. Đoka Jovanović, found a skull that he named "Neanderthal skull." Found in the outskirts of Belgrade, the skull was about two and a half meters below the surface. The finder indicated that a large tooth of *Elephas antiquus* was found next to it. Experts from France confirmed this. In his report dated 1892, Đ. Jovanović (1892) indicated that the skull belonged to the Neanderthal ancient man and described it as having very pronounced eyebrow ridges and back part, a strong saggital crest, etc., as well as having almost no forehead (Fig. 1). He added that it measured 504 mm long and 138 mm wide. If the quoted cranial measurements were exact, the length/width index would be 73.4.

In 1928, Niko Županić, an anthropologist who received his doctoral degree in Vienna in 1903, also briefly examined this skull (Županić 1928). He noted that he was unable to find it for a desired reexamination and it is assumed that the skull most likely vanished during the First World War. Is it lost to science forever? In the following seven decades of the 20th century there is no mention of it!

Interestingly, a similar mystery surrounds a fossil molar find in the Jerina Cave near Kragujevca (central Serbia). Diluvial fauna and numerous stone tools accompanied this find which H. Vallois assigned to a human of the *sapiens sapiens* type. The cave itself was excavated in 1951 and 1952 (Gavela 1988). Unfortunately, this find, too is lost and there are no indications as to where it may be today.

In summary, where the Paleolithic period is concerned, there is a huge imbalance in the investigation and knowledge of the material culture on one side and the human remains, i.e., the populations that made this material culture, on the other. We could say that at the present stage of research that there are practically no anthropological remains of the Paleolithic period in the central Balkans.

The Mesolithic era in this area spans a period from 9000 to 6000 BC and osteological remains are known only from the micro region of the Iron Gates Gorge (Fig. 2). The Iron Gates anthropological remains were found at sites on both banks of the Danube; however, compared to the Rumanian sites, the Serbian sites contain far more human skeletal remains. The six sites on the Serbian side, namely, Vlasac, Lepenski Vir, Padina, Kula, Ajmana and Velesnica, have around 500 individual burials. Chronologically, these can be attributed to the Late Mesolithic and Early Neolithic, i.e., they are considered to reflect the transition from the Mesolithic to the Neolithic anthropologically (Mikić 1992). Here, as in the contemporaneous sites in Rumania, the human skeletal remains were found not in necropolises but within the settlement areas. On the basis of demographic structure, it is clear that only selected individuals – regardless of their sex and individual ages – were buried within the settlements, and very often even inside habitation structures. The questions they pose for anthropology as well as for archaeology then are: who are these people and what is the bioanthropological structure of this Late Mesolithic population? However, up to today, some 30 years after the archaeological excavations of Vlasac, Lepenski Vir and Padina and other sites, physi-

Human skeletal remains from the central Balkans

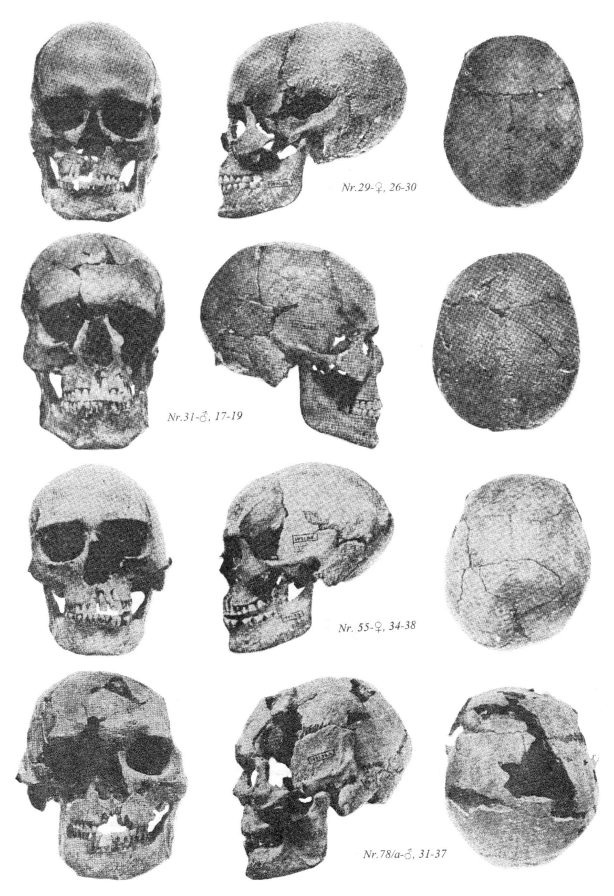

Fig. 2: Skulls from the Iron Gates/Vlasac.

cal anthropology has not found an answer. Classic anthropological methods have not helped to confine the interpretation to one of the two possible theories: autochthonous development of the population, i.e., microevolution in situ, or immigration from other regions. An answer is expected from some laboratory research that is in progress. This is exactly the mission for the present joint cooperation between Munich and Belgrade and it could determine if the individuals buried at the Iron Gate sites were immigrants or members (on the basis of a criterion or criteria yet unknown to us) of an indigenous population in transition from the Mesolithic to the Neolithic. Certainly in this context, questions of gracilisation (Schwidetzky & Mikić 1988), neolithisation (Mikić 1990) and all other bioanthropological processes are still unanswered.

Neolithic human remains come from 18 sites in the region: Lepenski Vir-layers III/a and III/b, Ajmana, Velesnica, Divostin, Kamenjar/Niš, Rudnik Kosovski, Padina b/III, Zlatara/Ruma, Starčevo, Tečić, Vinča, Obrež, Ušće Kameničkog potoka, Vizić/Golokut, Donja Branjevina, Gomolava and Mostonga/Moštanica i Bogojevo/Bancelapas. The total number of individual burials is around one hundred. Anthropological analyses of human skeletal remains from these sites have demonstrated that the Neolithic population is different from both older and also Late Neolithic populations. These individuals are far more gracile and considerably shorter. The average height of the male was around 163 cm, and the females around 156 cm (Mikić 2002).

Human skeletal remains from the Eneolithic Period, which globally lasted through the 3rd millennium BC, were found in several sites in the Pannonian part of Serbia (Nosa, Vajska, Vojlovica, Gomolava). At these sites one can identify anthropological types such as Proto-Nordic, Nordic, Atlantic, and Mediterranean as well as individuals of Eastern European origin. Thus, one could say that this is a period when the movement of populations can be documented anthropologically (Mikić 1981).

From the sites of Belotić and Bela Crkva in western Serbia and Mokrin in the Panonnian Plain where over 300 individual burials were excavated (Farkas 1971), human remains dating to the later part of the Metal Age, the Bronze Age (2nd millennium BC) have been recovered.

On the whole, we can conclude that the nonintegration of archaeological and anthropological research continues with damaging consequences for physical anthropology. A very pronounced heterogeneous population structure during this period was noted, and for the first time, skulls with trepanations made with metal implements appear in this area (Mikić 1998).

Human remains were found at 10 sites that have been dated to the Iron Age (covering most of the 1st millennium BC). One should emphasize the fact that in the Balkans, during the Late Iron Age, a core of brachycephalisation can be established, and consequent dinarisation of this late prehistoric population. The number of brachycephalic skulls increases and we already have an anthropological series with a mean index value on the boundary of brachycepalic and dolichocephalic. Certainly, the presence of a mostly dolichocephalic population has also been attested (Mikić 1987).

The Roman period lasted around 500 years in this region and is, when compared with other periods, the least known anthropologically. Frankly speaking, there is only one grand series – the Roman Viminacium.

At Viminacium there are around 12,000 excavated burials, out of which around 8,000 are inhumations and over 3,000 are cremations. There is here, a spectrum of anthropological types and variations, and due to the fact that both civil and military populations used the same necropolises, it is still not possible to establish either a demographic or an anthropological model for this series. In reality, since we have no other human osteological series at present, the only thing we know is what the so-called "Roman rethorta" looked like in this area (Mikić 1995).

The Early Middle Ages, i.e., the Migration Period, brought an influx of different groups of Germanic, Slavonic and Mongolian ethnic origins to the Balkan Peninsula. There are numerous anthropological traces of these populations and several thousand skeletons have been analyzed and published. In the central Balkans several groups have been found to have differed in size – the Gots, Gepids, different Slavonic and Mongolian groups (especially north of the Sava and the Danube). In addition, anthropological analyses indicate that the Germanic and Mongolian groups were isolated while Slavonic groups were much more integrated into the autochthonous population. The process of brachycephalisation, with its roots already established in the Iron Age, was not weakened during the 1st millennium A.D (Fig. 3). Anthropological results show that this process was only slightly altered by the influx of populations into the Balkans. The autochthonous population was already in the process of dinarisation and this is directly related to the brachycephalisation (Mikić 1997).

In spite of my introductory remark that only 70 out of around 400 excavated necropolises have been anthropologically analyzed and published, the last millennium of our era is anthropologically well established. We have around 10,000 individual skeletons that we can study and since the more recent necropolises (after the 16th century A.D.) are very rarely archaeologically investigated, they offer tremendous anthropological research opportunities.

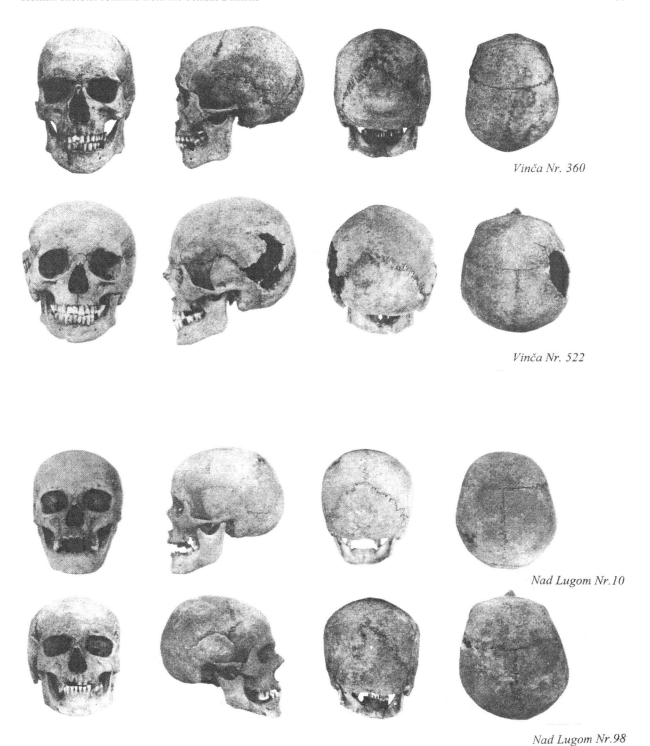

Vinča Nr. 360

Vinča Nr. 522

Nad Lugom Nr.10

Nad Lugom Nr.98

Fig. 3: Medieval dolichocranous and brachycranous skulls.

One should also note that the medieval population buried under standing tombs (13th to 16th century A.D.) in the central Balkans occupies a special place anthropologically (Fig. 4). They are characterized by a Dinaric anthropological type and are assumed to contain around 100 individuals (Mikić 2002). In contrast to these, most of the medieval population were buried in usual necropolises close to urban complexes or next to monasteries.

To conclude, we could say that the brachycephalisation process has not stopped in this millennium either. To the contrary, this process follows further dinarisation of the population, which is confirmed in the anthropological profile of the present-day population in this region (Mikić 2000/2001).

Radimlja-Medieval Stećak necropolis

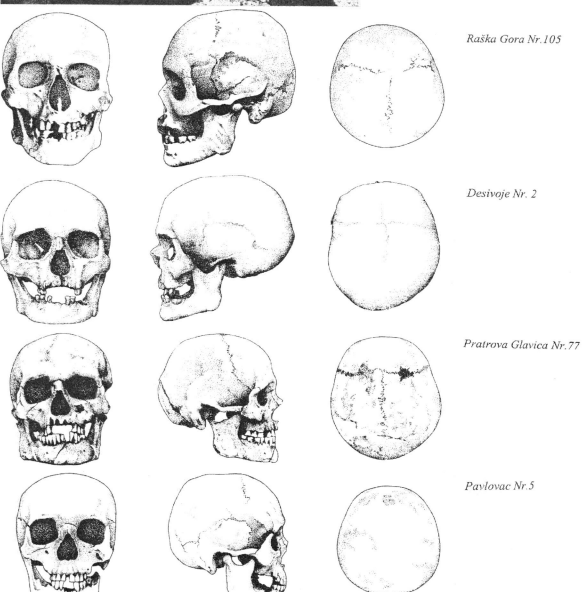

Raška Gora Nr.105

Desivoje Nr. 2

Pratrova Glavica Nr.77

Pavlovac Nr.5

Fig. 4

References

Farkas Gy. & Liptak, P., 1971.
Antropološka istraživanja nekropole u Mokrinu. Dissertationes et Monographie XI. Beograd: Smithsonian Institution and Narodni muzej Kikinda.

Gavela B., 1988.
Paleolit Srbije, Muzej u Aranđelovcu, Aranđelovac-Beograd: Napredak.

Jovanović Đ., 1892.
Prilozi za Paleontologiju Srpskih Zemalja. *Starinar* **IX/1**: 24-34.

Mikić Ž., 1981.
Stanje i problemi fizičke antropologije u Jugoslaviji-praistorijski periodi. *Centar za balkanološka ispitivanja* ABU BiH **LIII/9**: 92-106.

Mikić Ž., 1987.
Prilog antropologiji gvozdenog doba na tlu Jugoslavije. *Godišnjak Centra za balkanološka ispitivanja* ANU **BiH/23**: 37-50.

Mikić Ž., 1990.
Lepenski Vir und das Neolithisationproblem in der Anthropologie. Vinča and its World. Serbian Academy of Sciences and Arts. *Dept. of Historical Sciences*. Vol.**LI/14**: 61-65.

Mikić Ž., 1992.
Mesolithic Population of the Iron Gates Region. *Balcanica* **XXIII**: 33-45.

Mikić Ž., 1997.
Ein Beitrag zur anthropologischen Bevölkerungsgeschichte des Zentral- und Westbalkangebietes im 1. nachchristlichen Jahrtausend. *Starinar* **XLVIII**: 109-121.

Mikić Ž., 1998.
Über urgeschichtliche Trepanationen im Gebiet von der Adria bis zum Schwarzen Meer. *Praehistorische Zeitschrift* **73/2**: 145-150.

Mikić, Ž., 1999.
Die Gepiden von Viminacium-Anthropologischer Beitrag. *Anthropologischer Anzeiger* **57/3**: 257-268.

Mikić Ž., 2000/2001.
Bevölkerungsbewegungen im letzen nachchristlichen Jahrtausend auf dem Boden Jugoslawiens (1991). *Starinar* **L**: 221-235.

Mikić Ž., 2002a.
Istraživanje visine tela ljudi na tlu Jugoslavije od poznog mezolita do kasnog srednjeg veka. *Istorijski časopis* **XLVII**: 25-34.

Mikić Ž., 2002b.
Nekropola u Ćatićima i antropološka problematika stećaka. Zbornik za istoriju Bosne i Hercegovine **3**: 17-27.

Mikić Ž. & Hošovski E., 1995.
First Palaeodemographic Results of Antique Viminacium. *Etnoantropološki problemi* **10**: 103-112.

Schwidetzky I. & Mikić Ž., 1988.
Lepenski Vir und das Grazilisationsproblem in der Anthropologie. *Godišnjak Centra za balkanološka ispitivanja* ANU **XXVI/24**: 113-120.

Stefanović S., 2000.
Unpub. Magistar Thesis, University of Belgrade.

Županić N., 1928.
Paleolitsko Ljudstvo na tlu Južnih Slovena. *Narodna enciklopedija* **III**: 331-335.

Evaluating human fossil finds

Winfried Henke, Institut für Anthropologie, Universität Mainz

Abstract / Zusammenfassung

This contribution provides a comprehensive review of the research potential of human fossils. Palaeoanthropological research aims to explain the process of hominization. The biological approach to understanding human evolution as a self-organising process focuses on the structural and functional adaptations within the order of primates. During the past decades, our biological research concepts have advanced significantly, enabling today a better understanding of the complexity of the origin of *Homo sapiens*.

The issue is introduced by a brief discussion of the main subjects involved in the reconstruction of the evolutionary processes that have given rise to our uniqueness *sensu* Foley. Humans are primates, and thus it is the challenge of evolutionary research to explain the emergence of humans exclusively through the mechanisms of natural, inter-, as well as intrasexual selection. Using a multifaceted, multidisciplinary approach, palaeoanthropology tries to reconstruct the adaptive evolutionary patterns that have played an important role in human evolution. This contribution tries to outline some important evolutionary aspects, focusing on methodological problems, such as form-function-complexes, the concept of ecological niches, the middle range theory and the life history concept. Furthermore, three specific evolutionary phenomena are addressed: the origins of bipedalism, the Eurasian dispersal of early *Homo*, and finally, discussion on speciation processes in the genus *Homo* from a taxonomical perspective. The final conclusion can be understood as a plea for an increase in theory-oriented research in palaeoanthropology and a strong rejection of narrative approaches.

Der Beitrag liefert einen komprimierten Überblick zum Forschungspotential menschlicher Fossilien. Palaeoanthropologische Forschung zielt darauf ab, den Prozess der Menschwerdung zu erklären, wobei der evolutionsbiologische Ansatz die Humanevolution als einen selbstorganisatorischen Prozess strukturaler und funktionaler Adaptationen innerhalb der Ordo Primates versteht. Während der letzten Dekaden führten signifikante Fortschritte in den biologischen Forschungskonzepten zu wachsenden Erkenntnissen hinsichtlich der Komplexität des Ursprungs und der Entwicklung von *Homo sapiens* aus frühen Primatenvorfahren.

Der Abriß beginnt mit einer kurzen Darstellung der Hauptdisziplinen, die in die Rekonstruktion des einfachen Evolutionsprozesses, der unsere 'Einzigartigkeit' ('uniqueness' *sensu* Foley) hervorbrachte, involviert sind. Menschen sind Primaten und deshalb besteht die Herausforderung der Evolutionsbiologie darin, die Entstehung des Menschen ausschließlich durch Mechanismen der natürlichen sowie der inter- und intrasexuellen Selektion zu erklären. Die Palaeoanthropologie rekonstruiert mittels eines facettenreichen multidisziplinären Ansatzes die evolutiven Adaptationsmuster, die ausschlaggebend für die Hominisation waren. Der vorliegende Beitrag zeigt ausgewählte Aspekte der Humanevolution, insbesondere die methodologischen, z. B. Form-Funktion-Komplexe, ökologisches Nischenkonzept, Mittlertheorie, 'Life history'-Konzept. Drei Fallstudien wurden ausgewählt: die Entstehung der Bipedie, die Migration des frühen *Homo* in die außerafrikanische Alte Welt und schließlich eine taxonomische Diskussion zum Speziationsprozess des Genus *Homo*. Die abschließende Zusammenfassung ist ein Plädoyer für theoriegeleitete Forschung in der Palaeoanthropologie und eine strikte Absage an narrative, rein deskriptive Ansätze.

Keywords: Palaeoanthropology, adaptation, bipedalism, migration, speciation
Palaeoanthropologie, Adaptation, Bipedie, Migration, Speziation

Introduction

The present paper aims to give insight into the research potential of bioarchaeological collections, particularly of human fossil finds. Palaeoanthropology has changed during the last century from a more or less subjective, narrative discipline to a theory-oriented, deductive science (Aiello & Dean 1990, Bilsborough 1986, 1992, Delson et al. 2000, Foley 1987, 1995, Helmuth & Henke 1999, Henke & Rothe 1994, 1999a, 2003, Jones et al. 1992, Larsen 1997, Lewin 1993, Ullrich 1999, Wolpoff 1996-1997).

"*Fossils tell*" or statements of a similar populist nature have been used as titles for dozens of publications and

Scheme of hierarchy and relationships between the components of organisms and environment

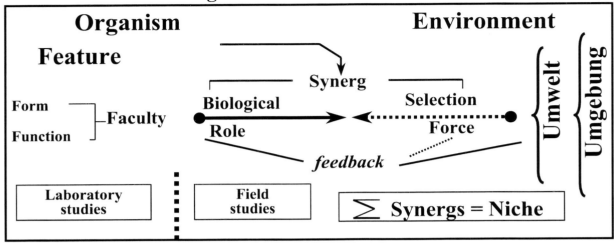

Fig. 1: Simplified scheme for the illustration of the hierarchy and cross-relationships between components of the organism and its "Lebensraum" (live space) (after Bock & von Wahlert 1965, modified, see Henke & Rothe 1994).

exhibitions. Of course, fossils are pathways to the past, but Milford Wolpoff (Univ. of Michigan, Ann Arbor, personal communication) is absolutely right in saying: *"Fossils don't tell – they only give silent witness."*

Robert Foley (Oxford University) adds another important point: *"The past cannot just be invented or imagined, nor reconstructed solely from observations of the way the world is at the present."* Or as L. P. Hartley put it: *"... the past is a foreign country, they do things differently there."* (Foley 1987, p. 78). So what can we conclude if we take both of these approaches into account? All we can do is build a model of our phylogeny! Since we are just modelling, any model will only be useful as long as it is not falsified by another (Darwin 1859, Popper 1994, see also Henke & Rothe 1999a, 2003, Vogel 1999, Wiesemüller et al. 2003).

Methodological aspects

The insight that *"form follows function"* (Fig. 1) is based on the observation that features are form-function complexes, so-called faculties. This simplified scheme by the evolutionary biologists Bock and v. Wahlert (1965) illustrates the hierarchy of and the relationship between the components of organisms and their environment that are pertinent to the understanding of biological adaptation. The bond between organism and environment constitutes the "synerge". It is formed by the biological role and the selectional force. The adaptation of a form-function-complex – morphological or behavioural – is the "faculty," and the sum of all synerges makes up the "ecological niche." For this reason, each adaptation is a compromise between the demands of all synergical relationships it is part of (Eckhardt 2000, Etter 1994, Fleagle 1999, Henke & Rothe 1999a, 2003). Undoubtedly, there are other explanatory models available, e.g., that of optimization. However, because it enables the testing of hypotheses on different levels, this adaptational concept seems to be of special value to palaeoanthropological research. Beside postadaptations, there are predispositions and/or preadaptations (Osche 1983, Vogel 1975, 1999): a trait is considered preadapted if its present form and functions allow for the acquisition of a new biological role, and hence a new synergetic relationship to its environment, whenever this becomes necessary due to the presence of new selective pressures favouring a new adaptation (see Aiello 1990, 1996, Aiello & Dean 1990, Bock & v. Wahlert 1965, Henke & Rothe 1994, 1997a, b, 1999a, 2001, 2003, Martin 1990, Preuschoft & Witte 1991, Rogers et al. 1996, Schmidt-Kittler & Vogel 1991).

Two aspects are relevant when dealing with hominid fossils:
- Firstly, we have to recognise the scarcity of hominine fossils i.e., all attempts to reconstruct the process of hominization are based on tiny samples.
- Secondly, we have to note that evolution displays a mosaic pattern, a continuous, successive change of different organic structures of a given species.

The fact that the analysis of skeletal remains may be carried out on several levels in this context is highlighted in Fig. 2. Some aspects of this table are especially interesting: the leftmost column shows that (as proposed by White 1988, p. 450) any mammal's biology may be classified under at least five basic overlapping categories: anatomy, ecology, demography, behaviour and phy-

	Skeletal	**Contextual**	**Speculation**
Anatomy	Ontogeny Body size Sexual dimorphism		Soft tissue
Behaviour	Locomotion	Sleeping habits Material culture	Communication Mating - Parenting Foraging Social structure Inter- und intraspecific relations
Demography	Disease Longevity	Predation Relative abundance Population density	Natality Mortality Group size Group composition
Ecology	Diet	Distribution Habitat	"Home Range" Territoriality Core area
Phylogeny	Number of species Relation between species	Appearance Extinction	after White 1988, modified

Fig. 2: Illustration of how skeletal and contextual data from the fossil record might be used to reconstruct life styles of fossil hominines. See text for details (after White 1988, modified, see Henke & Rothe 1994).

logeny. The further columns demonstrate that the study of skeletal traits alone can yield information about the ontogeny, body size and sexual dimorphism of a species.

Various aspects of our ancestors' behaviour can be inferred from the study of fossils as well (Chivers et al. 1984, Rogers et al. 1996, White 1988). For example, one can attempt to answer questions about locomotive patterns, sleeping habits, ways of food acquisition and consumption, manipulative abilities, and social interactions of early hominines. As far as conclusions about skin, hair, eyes and comparable features of our ancestors are concerned, the data permit nothing but speculations (Henke et al. 1996). Other topics which may also be addressed on the basis of skeletal analyses of fossil material are demographical, ecological and taxonomical (Jones et al. 1992, Martin 1990).

If we intend to link the past to the present, we need a "middle range theory" *sensu* Binford (1977, 1981) and Foley (1987) (Fig. 3). To build a plausible model on the basis of the interpretation of available fossil data, it is absolutely essential to incorporate our knowledge about comparable living systems and the skeletal remains they leave behind. Unfortunately, palaeoanthropologists ignored this theoretical approach until the second half of the last century. Rather than build scientific models, palaeoanthropology back then mostly produced "palaeopoetry," as Brigitte Senut (Centre National de Recherche Scientific, Paris) once put it.

The development of the "life history concept" was a big step forward on the path towards deciphering ancient bones. Research motivated by hypotheses centred on the "life history concept" (Fig. 4) focuses on the processes that take place throughout the ontogenetical development of members of different species, and seeks to answer questions about their evolutionary significance. Even though this idea is not completely new, as is often claimed (perhaps the reader is still familiar with the work of Jakob von Uexküll), the "life history concept" can be seen as a powerful tool in its own right. Amongst other things, it can be used as a heuristic device that aids in the selection of variables for population studies (DeRousseau 1990). Fig. 4 demonstrates differences between the ontogeny of nonhuman and human primates. Prolonged preadult ontogenetical phases and lifespans can be considered a general evolutionary trend in the order of primates (Schultz 1969). Bearing this trend in mind, taking fossil, primatological and anthropological data into account, one can make use of the "life history concept" to speculate on the life cycles of early hominines (e.g., Jones et al. 1992, Bromage 1990). Still, we have to remind ourselves that attempts to come up with at least approximately correct

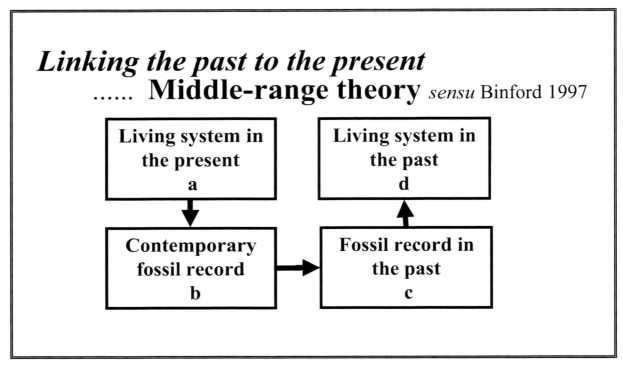

Fig. 3: The route of inference for knowledge about life in the past (d) depends upon observations of the fossil record (c). An interpretation of the fossil sources depends on oberservations of the contemporary fossil formation (b) and of living systems in the present (a) (see Binford 1977, 1981, redrawn).

regressions to calculate the individual ages of the Taung child, the boy from Nariokotome or the Old Man from La Chapelle aux Saints face numerous methodological problems.

These brief preliminary remarks on different palaeoanthropological approaches are meant to emphasize the fact that palaeoanthropology has but one chance to soundly reconstruct the past, and that is: cooperate with all scientific disciplines that are able to contribute to this endeavour (Fig. 5). Not only must our research be firmly rooted in the biological concepts of the system theory of evolution, but it must also integrate all available information on hominization. The challenge is to reconstruct the process of anthropogenesis as the tremendously complex historical and genetical single event that it is.

Palaeoanthropological case studies – major steps in human evolution

From here on we will focus on three issues of special interest in the vast field of palaeoanthropological research.
- Firstly, we will draw attention to bipedalism, both as a human locomotive pattern and as an essential hominine adaptation;
- Secondly, we are going to critically discuss the earliest dispersal of the genus *Homo*, and
- Thirdly, we will raise the question about the number of species within the genus *Homo*.

Bipedalism – a comparative morphological approach

The study of primate fossils is largely a comparative undertaking that relies heavily on comparison with extant primates. In this context, descriptive analyses are just as valid a tool as say, univariate analyses, multivariate analyses, x-ray analyses, computer tomography, or SEM-analyses. Integrated analyses of field and laboratory data enable us to draw conclusions about our ancestors' locomotive abilities. Experimental morphological analyses compare fossil taxa and modern, extant primates, taking experimental studies on locomotive behaviour during ontogenetic stages as well as social-, subsistence- and reproductive behaviour into account. We can arrive at likely palaeoecological scenarios in much the same fashion, but we always have to remind ourselves that the complexities of life only permit uncertain reductionistic reconstructions.

When Australopithecines were first discovered in South Africa, there was little doubt that these early hominines were habitual bipeds (Dart 1925). Soon East African sites were yielding more and more Plio-Pleistocene fossils confirming this opinion. In view of this evidence, most palaeoanthropologists were convinced of the correctness of the "Eastside story" hypothesis, i.e., the idea that a savannah-like East African biotop was conquered by bipedal Australopithecines. The discovery of "Lucy" at the Afar Locality in Ethiopia and the Laetoli

Life history concept

Ranges of normalcy in the eruption of permanent teeth

Panini	M1	M3
Pan troglodytes	3.26 y	11.35 y
Gorilla gorilla	3.50 y	11.40 y
Hominini		
Homo sapiens	6.24 y	20.50 y

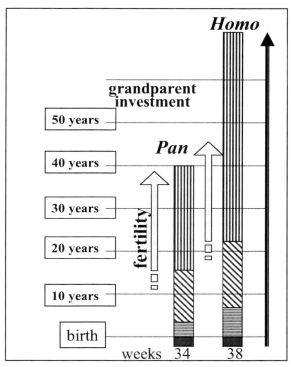

Fig. 4: Compared with other primates humans live slowly. The "life history concept" allows to reconstruct the evolutionary trend which increased the life spans and timing of life-history events (data from Schultz 1969, see Henke & Rothe 1999a).

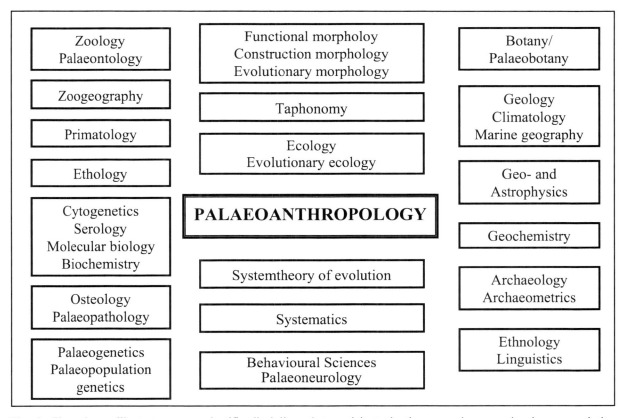

Fig. 5: The scheme illustrates some scientific disciplines that participate in the research concerning human evolution (Henke & Rothe 1994).

footprints, especially, were in accordance with this hypothesis (Johanson & Edey 1981, Johanson & Edgar 1996, Johanson et al. 1978). In spite of this evidence, Rak (1991, p. 283) argued that Australopithecines probably had very original locomotive habits, or as he put it, *"Lucy evidently achieved this mode of locomotion through a solution all her own"*.

The four-million-year-old Allia Bay fossils from the heart of the palaeobiotop of the East African Rift Valley finally proved Yoel Rak right, and profoundly changed the image of early hominines. In particular, the structure of the Kanapoi hominine's wrist (e.g., the oversized hamate and the form of the carpal tunnel) suggested that *Australopithecus anamensis* was probably still a very able climber, since his grip was extremely powerful in comparison to ours (Leakey et al. 1995, 1998, 1999, Stearns 1997, Stern 2000).

Resolving the controversy concerning Lucy's locomotive adaptations is crucial for the palaeoecological discussion. One of several points of disagreement is whether "Lucy" was partially arboreal. There are disagreements as to whether Lucy's curved phalanges represent the bone's response to mechanical stresses in the environment, or are as others suggest, a "hold over" due to a shared ancestry of apes. Paciulli (1995) was able to demonstrate by an ontogenetical analysis of the phalanges of *Pan* that chimpanzees are born with slightly curved phalanges which become more curved in early infancy, while curvature decreases in adulthood. His interpretation is *"....therefore, curvature appears to have a heritable component that is modified by changes in mechanical function."* (Paciulli 1995, p. 165). These findings imply that phalangeal curvature in *Australopithecus afarensis* is probably also a reflection of mechanical, i.e., locomotive functions, and this may be interpreted as an indication of an adaptation to an arboreal habitat.

Any one of a number of recent finds, e.g., *Sahelanthropus tchadensis, Orrorin tugenensis, Kenyanthropus platyops, or Ardipithecus ramidus kadabba* (see phylogenetic model Fig. 6) could possibly be the oldest hominine. Regardless of these discoveries, enough good reasons remain to continue to assume that *Homo* was the first savannah-dwelling hominine and that Australopithecines and earlier hominines still mainly inhabited gallery forests (Henke & Rothe 2003, Rothe et al. 1997, White et al. 1994, 1995).

In comparison with earlier hominids, i.e., fossil African apes and Mio-, Plio- and Pleistocene hominines, a much younger and undoubtedly fully bipedal human ancestor, *Homo ergaster*, displays a number of characteristics which make him a very likely direct ancestor of ours. Like other finds assigned to this taxon, the Turkana Boy had narrow hips and a narrow birth canal, an enlarged brain, an increased body size, small jaws and teeth. What are the effects of these morphological adaptations? Some possible explanations concerning the increase in body size are given in Fig. 7 (Aiello 1990, Larick & Ciochon 1996, Rogers et al. 1996, Walker & Leakey 1993).

Brain and body size seem to have increased simultaneously in *Homo ergaster*. The increased body size was most likely connected with several physiological and behavioural changes, one of which must have been a longer maturation period that was surely related to more intensive parental investment. A larger body needs larger amounts of high quality food, and to satisfy this need more time had to be spent acquiring it. Moreover, a larger body not only meant increased physical strength and improved locomotive abilities (especially to master long distances), but also greater resistance to heat stress. Both improved locomotive abilities and increased heat resistance indicate larger home ranges, prolonged diurnal activity and lower population densities. Accordingly, an increase in body size can be considered an elementary adaptation for the dispersal of *Homo* (Henke & Rothe 1999b, 2003), but the main trigger of hominization was, finally, the dramatic increase in brain functions within the genus *Homo* (see Fig. 8).

Dispersal out of Africa – when and how often?

The second case study concerns the dispersal of early *Homo*. There is no doubt that *Homo* emerged in Africa, but until recently, evidence for early dispersals out of Africa was very scarce. Finds dated to 1.9 – 1.8 million years BP from many different archaeological and palaeoanthropological sites in Asia indicate that the earliest dispersal out of Africa must have started at least two million years ago (Fig. 9). In 1991, the most surprising news arrived from Dmanisi in the Republic of Georgia. A hominine mandible (Fig. 10) had been found at this site, 60 kilometers east of Tbilisi, at the gates of Europe. The hominine fossil, associated with a Villafranchian fauna, has been dated with different methods to approximately 1.8 million years BP. Since the excavation campaign was a joint venture of the Georgian Academy of Sciences and the Römisch-Germanisches Zentralmuseum Mainz, I had – thanks to Prof. Gerhard Bosinski, the former head of the Department of Palaeolithic Archaeology – the chance to analyze the fossil together with other colleagues (Bräuer et al. 1995, Henke 1995, Henke et al. 1999).

One of the results, the output of a Principal Component Analysis (PCA), revealed strong affinities between the Dmanisi mandible and early African hominines, especially early African *Homo erectus,* and even older African specimens (Fig. 10). Cluster analyses pointed towards an affiliation between the Georgian fossil and

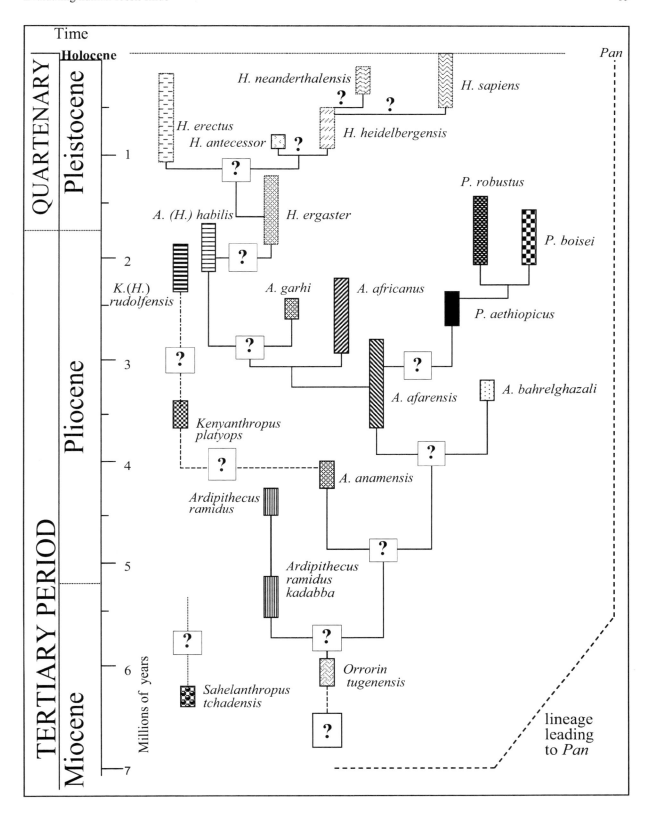

Fig. 6: Hypothesis of evolutionary relationships between recently suggested early hominine taxa (*Tchadanthropus*, *Orrorin*, *Kenyanthropus*, *Ardipithecus*), australopithecine species and the human lineage (genus *Homo*) (Henke & Rothe 2003).

Anatomical Change	Physiological Change	Behavioural Implications
Increase in body size	Longer period of maturation	Greater parental investment
	Increase in need for food quantity and quality	More time spent on food procurement
	Increase in strength and long-distance locomotor efficiency	Larger home range, longer day range, and lower population density
	Greater resistance to heat stress	

Fig. 7: Anatomical and physiological changes reflected in the *Homo ergaster* specimens (adapted from Rogers et al. 1996, see Henke & Rothe 1999a).

Cranial capacity and evolutionary success

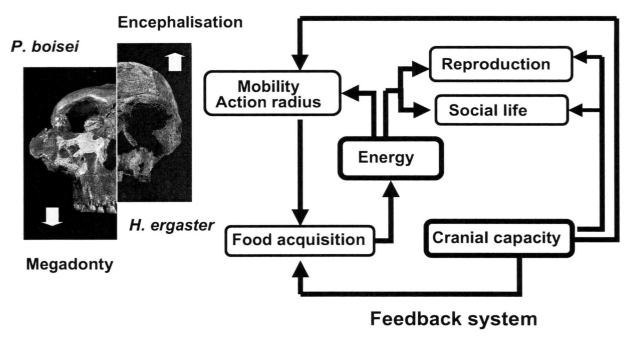

Fig. 8: Energy costs of brain size increase: feedback system (adapted from Martin 1996, modified).

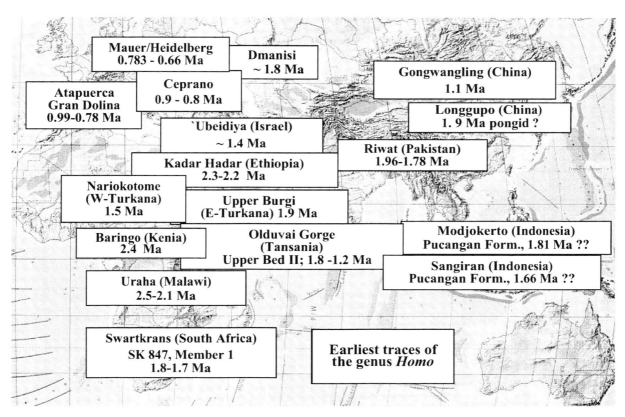

Fig. 9: Dispersal of the genus *Homo*: earliest traces.

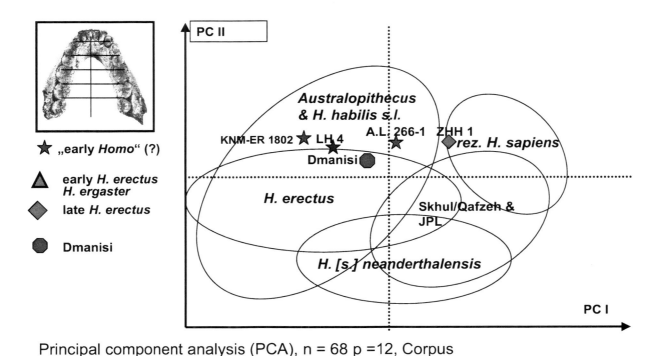

Principal component analysis (PCA), n = 68 p =12, Corpus

Fig. 10: Principal component analysis (PCA), plott diagram illustrating the morphological affinities of the mandible Dmanisi 211 in comparison with other hominine fossils (see Henke et al. 1999b).

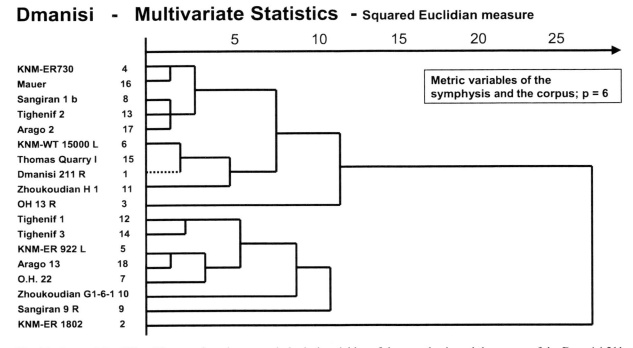

Fig. 11: Squared Euclidian Distances based on morphological variables of the symphysis and the corpus of the Dmanisi 211 mandible (from Henke et al. 1999).

the mandibles from Nariokotome (KNM-WT 15000 L) and the Thomas Quarry (Fig. 11). It also revealed affinities with the finds from Zhoukoudian. However, as the fossil was considered to have been rather insecurely dated, most anthropologists doubted that Dmanisi 211 could be seen as proof of an early dispersal of *Homo* (Henke 1995).

The morphometrical analysis could not convince the scientific community either, but even the most critically minded had to reconsider when two excellently preserved skulls were found in 1999 at the Dmanisi site. Both skulls (D 2280 and D 2282) had very small cranial capacities, resembled *Homo ergaster* in most respects, and finally not only erased the doubts about an early dispersal out of Africa, but proved in fact that the dispersal was a lot earlier than previously thought (Gabunia et al. 1999).

On the fifth of July 2002, in volume 297 of Science Magazine, Vekua et al. published the discovery of yet another hominine skull (D 2700) at the Dmanisi site. This third skull obviously exhibits so many archaic traits that Vekua and his co-workers "...*suggest that the ancestors of the Dmanisi population dispersed from Africa before the emergence of humans identified broadly with the* H. erectus *grade*" (Vekua et al. 2002, p. 85).

The extraordinarily small cranial capacity of 600 cm³ demonstrates that only a comparatively small brain was required to reach new continents and to cope with new habitats. Since that is so, it becomes more and more apparent that we have to thoroughly revise our models concerning the abilities of early *Homo*. A fourth skull and other bones of a postcranial skeleton from the Dmanisi site demonstrate that some 100,000 years after the first appearance of *Homo* in Africa, this genus conquered the non-African Old World. The main question which palaeoanthropologists have to answer is whether the early out-of-Africa dispersal was followed by a second and even a third wave, as the proponents of a recent African Origin of anatomically modern man suggest, or whether the multiregionalists, such as Milford Wolpoff, David Frayer, Jan Jelinek, Alan Thorne, Fred Smith and John Relethford are correct in assuming that the dispersal was a single event without further speciations (Frayer et al. 1993, Relethford 1998, 2001, Smith 1992, Smith et al. 1989, Thorne & Wolpoff 1992, Wolpoff 1984, 1989, 1992, 1996, Wolpoff 1996-1997, Wolpoff et al. 1994, Wolpoff & Caspari 1997a, b).

How many Homo species were there?

The third palaeoanthropological discussion focuses on the question of how many speciation processes took place within the genus *Homo*. In the last few decades, an inflationary number of new hominine species has been proposed. Reinterpretations of species like *Homo heidelbergensis*, *Homo rhodesiensis*, *Homo soloensis*, *Homo helmei* and newly established taxa like *Homo antecessor* were placed alongside wellknown and seemingly likewise accepted species like *Homo habilis*, *Homo rudolfensis* and *Homo erectus* (Table 1, Collard 2002, Henke & Rothe 2001, 2003, Tobias 1989, 1991).

Table 1 Hominine taxonomy: Genera and species designations of former and current taxa; temporal and geographic ranges.
Except *Homo sapiens* all the other taxa are extinct (adapted from Collard 2002, modified; sources of citations see Henke & Rothe 1994, 2001, 2003).

Genus *Homo* Linneaus, 1758 [including the following genera: *Anthropopithecus* Dubois, 1892; *Pithecanthropus* Dubois, 1894; *Protanthropus* Haeckel, 1895; *Sinanthropus* Black, 1927; *Cyphanthropus* Pycraft, 1928; *Meganthropus* Weidenreich, 1945; *Atlanthropus* Arambourg, 1954; *Telanthropus* Broom u. Robinson, 1949]; earliest appearence **in Pliocene, world-wide**.

Homo sapiens Linneaus, 1758. Pleistocene to present, worldwide
Homo neanderthalensis King, 1864. Pleistocene, western Eurasia
Homo erectus (Dubois 1892), Weidenreich, 1940. Pleistocene, Africa and Eurasia
Homo soloensis (Dubois 1940). Pleistocene. SO-Asia
Homo heidelbergensis Schoetensack, 1908. Pleistocene, Africa and Eurasia
Homo helmei Dreyer, 1935. Pleistocene, northern and East Africa
Homo habilis L.S.B. Leakey et al., 1964. Pliocene-Pleistocene, Africa
Homo ergaster Groves u. Mazák, 1975. Pleistocene, Africa and Eurasia
Homo rudolfensis (Alexeev 1986) Wood, 1992. Pliocene-Pleistocene, East Africa and Malawi
Homo antecessor Bermudez de Castro et al., 1997. Pleistocene, western Europe

Genus *Australopithecus* Dart, 1925 [includes the Genus *Plesianthropus* Broom, 1938]. Pliocene, Africa.
Australopithecus africanus Dart, 1925. Pliocene, Africa
Australopithecus afarensis Johanson et al., 1978. Pliocene, East Africa
Australopithecus anamensis M.G. Leakey et al., 1995. Pliocene, East Africa
Australopithecus bahrelghazali Brunet et al, 1996. Pliocene, northern Africa
Australopithecus garhi Asfaw et al., 1999. Pliocene, East Africa

Genus *Paranthropus* Broom, 1938 [includes the Genera *Zinanthropus* L.S.B. Leakey, 1959, *Paraustralopithecus* Arambourg u. Coppens, 1967.] Pliocene-Pleistocene, Africa
Paranthropus robustus Broom 1938. Pleistocene, South Africa
Paranthropus boisei (L.S.B Leakey 1959) Robinson, 1969. Pliocene-Pleistocene, East Africa
Paranthropus aethiopicus (Arambourg u. Coppens 1968) Kimbel et al., 1988. Pliocene, East Africa

Genus *Ardipithecus* White *et al.*, 1995. Pliocene, East Africa
Ardipithecus ramidus (White *et al.* 1994) White *et al.*, 1994, Pliocene, East Africa

Genus *Kenyanthropus* M.G. Leakey *et al.*, 2001. Pliocene, East Africa
Kenyanthropus platyops M. G. Leakey et al., 2001. Pliocene, East Africa

Genus *Orrorin* Senut *et al.*, 2001. Miocene, East Africa
Orrorin tugenensis Senut et al. 2001. Pliocene, East Africa

Genus *Sahelanthropus* Brunet *et al.*, 2002. Miocene, northern East Africa
Sahelanthropus tchadensis Brunet *et al.*, 2002, Pliocene, northern East Africa

Without a doubt, this questionable development merely reflects the different methods and taxonomical principles applied to available data by different palaeoanthropologists and their attitudes towards species recognition (Rothe & Henke 2001, Wägele 2000, Wiesemüller et al. 2003). However, reinterpretations and newly proposed species within the genus *Homo* are not the only noteworthy developments in palaeoanthropology relevant in this context. On the basis of a cladistic analysis, Wood & Collard (1999) have recently reached the conclusion that neither *Homo habilis* nor *Homo rudolfensis* meet the necessary criteria to be included in the genus *Homo* (Fig. 12). Hence, both palaeoanthropologists suggest that the earliest species in the genus *Homo* could be *Homo ergaster* i.e., early African *Homo erectus*. Recent research suggests that *Homo habilis* originated from *Australopithecus afarensis* or a later taxon, while the ancestor of *Homo rudolfensis* could be *Kenyanthropus platyops* (see Fig. 6).

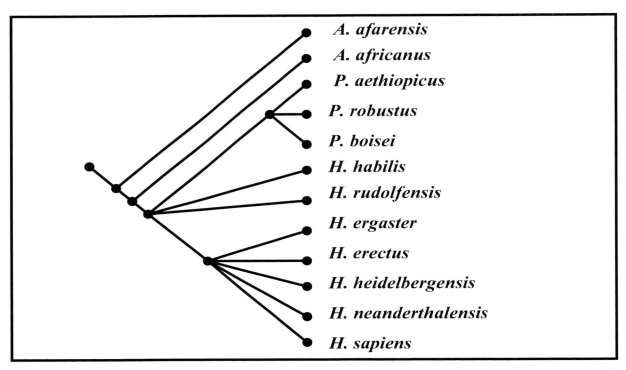

Fig. 12: Consensus Cladogram showing the relationships between hominine species demonstrating that the taxa *Homo habilis* and *Homo rudolfensis* lumper with the australopithecines (after Wood & Collard 1999; redrawn).

Anybody concerned with palaeoanthropology is aware that the nomenclatural history of the *Homo erectus* hypodigm is rather complicated. Originally named *Anthropopithecus* by Dubois in 1892, it was first renamed *Pithecanthropus*, and then later extended to include *Sinanthropus* (Rightmire 1990, 1992, 1998, Henke & Rothe 1994, 2003).

Furthermore, the question whether close, almost contemporary African and European fossils should be placed in the same hypodigm as Asian *Homo erectus* fossils is a matter of controversy to this day. While comparable Asian and African *Homo* fossils are considered to belong to the same hypodigm by many colleagues (Bräuer & Mbua 1992, Kennedy 1991, Rightmire 1992), the answer to the question as to whether *Homo erectus* is a valid European species is still unanswered.

At the same time, even though there is a noticable tendency to accept *Homo heidelbergensis* as a European and African taxon that developed from African *Homo erectus*, i.e., *Homo ergaster* (Rightmire 1999, Tattersall & Schwartz 2000), a recently discovered fossil from Ceprano in Italy, which may represent the earliest specimen of this taxon in Europe, raises certain doubts concerning this opinion (Manzi 1999). The Ceprano find resembles a new *Homo erectus* fossil from Bouri (Ethiopia) that has been dated to 1.1 million years. In the light of this evidence, an early *Homo erectus* expansion along the northern Mediterranean coast or *via* Great Syrte, Sicily, or other possible "land bridges" does not seem too unlikely. Arsuaga et al. (1999) and Carbonell et al. (1999) suggest *Homo antecessor*, a newly described species from the Gran Dolina of Atapuerca (Spain), as the first immigrant.

Whether these early European immigrants were ancestral to *Homo heidelbergensis* or whether another later migration to Europe followed an independent development of *Homo heidelbergensis* in Africa is an unsolved problem. Considering that the different European specimens in question have been variously assigned to *Homo erectus tautavelensis*, *Homo erectus petraloniensis*, *Homo rhodesiensis*, *Homo sapiens* and archaic *Homo sapiens*, one might also wonder if *Homo heidelbergensis* simply constituted an [archaic] *Homo sapiens* (Henke & Rothe 1995, 2001, Henke et al. 1999).

Assuming that the later part of human evolution was a gradual transformation without speciation which began around 500,000 to 400,000 years ago in Africa (as suggested by recent-African-origin proponents) rather than a succession of speciation events, one might as well seriously think about drastically reducing the number of species within the taxon *Homo* (Fig. 13).

If Wolpoff and other multiregionalists are right in supposing that there was no speciation in the evolutionary line leading to "anatomically modern" humans since the emergence of *Homo ergaster* i.e., early African *Homo erectus*, then all *Homo* species, except for *Homo habilis* and *Homo rudolfensis*, should be lumped together (Fig. 14). If that is the case, what about the Neanderthals? The contro-

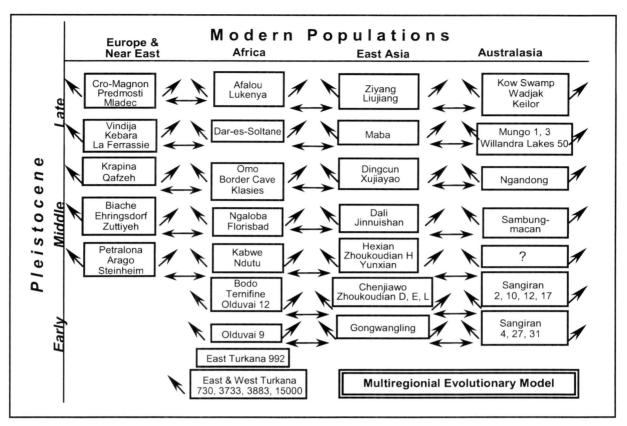

Fig. 13: Multiregional model (after Frayer et al. 1993, Wolpoff 1996-1997, modified).

Fig. 14: Palaeoanthropologists should seriously think about drastically reducing the number of species within the taxon *Homo*.

versy about the phylogenetic role of Neanderthals is as old as palaeoanthropology itself. The phylogenetical status of the Neanderthals is also one of several crucial issues for the multiregional theory. Neanderthals undoubtedly differed morphologically from "anatomically modern" humans, but the morphological data alone do not enable us to make a final decision concerning their taxonomic status (Churchill 1998, Clark & Willermet 1997, Crow 2002, Henke et al. 1996, Henke & Rothe 1999c, Hublin 1998, Tattersall 1995a, b 1997, 2000, Trinkaus 1989). Many, though, think results of recent aDNA analyses do provide us with sufficient additional evidence to make that very decision (Krings et al. 1997). Conclusions drawn on the basis of the results of mtDNA-based aDNA analyses of the fossils from the Feldhofer Grotte, Mezmaiskaya and Vindija seemed to support the now widespread opinion that Neanderthals and "anatomically modern" humans were members of different species (Tattersall & Schwartz 2000).

This interpretation of the results of the just mentioned aDNA analyses had already been repeatedly criticized (cf. e.g., Templeton 1997, Relethford 1998) when a new find from Portugal added yet another twist to the still ongoing debate (Duarte et al. 1999). In 1999, the skeleton of a circa four-year-old child was recovered from the Abrigo do Lagar Velho in Portugal. The find was dated to approximately 24,500 years BP and exhibits a number of traits which seem to prove it to be the result of the hybridisation of Neanderthals and "anatomically modern" humans. This evidence matches new interpretations of results of modern DNA analyses that have been conducted by population geneticists, and which would be in agreement with a "mostly out-of-Africa" model (cf. Relethford 2001, Fig. 15).

Even geneticists who were successful in sequencing mitochrondrial aDNA of different Neandertal specimens have changed their views on this issue and now liken the genetic relationship between Neanderthals and "anatomically modern" humans to that of subspecies of contemporary ape-species (cf. Hofreiter et al. 2001). However, popular as such a "mostly out-of-Africa" model *sensu* Relethford may become in the future, one should not fail to recognize that it is clearly a multiregional and not an "out-of-Africa" model. I am afraid, though, that we will need more sophisticated research methods to produce the type of information which will allow us to conclusively reconstruct our own evolutionary history and come to a final conclusion concerning the phylogenetical status of Neanderthals. Especially with regard to the Neanderthals, we have to ask ourselves whether our scenarios are realistic or merely the product of a "beloved" stereotype. This fictitious dialogue (adapted from João Zilhão, personal communication) between a Neanderthal and a Cro-Magnon man may serve to illustrate this problem:

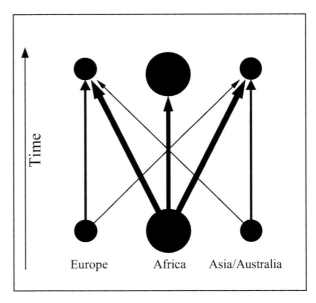

Fig. 15: The "Mostly Out of Africa" model is a multiregional model (!) in which Africa contributes the most to accumulated ancestry in all regions. The width of the lines indicates possible relative contributions in terms of accumulated ancestry (Relethford 2001, redrawn).

Neanderthal: *"I am told you have been fooling around with my sister..."*

Cro-Magnon man: *"Are you crazy? You know very well I am not interested in dating women with suprainiac fossa..."*

Conclusion

The given characterization of the principles of palaeoanthropogical research and the three case studies presented in the course of this presentation imply the following conclusions:
– As matters stand, we have to be very aware of the fact that most of our views on human evolution have a very short half-life.
– We are just modelling.
– Our models should be constructed in accordance with the highest methodological standards and general biological principles.
– Story-telling should be frowned on.

To bring my presentation to a close, finally a quote taken from McHenry (1996, p. 86): *"One needs to make the best of our tiny sample of life in the past, to be open to new discoveries and ideas, and to enjoy the pleasure of learning and changing"*.

Ackowledgements

I am grateful for the opportunity to contribute to these proceedings, especially in the face of the excellent efforts which have been made during the last decades to increase the scientific output of the Collection of Anthropology and Palaeoanatomy in Munich. I would like to thank the organisers of this anniversary, my dear colleagues Prof. Gisela Grupe and Prof. Joris Peters, for the invitation to this event which proved as a really multi- and interdisciplinary approach, according to Glynn Isaac's definition, that *"prehistoric archaeology is [...] in its total aims not a natural science, a social science or a branch of the humanities; rather it is a distinctive pursuit in which all of these meet.* (G. Isaac 1971, see B. Isaac 1989, p. 397)

I am especially grateful to Erik Becker for improving the English version of the manuscript.

Bibliography

Aiello L. C., 1990.
Patterns of stature and weight in human evolution. *American Journal of Physical Anthropology* **81**, 186-187.

Aiello L.C., 1996.
Terrestriality, bipedalism and the origin of language. In: Runciman W.G., Maynard Smith J. & Dunbar R. I. M. (eds.). *Evolution of Social Patterns in Primates and Man*. The British Academy, 269-289.

Aiello L.C. & Dean C., 1990.
An Introduction to Human Evolutionary Anatomy. San Diego, Academic Press.

Arsuaga J.-L., Martínez I., Lorenzo C., Garcia A., Muñoz A., Alonso O. & Gallego J., 1999.
The human cranial remains from Gran Dolina Lower Pleistocene site (Sierra de Atapuerca, Spain). *Journal of Human Evolution* **37**: 431-457.

Bilsborough, A., 1986.
Diversity, evolution and adaptation in early hominids. In: Baily G. N. & Callow P. (eds.). *Stone Age Prehistory*: 197-220. Cambridge, Cambridge University Press.

Bilsborough A., 1992.
Human Evolution. London, Blackie Academic & Professional.

Binford L. R., 1977.
For Theory Building in Archaeology. London & New York, Academic Press.

Binford L. R., 1981.
Bones: Ancient Man and Modern Myths. New York, Academic Press.

Bock W. J. & von Wahlert G., 1965.
Adaptation and the form function complex. *Evolution* **19**: 269-299.

Bräuer G., Henke W. & Schultz M., 1995.
Der hominide Unterkiefer von Dmanisi: Morphologie, Pathologie und Analysen zur Klassifikation. *Jahrbuch des Römisch-Germanischen Zentralmuseums Mainz* **42**: 183-203.

Bräuer G. & Mbua E., 1992.
Homo erectus features used in cladistics and their variability in Asian and African hominids. *Journal of Human Evolution* **22**: 79-108.

Bromage T. G., 1990.
Early Hominid Development and Life History. In: DeRousseau C. J. (ed.). *Primate Life History and Evolution*:105-113. New York, Wiley-Liss..

Carbonell E., Bermúdez J.- M. & Arsuaga J. L., 1999.
Preface. *Journal of Human Evolution* **37**, 309-311.

Chivers D. J., Wood B. A. & Bilsborough A., (eds.) 1984.
Food Acquistion and Processing in Primates. New York, Plenum Press.

Churchill S. M., 1998.
Cold Adaptation, Heterochrony, and Neandertals. *Evolutionary Anthropology* **7**: 46-61.

Clark G. A. & Willermet C. M., (eds.) 1997.
Conceptual Issues in Modern Human Origins Research. New York, Aldine de Gruyter.

Collard M., 2002.
Grades and Transitions in Human Evolution. In Crow T.J. (ed.). *The Speciation of Modern Homo sapiens. Proceedings of the British Academy* **106**: 61-100. Oxford, Oxford University Press.

Crow T.J., (ed.) 2002.
The Speciation of Modern Homo sapiens. Oxford, Oxford University Press.

Dart R., 1925.
Australopithecus africanus: the man-ape of South Africa. *Nature* **115**: 195-199.

Darwin, Ch., 1859.
On the Origin of Species by Means of Natural Selection. London, Murray, Dt. Übers. H. Schmidt 1982, 4. Aufl.

Delson E., Tattersall I., Van Couvering J.A. & Brooks A.S., (eds.) 2000.
Encyclopedia of Human Evolution and Prehistory, 2nd ed. New York, Cambridge University Press, Garland Publishing.

DeRousseau C. J., (ed.) 1990.
Primate Life History and Evolution. New York, Wiley-Liss.

Duarte, C., Maurício J., Pettitt, P. B., Souto P., Trinkaus E., Van Der Plicht H. & Zilhão J., 1999.
The early Upper Paleolithic human skeleton from the Abrigo do Lagar Velho (Portugal) and modern human emergence in Iberia. *Proceedings National Academy of Science, USA, Anthropology*, **96**: 7604-7609.

Eckhardt R. B., 2000.
Human Paleobiology. Cambridge, Cambridge University Press

Etter W., 1994.
Palökologie. Eine methodische Einführung. Basel, Birkhäuser.

Fleagle J. G., 1999.
Primate Adaptation and Evolution. 2nd ed., San Diego, Academic Press.

Foley R. A., 1987.
Another Unique Species. Patterns in Human Evolutionary Ecology. Harlow, Longman.

Foley R. A., 1995.
Humans before Humanity. Cambridge, Mass., Blackwell Publishers Ltd..

Frayer D. W., Wolpoff M. H., Smith F. H., Thorne, A. G. & Pope, G. G., 1993.
The fossil evidence for modern human origins. *American Anthropologist* **95**: 14-50.

Gabunia L., Vekua A., Lordkipanidze D., Justus A., Nioradze M. & Bosinski, G., 1999.
Neue Urmenschenfunde von Dmanisi (Ost-Georgien). *Jahrbuch des Römisch-Germanischen Zentralmuseums Mainz* **46**: 23-38, Tafel VII-XI

Helmuth H. & Henke W., 1999.
The Path to Humanity. Tornoto, Canadian Scholars' Press Inc..

Henke W. in Zusammenarbeit mit Helga Roth und Christian Simon, 1995.
Qualitative and quantitative analysis of the Dmanisi mandible. In: Radlanski R.J. & Renz H. (eds.). *Proceedings of the 10th International Symposium on Dental Morphology*, Berlin 1995: 6-10. Berlin, C. & M. Brünne GbR.

Henke W. & Rothe H., 1994.
Paläoanthropologie. Berlin, Heidelberg, New York, Springer-Verlag.

Henke W. & Rothe H., 1995.
Homo erectus – valides Taxon der europäischen Hominiden? *Bulletin Société Suisse d'Anthropologie* **1**: 15-26.

Henke W. & Rothe H., 1997a.
Zahnphylogenese der nicht-menschlichen Primaten. In: Alt K. W. & Türp J. C. (eds.). *Die Evolution der Zähne. Phylogenie, Ontogenie, Variation*: 229-278. Berlin, Quintessenz.

Henke W. & Rothe H., 1997b.
Zahnphylogenese der Hominiden. In: Alt K. W. & Türp J. C. (eds.). *Die Evolution der Zähne. Phylogenie, Ontogenie, Variation*: 279-360. Berlin, Quintessenz.

Henke W. & Rothe H., 1999a.
Stammesgeschichte des Menschen. Eine Einführung. Heidelberg, Berlin, New York, Springer.

Henke W. & Rothe H., 1999b.
Migrationen früher Hominini – Überlegungen zu Eurytopie, Exogenie und Expansion in Verbindung mit tiergeographischen Befunden. *Beiträge zur Archäozoologie und Prähististorischen Anthropologie* **II**: 28- 35.

Henke W. & Rothe H., 1999c.
Die phylogenetische Stellung des Neandertalers. *Biologie in unserer Zeit* **29**: 320-329.

Henke W. & Rothe H., 2001.
Entstehung des Genus *Homo*. Ein Quantensprung in der Evolution des Menschen. *Praxis der Naturwissenschaften, Biologie in der Schule* **2/50**: 29-33.

Henke W. & Rothe H., 2003.
Menschwerdung. Frankfurt, S. Fischer.

Henke W., Kieser N. & Schnaubelt W., 1996.
Die Neandertalerin. Botschafterin der Vorzeit. Gelsenkrichen, Schwelm, Edition Archaea.

Henke W., Rothe H. & Alt K. W., 1999.
Dmanisi and the early Eurasian dispersal of the genus *Homo*. In: Ullrich H. (ed.). *Lifestyles and Survival Strategies in Pliocene and Pleistocene Hominids*: 138-155. Gelsenkirchen, Schwelm, Edition Archaea.

Hofreiter M., Serre D., Pinar H. N., Kuch M. & Pääbo S., 2001.
Ancient DNA. Nature Reviews. *Genetics* **2**: 253-258.

Hublin J. J., 1998.
Die Sonderevolution der Neandertaler. *Spektrum der Wissenschaft* **7**: 56-63.

Isaac B., (ed.) 1989.
The Archaeology of Human Origins. Cambridge, Cambridge University Press.

Johanson D. C. & Edey M., 1981.
Lucy. Die Anfänge der Menschheit. München, Zürich, R. Piper & Co. Verlag .

Johanson D. C. & Edgar B., 1996.
From Lucy to Language. New York, George Weidenfeld & Nicholson Ltd., A Peter N. Nevraumont Book.

Johanson D. C., White T. D. & Coppens Y., 1978.
A new species of the genus *Australopithecus* (Primates; Hominidae) from the Pliocene of Eastern Africa. *Kirtlandia* **28**, 1-14.

Jones S., Martin R. D. & Pilbeam D., (eds.) 1992.
The Cambridge Encyclopedia of Human Evolution. Cambridge, Cambridge University Press.

Kennedy G. E., 1991.
On the autapomorphic traits of *Homo erectus*. *Journal of Human Evolution* **20**: 375-412.

Krings M., Stone A., Schmitz R. W., Krainitzki H., Stoneking, M. & Pääbo S., 1997.
Neandertal DNA Sequences and the Origin of Modern Humans. *Cell* **90**: 19-30.

Larick R. & Ciochon R. L., 1996.
The African emergence and early Asian dispersal of the genus *Homo*. *American Scientist* **84**: 538-551.

Larsen C.S., 1997.
Bioarchaeology. Interpreting Behavior from the Human Skeleton. Cambridge, Cambridge University Press.

Leakey M. G., Feibel C. S., McDougall, I. & Walker A., 1995.
New four-million-year-old hominid species from Kanapoi and Allia Bay, Kenya. *Nature* **376**: 565-571.

Leakey M. G., Feibel C. S., McDougall I., Ward C. & Walker A., 1998.
New specimens and confirmation of an early age for *Australopithecus anamensis*. *Nature* **393**: 62-66.

Leakey M. G., Ward C. V. & Walker A. C., 1999.
Australopithecus anamensis – a new hominid species from Kanapoi. In: Schultz M. et al. (ed.). *"Internationale Anthropologie"*, Proceedings der Ges. für Anthropologie, Tagung 1998, Göttingen, Cuvillier Verlag, S. 5-9.

Lewin R., 1993.
Human Evolution. An Illustrated Introduction. 3rd Edition. Cambridge, Mass., Blackwell Sientific Publications.

Manzi G., 1999.
The Earliest Diffusion of the Genus *Homo*. Toward Asia and Europe: a Brief Overview. In: Tobias Ph. V., Raath M. A., Moggi-Cecchi J. & Doyle, G. A. (eds.). *Humanity from African Naissance to Coming Millennia*: 117-124. Firenze, Firenze University Press.

Martin R. D., 1990.
Primate Origin and Evolution. A Phylogenetic Reconstruction. London, Chapman and Hall.

Martin R. D., 1996.
Hirngröße und menschliche Evolution. In: Sommer V. (ed.). *Biologie des Menschen. Verständliche Forschung*: 2-9. Spektrum Akademischer Verlag.

McHenry H. M., 1996.
Homoplasy, clades, and hominid phylogeny. In: Meikle W. E., Howell F. C. & Jablonski N. G. (eds.). *Contemporary Issues in Human Evolution*: 77-92. San Francisco, California Academy of Sciences.

Osche G., 1983.
Die Sonderstellung des Menschen aus evolutionsökologischer Sicht. *Nova Acta Leopoldina NF* **55**, Nr. 253: 57-72.

Paciulli L. M., 1995.
Ontogeny and phalangeal curvature and positional behavior in chimpanzees. *American Journal of Physical Anthropology*, Supplement **20**: 165.

Popper K.R., 1994.
Objektive Erkenntnis: Ein evolutionärer Entwurf. 2. Aufl. Hamburg, Parey.

Preuschoft H. & Witte H., 1991.
Biomechanical reasons for the evolution of hominid body shape. In: Coppens Y. & Senut B. (eds.). *Origine(s). De La Bipedie chez les Hominidés*: 59-78. Édition CNRS, Paris.

Rak Y., 1991.
Lucy's pelvic anatomy: its role in bipedal gait. *Journal of Human Evolution* **20**: 283-290.

Relethford J. H., 1998.
Genetics of Modern Human Origins and Diversity. *Annual Review of Anthropolology* **27**: 1-23.

Relethford J. H., 2001.
Genetics and the Search for Modern Human Origins. New York, Wiley-Liss.

Rightmire G. P., 1990.
The Evolution of Homo erectus. Comparative Anatomical Studies of an Extinct Human Species. Cambridge, Cambridge University Press.

Rightmire G. P., 1992.
Homo erectus: Ancestor or Evolutionary Side Branch? *Evolutionary Anthropology* **2**: 43-49.

Rightmire G. P., 1998.
Evidence from facial morphology for similarity of Asian and African representatives of *Homo erectus*. *American Journal of Physical Anthropology* **106**: 61-85.

Rightmire G. P., 1999.
Diversity in the Earliest 'Modern' Populations from South Africa, Northern Africa and Southwest Africa. In: Tobias Ph. V., Raath M. A., Moggi-Cecchi J. & Doyle, G. A. (eds.). *Humanity from African Naissance to Coming Millennia*: 231-236. Firenze, Firenze University Press.

Rogers M. J., Feibel, C. S. & Harris J. W. K., 1996.
Deciphering early hominid land use and behaviour: A multidisciplinary approach from the Lake Turkana basin. In: Magori C. C., Saanane C. B. & Schrenk F. (eds.). Four Million Years of Hominid Evolution in Africa: Papers in Honour of Dr. Mary Douglas Leakey's Outstanding Contribution in Palaeoanthropology. *Kaupia* **6**: 9-19.

Rothe H. & Henke W., 2001.
Methoden der Klassifikation und Systematisierung. *Naturwissenschaften, Biologie in der Schule* **2/50**: 1-7.

Rothe H., Wiesemüller B. & Henke W., 1997.
Phylogenetischer Status des fossilen Neulings *Ardipithecus ramidus*: eine kritische Evaluation gegenwärtiger Konzepte. In: Überseemuseum Bremen – König V. & Hohmann H. (eds.). *Bausteine der Evolution*: 159-168. Gelsenkirchen, Schwelm (Edition Archaea).

Schmidt-Kittler N. & Vogel K. (eds.) 1991.
Constructional Morphology and Evolution. Berlin, Heidelberg, New York, Springer-Verlag.

Schultz A. H., 1969.
The Life of Primates. London, Weidenfeld & Nicolson.

Smith F. H., 1992.
The role of continuity in modern human origins. In: Bräuer G. & Smith F. H. (eds.). *Continuity or Replacement. Controversies in Homo sapiens Evolution*: 145-156. Rotterdam, A.A. Balkema.

Smith F. H., Falsetti A. B. & Donnelly S. M., 1989.
Modern human origins. *Yearbook of Physical Anthropology* **32**: 35-68.

Stearns S.C., 1997.
The Evolution of Life Histories. Oxford.

Stern J. T., 2000.
Climbing to the Top: A Personal Memoir of *Australopithcus afarensis*. *Evolutionary Anthropology* **9**: 113-133.

Tattersall I., 1995a.
The Fossil Trail. How We Know What We Think We Know about Human Evolution. New York, Oxford, Oxford University Press.

Tattersall I., 1995b.
The Last Neanderthal. The Rise, Success, and Mysterious Extinction of Our Closest Relatives. New York, Macmillian, A Peter N. Nevraumont Book.

Tattersall I., 1997.
Puzzle Menschwerdung. Auf der Spur der menschlichen Evolution. Berlin, Heidelberg, Spektrum Akademischer Verlag.

Tattersall I., 2000.
Once We Were Not Alone. *Scientific American* **2**: 56-63.

Tattersall I. & Schwartz J., 2000.
Extinct Humans. New York, Westview Press, A Peter N. Nevraumont Book.

Templeton A. R., 1997.
Out of Africa? What do genes tell us? *Current Opinions Genetical Development* **7**: 841-847.

Thorne A. G. & Wolpoff M. H., 1992.
Multiregionaler Ursprung des modernen Menschen. *Spektrum der Wissenschaft* **6**: 80-87.

Tobias P. V., 1989.
The status of *Homo habilis* in 1987 and some outstanding problems. In: Giacobini G. (ed.). *Hominidae: Proc 2nd Int Congr Hum Paleont, Turin 1987*: 141-149. Mailand, Jaca Book.

Tobias P. V., 1991.
Olduvai Gorge, Vol. 4, parts V-IX. The Skulls, Endocasts and Teeth of Homo habilis. Cambridge, Cambridge University Press.

Trinkaus E., (ed.) 1989.
The Emergence of Modern Humans. Biocultural Adaptations in the Later Pleistocene. Cambridge, Cambridge University Press.

Ullrich H., (ed.) 1999.
Hominid Evolution. Lifestyles and Survival Strategies. Gelsenkirchen, Schwelm. Edition Archaea.

Vekua A., Lordkipanidze D, Rightmire G. P., Agusti J., Ferring R., Maisuradze G., Mouskhelishvili A., Nioradze M., Ponce de León M., Tappen M., Tvalchrelidze M. & Zollikofer, C., 2002.
A New Skull of Early *Homo* from Dmanisi, Georgia. *Science* **297**: 85-89.

Vogel C., 1975.
Praedispositionen und Praeadaptationen der Primaten-Evolution im Hinblick auf die Hominisation. In: Kurth G. & Eibl-Eibesfeldt I. (eds.). *Hominisation und Verhalten*: 1-31. Stuttgart, Gustav Fischer Verlag.

Vogel C., 1999.
Anthropologische Spuren. Zur Natur des Menschen / Christian Vogel, hrsg. von Volker Sommer. Stuttgart, Leipzig, S. Hirzel.

Wägele J.-W., 2000.
Grundlagen der phylogenetischen Systematik. München, Dr. Pfeil Verlag.

Walker A. C. & Leakey R. E. F. (eds.) 1993.
The Nariokotome Homo erectus skeleton. Cambridge, Massachusets, Harvard University Press.

White T., 1988.
The comparative biology of "robust" *Australopithecus*: clues from the context. In: Grine F. (ed.). *Evolutionary History of the "Robust" Australopithecines*: 449-483. New York, Aldine de Gruyter.

White T. D, Suwa G, & Asfaw B., 1994.
Australopithecus ramidus, a new species of early hominid from Aramis, Ethiopia. *Nature* **371**: 306-312.

White T. D., Suwa, G. & Asfaw B., 1995.
Corrigendum. *Nature* **375**: 88.

Wiesemüller B., Rothe H. & Henke W. 2003.
Phylogenetische Systematik. Eine Einführung. Berlin und Heidelberg, Springer-Verlag.

Wolpoff M. H., 1984.
Evolution of *Homo erectus*: The question of stasis. *Palaeobiology* **10**: 389-406.

Wolpoff M. H., 1989.
Multiregional evolution: the fossil alternative to Eden. In: Mellars, P. & Stringer C. (eds.). *The Human Revolution: Behavioural and Biological Perspectives on the Origins of Modern Humans*: 62-108. Edinburgh, University of Edinburgh Press.

Wolpoff M. H., 1992.
Theories of modern human origins. In: Bräuer G. & Smith F. H. (eds.). *Continuity or Replacement. Controversies in Homo sapiens Evolution*: 25-63. Rotterdam, A.A. Balkema.

Wolpoff M. H., 1996.
Neandertals are a race of *Homo sapiens*. *Journal of Human Evolution* **32**: A25.

Wolpoff M. H., 1996-1997.
Human Evolution. New York, The McGraw-Hill Company.

Wolpoff M. H. & Caspari R., 1997a.
Race and Human Evolution. A Fatal Attraction. New York, Simon & Schuster.

Wolpoff M. H. & Caspari R., 1997b.
What Does It Mean To Be Modern? In: Clark G. A. & Willermet C. M. (eds.). *Conceptual Issues in Modern Human Origins Research*: 28-44. New York, Aldine de Gruyter.

Wolpoff M. H., Thorne A. G., Jelinek J. & Yinyun Z., 1994.
The case for sinking *Homo erectus*. 100 years of *Pithecanthropus* is enough! In: Franzen J. L. (ed.). *100 Years of Pithecanthropus. The Homo erectus Problem*. Courier Forschungsinstitut Senckenberg **171**: 341-361.

Wood B. A. & Collard M. 1999.
The changing face of genus *Homo*. *Evolutionary Anthropology* **8**: 195-207.

Contributions of Primatological Collections to modern biodiversity research

Bernhard Wiesemüller, Hartmut Rothe, Institut für Zoologie und Anthropologie, Universität Göttingen

Abstract / Zusammenfassung

Since biodiversity research focuses on past events, the importance of primatological collections, which consist mainly of recent taxa, might seem subordinate. However, the knowledge of recent taxa is indispensable in the understanding of fossil evidence. Therefore, multiple primatological collections are needed for primate biodiversity research.

Zumal sich die Biodiversitätsforschung auf Ereignisse in der Vergangenheit konzentriert, mag die Bedeutung primatologischer Sammlungen, die überwiegend aus rezenten Taxa bestehen, untergeordnet erscheinen. Die Kenntnis rezenter Taxa ist indes unverzichtbar für das Verständnis der fossilen Überlieferung. Daher sind viele primatologische Sammlungen zur Erforschung der Biodiversität der Primaten erforderlich.

Keywords: Biodiversity, Primates, Collections, Phylogeny
Biodiversität, Primaten, Sammlungen, Phylogenese

Introduction

Since biodiversity research mainly deals with the historical reconstruction of diversification of taxa, it is obvious that fossils, as preserved evidence from the past, are important research materials. However, primatological collections, or more generally speaking, biological collections, consist, for the most part, of dead specimens of recent species. There are many large collections of these recent samples all over the world, and their importance at first glance might be less obvious than that of the comparatively small fossil collections. The aim of this paper is to explain why biodiversity research is so highly dependent on these extant or recent species collections.

The importance of studying recent taxa for an understanding of natural history

The interpretation of fossil studies necessarily requires extensive knowledge and an understanding of living species. This is because there are frequently large gaps in fossil records, and therefore many taxic branches cannot be reconstructed from fossil evidence alone (overview in Wiesemüller et al. 2002, p. 129ff), but instead have to be established from data of recent organisms. The Theory of Phylogenetic Systematics (Hennig 1966, 1982) provides the necessary methods for this aim. Studies of taxa without hard tissue, or where the fossil record is extremely poor, in particular, depend on this approach. But also in the case of primates, an understanding of the evolution of soft parts, i.e., tissues and macromolecules, can usually only be observed in specimens of recent species.

However, it is not only the gaps in fossil records that make investigation of recent taxa so important: biodiversity is connected not only to phylogeny, but to other genetic phenomena and external, nonhereditary factors that influence the phenotype as well. Not all these aspects can be recorded from fossils directly, and therefore fossils have to be interpreted against a background of observations of recent taxa. An important aspect in this connection is that organisms change their form and structure during ontogeny (metamorphism *sensu* Hennig 1950). In some taxa, like holometabolic insects, echinoderms or amphibians, these changes can be very drastic (metamorphosis), and only a direct observation of ontogeny can show certain quite different forms to actually belong to the same species, or even to the same individual. Ontogenetic changes can also be extensive in primates, e.g., the development of the secondary dentition. Without a basic knowledge of primate ontogeny, it would be impossible to identify infant, juvenile and adult fossils, but this again, relies on extensive information about recent taxa.

Polymorphism, the ability of individuals of the same species to look quite different, causes another problem that makes investigation of recent species very important. In primates, for instance, the most prominent polymorphism is sexual dimorphism, and indeed, in some primate species, such as orang utans or gorillas, the males and females differ tremendously in size and shape. The hypothetical fossil taxon *Australopithecus afarensis* demonstates how many difficulties this phenomenon can cause. Johanson et al. (1978) interpreted two very distinct forms as males and females of *A. afarensis*, a species with strong sexual dimorphism, whereas Schmid (1989) concluded that they belonged to different species. This problem remains unsolved (Rothe et al. 1997), proving in general, that an extensive knowledge of sexual dimorphism, or other polymorphisms, is necessary for drawing conclusions in such difficult situations. This knowledge, however, can only be acquired through studies of recent taxa.

The aspect of function and adaptation is also very important to the understanding of biological diversification. Again, there is often not much information available about the external surroundings to which a given fossil was exposed. Functional morphology is also highly dependent on investigations of living species to construct models, which can then be applied to fossils.

Finally, the study of recent species is vital for the reconstruction of phylogenetic trees. Living taxa are complete sources of information and therefore needed as starting points for phylogenetic reconstructions. Fossils can be integrated into the reconstruction after the fact (Fig. 1, see also Ax 1989), and the integration of fossils can sometimes correct findings from recent taxa, such as in the case of the thickness of tooth enamel in large hominoids (e.g., Wiesemüller et al 2000; 2002, p. 138f). Consequently, biodiversity research and palaeontological studies would be impossible without background knowledge acquired from recent taxa.

The necessity of collections

As stated, investigation of living or recent species are indispensable for the understanding of biodiversity, but the question as to why large collections of dead specimens from these recent species are needed may still arise. For rare species, the answer is very simple. Individuals of these species are hard to find, and every invasive study of such individuals endangers the survival of an already endangered taxon. Consequently, all dead specimens of rare, endangered species are very important for studies of extant taxa. On the other hand, dead specimens of more common species are of equal importance for certain studies. For instance, Wiesemüller & Rothe (in review) performed a brief study on the distribution of the entepicondylar foramen on the humeri of callithrichids. The 168 postcranial skeletons studied were derived from one and the same collection, thus

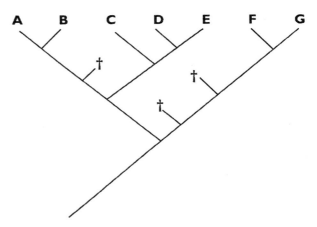

Fig.1: A scientific reconstruction of phylogenetic relationships between species usually starts with an analysis with data from recent taxa. Fossils, symbolized by crosses, are integrated into the reconstructed tree afterwards.

enabling the study of the left and right humeri to be completed in a relatively short time of several hours. Without primatological collections, this study would have required the catching, anaesthetizing and x-raying of 168 monkeys, partially from zoos and partially from the wild. Then again, some of the x-ray photos may still not have yielded clear results, simply because it is difficult to position a living, anaesthetized monkey with both humeral trochleae in a plane position. Consequently, studies like this would cost too much effort to be ever performed, and hence, emphasizes the necessity for many biodiversity studies to have collections of dead specimens available.

Many collections are needed

The final question that needs to be addressed is why we need to have many *and* globally dispersed collections. The first and most obvious reason is that rare taxa are only represented in certain collections. At this point it might seem sensible to group all these collections together at one place to facilitate access. However, there is at least one important reason why this is counterproductive: the risk of curating such rare material that can neither be duplicated or replaced in one place and having it destroyed in a single chance catastrophe is unacceptable. Furthermore, a centralised collection would probably also hinder international research programmes because researchers from distant countries would have fewer opportunities to see the collection.

Conclusion

The reconstruction of biodiversity depends on the investigation of recent taxa. For this purpose, multiple, large collections of specimens of recent taxa are required.

Bibliography

Ax P., 1989.
 The integration of fossils in the phylogenetic system of organisms. *Abhandlungen des Naturwissenschaftlichen Vereins in Hamburg (NF)* **28**: 27-43.

Hennig W., 1950.
 Grundzüge einer Theorie der phylogenetischen Systematik. Berlin: Deutscher Zentralverlag.

Hennig W., 1966.
 Phylogenetic Systematics. Urbana: University of Illinois Press.

Hennig W., 1982.
 Phylogenetische Systematik. Berlin: Verlag Paul Parey.

Johanson D.C., White T.D., Coppens Y., 1978.
 A new species of the genus *Australopithecus* (Primates: Hominidae) from the Pliocene of Eastern Africa. *Kirtlandia* **28**: 1-14.

Rothe H., Wiesemüller B., Henke W., 1997.
 Phylogenetischer Status des fossilen Neulings *Ardipithecus ramidus*: eine kritische Evaluation gegenwärtiger Konzepte. *Symposium Übersee-Museum Bremen*: 159-168. Gelsenkirchen: Edition Archaea.

Schmid P., 1989.
 How different is Lucy? In: Giacobini G. (ed). *Proceedings of the 2nd International Congress of Human Paleontology*, Turin 1987: 109-114. Milano: Yaca Book.

Wiesemüller B., Rothe H., (in review).
 Taxonomic Distribution of the Entepicondylar Foramen in Callithrichids. *American Journal of Physical Anthropology*.

Wiesemüller B., Rothe H. Henke W., 2000.
 Good Characters and Homology. *Evolutionary Anthropology* **9**: 151-152.

Wiesemüller B., Rothe H., Henke W., 2002.
 Phylogenetische Systematik. Eine Einführung. Berlin: Springer.

Part III:

Man and his Animal World

Bone artefacts and man – an attempt at a cultural synthesis

Cornelia Becker, Seminar für Ur- und Frühgeschichte, Freie Universität Berlin

Abstract / Zusammenfassung

This paper poses the question of how far we have come in our attempt to understand the significance of bone artefacts for the recognition of human behaviour and cultural development. It aims to provide insight into the relations between bone artefacts and man from the first use of simple items by early hominids and Neandertals to the sophisticated bone industry in the Roman era. During this long period, bone working fluctuated considerably, ranging from the manufacture of ad-hoc implements to highly elaborated objects of prestige, from musical instruments to artefacts of narrative and symbolic connotation. Fundamental questions as to the procurement of raw material, to the level of bone technology and to the mental abilities and skills of prehistoric people are raised and discussed according to different archaeological contexts (camp-sites, settlements, burials). Prominence is given to those data which characterise best the specific repertoire of bone artefacts in a particular period and which have to be seen in close association with technological achievements and changes such as the artistic "explosion" in the Upper Palaeolithic, specialised hunting techniques during the Mesolithic, spinning and weaving, mining and pottery production during the Neolithic or the harnessing of horses in the Bronze and Iron Ages. Furthermore, a methodological approach is presented to illustrate the vast spectrum of information that can be drawn from a trans-disciplinary analysis of bone artefacts. From that basis, a number of themes are pursued leading to a more complex description of the function and meaning of such items. Bone artefacts not only provide a window into subsistence strategies, exploitation of raw material, skill and inventive abilities, but also to social structures, interaction and ritual behaviour, as well as concepts of the after-life.

Der vorliegende Beitrag beschäftigt sich mit der Frage, inwieweit es bisher gelungen ist, die Bedeutung von Knochenartefakten im Hinblick auf Erkenntnisse über menschliches Verhalten und kulturelle Entwicklungen richtig einzuschätzen. Es wird versucht, einen Einblick in die Beziehung zwischen Knochenartefakten und Menschen zu geben, angefangen vom Gebrauch einfacher Knochenstücke durch frühe Hominiden und Neandertaler bis hin zum Knochenhandwerk in der Römerzeit. In dieser langen Abfolge unterliegt die Knochenverarbeitung starken Schwankungen; sie reichen von der Herstellung einfacher Gelegenheitsgeräte zu kunstvoll verarbeiteten Prestigeobjekten, von Musikinstrumenten bis zu Gegenständen mit erzählendem oder symbolischem Charakter. Es werden Fragen zur Beschaffung von Rohmaterial, dem Niveau der Bearbeitungstechnik und den mentalen wie handwerklichen Fähigkeiten prähistorischer Menschen gestellt und in Verbindung mit Befunden aus verschiedenen archäologischen Kontexten wie Jagdstationen, Siedlungen und Bestattungen diskutiert. Dabei wird solchen Daten Vorrang eingeräumt, die das spezifische Repertoire von Knochenartefakten einer Epoche am eindrucksvollsten widerspiegeln und die in enger Verknüpfung zu technologisch neuen Errungenschaften und Veränderungen zu sehen sind, wie das scheinbar schlagartige Aufkommen künstlerisch gestalteter Knochenobjekte im Jungpaläolithikum, spezialisierte Jagdtechniken des Mesolithikums, Spinnen und Weben, Bergbau und Keramikproduktion im Neolithikum oder die Anschirrung von Pferden in der Bronze- und Eisenzeit. In einer kurzen Erläuterung zur Methodik wird deutlich gemacht, dass sich das breite Spektrum an Informationen erst im Rahmen transdisziplinärer Analysen von Knochenartefakten erschließt. Erst auf dieser Basis kann deren Beschreibung zunehmend an Komplexität und die Interpretation von Funktion und Bedeutung an Plausibilität gewinnen. Knochenartefakte öffnen nicht nur ein Fenster des Wissens in Richtung Ernährungsstrategien, Ausbeute von Rohmaterial, Fähigkeiten und Innovationsvermögen, sondern auch zu sozialen Strukturen und Wechselwirkungen, zu Ritualen und Glaubenskonzepten.

Keywords: Bone artefacts and man, specific cultural contexts, early hominids to Roman craftsmen, methods, range of results and interpretation
Knochenartefakte und Mensch, kulturspezifische Kontexte, frühe Hominiden bis römische Knochenschnitzer, Methoden, Aussagepotential

Introduction

The relationship between man and bone artefacts spans many millennia from the early hominids who lived more than one million years ago to the people of the 21st century. Even today, a walk through a flea-market may provide us with useful or decorative items such as buttons, crochet hooks, curtain rings, spatulae, hairpins and spoons (Jünger 2002), not to mention beads and pendants, strung on necklaces and bracelets, all made of bone. On this occasion, I would like to focus on bone artefacts in the prehistoric period. When speaking of "bone artefacts" I include also antler, tooth and to a lesser extent ivory as well while recognizing that in a physiological term the latter are quite distinct.

Whenever a bone artefact is found in an archaeological context, an ambitious excavator will try to visualise the process of creating it and to share the aspirations of its maker. Seen from the perspective of an archaeozoologist, however, bone artefacts are items with somewhat fluctuating fortunes. Only occasionally, they take centre stage in archaeological research, are documented in detail and interpreted as carefully as any other archaeological find. In the course of studying the literature I became increasingly aware of the fact that comprehensive analyses featuring bone artefacts are the exception rather than the rule. I have the feeling that the enormous potential of information held by bone artefacts, consequently, is ignored. This potential can easily be acknowledged through the variety of artefact types identified from prehistoric sites in Europe to date (Tab. 1). Each type may be assigned to a particular aspect of ancient life such as hunting and fishing, reflected in implements like spear-throwers, projectiles, harpoons and gorges, the protection of the body against harsh environs with cloths and tents made of leather as mirrored by perforators and needles, and so forth (Tab. 2). We should be aware of the fact that some of these items can be placed under one of several classifications, even more so if the function is debatable. A pin may be a useful object for leather work, but it could also be a hairpin or a utensil for tattooing.

The more ancient and spectacular the find is, the more compelling a comprehensive deciphering becomes. To decipher means to identify through meticulous analysis how, by whom and for what purpose a piece of bone was processed. That sounds simple, but it involves a large number of different aspects and methodological problems that will be touched upon below.

We know that manipulative skills and a dependence on tools to exploit resources and make vital equipment are hallmarks of the development of the human species in general. It has been claimed that through worked bones, traces have been left on the world which are not only connected with certain techniques, but also with innovative thinking, imagination, communication, art and religion. For that very reason, an artefact may constitute one means of understanding human behaviour and cultural development. This is even more important for those periods for which our knowledge still is extremely meagre: the Lower and Middle Palaeolithic. M. Kokabi and his co-authors claim in their catalogue "Knochenarbeit – Artefakte aus tierischen Rohstoffen im Wandel der Zeit" (Kokabi et al. 1996: 29) that it is only from the Aurignacian onwards, that the processing of bone, antler and ivory can be ascertained. Evidence has been gathered, however, to show that such activities reached much further back.

The earliest evidence

One of the oldest pieces of evidence is an assemblage of 70 bone artefacts excavated from layers at the Swartkrans Cave in South Africa, dating to about 1.8 to 1 million years ago (Kuckenburg 2001: 38ff.). Six of these are redrawn in Figure 1. Microscopic analysis (cf. Fig. 1, bottom line) has revealed that these bone "spatulae" were used for digging, perhaps in search of termites and roots. As far as I am able to judge from the illustrations, processing of the bones at best involved a rough shaping. One might critically ask whether these implements deserve to be called "artefacts" in the proper sense of the word. I tend to think that they acquired their final shape through their utilisation as digging sticks and not through purposeful shaping. It has to be stressed that between true artefacts and bone fragments with unspecific traces (for example those produced during butchery), a "grey zone" exists in which one might place less extensively worked bones with only minor traces of use as well as previously misinterpreted finds such as "pseudo-tools" (e.g. Brain 1976; Olsen 1989; Lyman 1984) and maybe even items which have been utilised by inventive bonobos, chimpanzees and macaques (Paul 1998: 205ff.).

Other comparably famous sites with remains of early humans as well as considerable amounts of refuse from beasts of prey can be quoted which could also be examined according to bone artefacts: the Klasies River Mouth Cave on the Tsitsikamma coast in South Africa (for further information on "bones with parallel grooves" see Stringer & Gamble 1993: 161) and the Dmanisi site in Georgia (Gabunia et al. 1999).

In Europe, a camp-site which initiated further discussion on ancient bone manipulation is Bilzingsleben in Thuringia, Germany. The site dates to about 400,000 BP. According to D. Mania (1990a, b, 1998), it has yielded a rich faunal assemblage and a certain quantity of artefacts which were processed from red deer

adze	disc	polisher
amulet	drilled astragali	projectile
antler band	drum stick	punch
antler beaker	figurine	rake
antler head-gear	fish hook	rattle
antler pick	fish spear	receptacle
antler rake	flax hackle	retoucher
arm-guard	fleshing tool	ring
arrow-head	flute	*sagaie*
awl	gorge	seal
axe	gouge	scraper
axe sleeve	haft	scapula scoop
baguette	hairpin	skate
basketry pin	hammer	skinner
bâton de commandement	handle	shovel
bead	harpoon	skewer
belt-hook	hilt	smoother
blast horn	hinge	socket
bobbin	knife	spacer-bead
bone skate	limpet scoop	spade
boomerang	macehead	spatula
borer	mattock	spear-head
bow	mount	spear-thrower
bowl	music bow	spindle
bracelet	needle	spindle-whorl
brooch	needle case	spoon
bull-roarer	net needle	stamp
burnisher	net sinker	strap loop
button	ornaments*	strap distributor
cheek-piece	peg	thong smoother
chisel	pendant*	toggle
comb	percussion instrument	tweezer
container	perforator	weaving comb
counting bar	plaque	wedge
cup	plate	whistle
dagger	pin	worked beaver mandible
denticulated ribs	pipe	
digging stick	planting stick	

Tab. 1: Range of bone and antler artefacts from prehistoric periods (arranged in alphabetical order; ornaments* and pendants* occur in the form of perforated teeth, mandibles, vertebra, antler burrs and many other items).

antler (n >225), mammoth bones (n = 120) and ivory (n = 4). It is presumed that they were used as tools mainly for processing carcasses and other food sources, for leatherwork and woodworking. Personally I retain some doubts as to whether all the items illustrated in the publication under the heading of "Geräte" (Mania 1990b: 148ff.), can be characterised as "artefacts". Nonetheless, at least some of them exhibit traces of manipulation in the form of retouched edges (Fig. 2.1) and rounded tips (Fig. 2.2). Moreover, Mania presents comparable finds from other Middle Palaeolithic sites such as Malagrotta, Castel di Guido and Torralba (Mania 1990b: 172). S. Gaudzinski (1998b: 199) comments on the Bilzingsleben items that a manipulation by carnivores, especially hyenas, may yield very similar traces – an observation that should be brought to mind more often while considering Middle Palaeolithic "tools". The most debatable finds from Bilzingsleben, however, are those with seemingly regular cuts on their surfaces (Fig. 2. 3, 4). The excavator assumes these lines to be the oldest documentation of human thought (Mania 1998: 60). He is of the opinion that the *Homo erectus* from Bilzingsleben had the ability of abstract thinking and verbal communication as visualised in these marks. Mania rejects the idea that they could have been produced during the preparation of meat or the cutting of leather using these flat bones as pads (cf. Fig. 2. 5). In my personal experience, mundane activities are inherently more plausible. This may also be true for some other early bone finds with cutmarks, notably those believed to be of symbolic value (Stringer & Gamble 1993: 161).

Hunting – Gathering – Fishing

Preparation of food

Leatherwork – Textile production – Wattle work

Fieldwork (agriculture)

Woodworking

Flint-knapping

Ceramic production

Personal equipment – Prestige and social ranking

Household equipment

Activities in 'leisure times' – Music – Play

Transfer of information

Religion and ritual – Burial contexts

Horse harness

Warfare

Tab. 2: Activities of prehistoric men in which bone artefacts played a major role.

At Salzgitter-Lebenstedt, a Middle Palaeolithic site south of Hanover, Germany, dated to ca. 55,600 ± 900 years BP, an important collection of bone artefacts was excavated (Gaudzinski 1998a, b, 1999). Twenty-nine bone artefacts were unearthed, namely processed mammoth fibulae and ribs, as well as one processed antler (Fig. 3. 2-5); the most remarkable item is a carefully shaped point made of a long bone fragment (Fig. 3. 1). The function of these tools is questionable. The authoress could reasonably argue that Salzgitter-Lebenstedt is one of the oldest sites where a purposeful selection of elements for bone processing can be evidenced (Gaudzinski 1998a: 329). Her view on Palaeolithic bone artefacts in general has to be stressed, for she claims the Salzgitter-Lebenstedt items to be almost unique. Only from Predmostí in the Czech Republic (Valoch 1982) and from some sites in France (Vincent 1988) can comparably reliable finds be noted.

A. Vincent (1988, 1993) has analysed bone material from Middle Palaeolithic sites in France (Bois-Roche, Grotte Vaufrey). She was successful in deciphering traces of a variety of techniques in bone manipulation such as "pointe libérée par retouches alternes" and "racloir bifaces" in addition to perforating, sawing and scraping (Vincent 1988: 187ff.). In particular her mention of bone working by means of a flaking technique deserves major attention. This idea was already noted by A. Breuil in 1932 while Mania reached a basically similar view, concerning material from Bilzingsleben (see above). Two other examples from the Middle Palaeolithic can be quoted: first, the site of Clacton-on-Sea, England (Singer et al. 1973) with faunal material and stone tools in primary contexts and excellent conditions of preservation[1]. Again, evaluation of the bone technology is problematic, since only three implements were identified: "a bar-hammer of deer tibia, a polished deer radius and a stumpy, apparently flaked, piece of rhinoceros [bone] resembling a pebble chopper-core" (ibid.: 43). The second site worth mentioning is the Large Grotto of Blaubeuren, Germany, where a carefully shaped bone point from a Middle Palaeolithic context was recovered which resembles very much points from Upper Palaeolithic contexts (Veil 1999: 146f.).

Whether or not these and other rare items adequately represent the former situation, cannot be decided for the moment, since we presently lack the full picture. Despite some exceptional cases of bone manipulation, as already mentioned, it seems that a general exploitation of bone remains doubtful (d'Errico & Giacobini 1988; Mithen 1998). However, one might also agree with A. Vincent's estimation that "...nous sommes donc en mesure d'envisager une réelle technologie de l'os, aussi balbutiante qu'elle soit, pour ces époques" (Vincent 1988: 191). This scanty evidence of bone processing can perhaps be explained by a certain lack of development of human manipulative skills which were largely dependent on the manual dexterity (Trinkaus 1992; Jeannerod 1997). Undoubtedly, "manual intelligence" (Wilson 2000: 71ff.) was indeed already developed at that time, but the range of grip positions may still have been limited. A certain lack of fine control might additionally have reduced the level of technology which could be achieved. Or was it perhaps a missing "mental" ability? This pivotal question will be discussed later. It must be stressed that according to an increasing consensus among anthropologists, hominids inhabiting Europe at that time are considered to be predecessors of the Neandertals. They are currently labelled "Ante-" and "Pre-Neandertals" rather than *Homo erectus* (Henke et al. 1996: 89ff.; Henke & Rothe 1998: 203).

If we turn now to the era of the "classical" Neandertals who roamed Europe and Eurasia from ca. 100,000 to 35,000 BP, again the question arises: did they process bone in major quantities? Although the story of the Neandertals fills hundreds of scientific publications (see the long cast of characters involved in the discovery and interpretation of Neandertals, as compiled by Trinkaus and Shipman in 1993), the information on this particular field of research is extremely meagre. Stringer & Gamble's book entitled "In Search of the Neanderthals" (1993) sets an example: one finds not a single drawing nor any mention of a true bone artefact in the text, apart from a short chapter about "Middle Palaeolithic artifacts with apparently deliberate marking" and the question of

---

,1. as reflected, for example in the recovery of a wooden spear (Singer et al. 1973: 43).

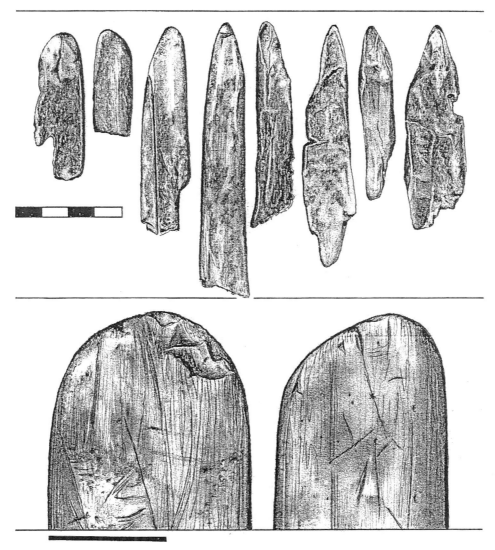

Fig. 1: Swartkrans Cave, South Africa (1,8 to 1 million years). Bone tools (top line) and close-up of tips (bottom line). Bottom line, right is an experimentally used modern bone tool (all redrawn from Kuckenberg 2001: 39 and 42 by H. Hähnl; scale in cm).

whether such traces could symbolise art or some sort of an abstract code (ibid.: 160f.). G. Bosinski summarises as follows (Bosinski 1991: 44): "Geräte aus Geweih, Elfenbein und Knochen sind in der Zeit der Neandertaler außerordentlich selten. Dies kann nicht durch die Erhaltungsmöglichkeiten erklärt werden, denn deren Materialien sind in kalkhaltigen Schichten, zum Beispiel in Höhlen oder im Löß zahlreich überliefert." Is there any plausible explanation for this surprising lack of bone artefacts?

Even until the 1990s, some anthropologists viewed the Neandertals as subhumans of low intelligence; the remains found at several Neandertal camp-sites lead to controversial discussions among anthropologists and archaeologists, as reflected for example in the many contributions at the Liège conferences in 1986 and 1993 (Otte 1988, 1996). But, as recent excavations and research have demonstrated, the former simplistic view has had to be corrected (Henke et al. 1996: 100; Henke & Rothe 1998; Zilhão & d'Errico 2000). In general, the mental skill of the Neandertals must have been of a high order, for they were able to adapt to a variety of extremely diverse environments within the vast area of their distribution from Spain to Eurasia and the Middle East. Moreover, during their long period of existence they coped with varying climatic changes. Within the Neandertal populations, the existence of a quite complex social organisation can be proposed including particular burial practices[2] (Bosinski & Henke 1993; Henke et

---

2. Burials are known, e.g., from La Chapelle-aux-Saints, La Ferrasie, Le Moustier, La Quina, Saint Césare, Kebara, Amud, Teshik Tash, Shanidar (Orschiedt et al. 1999).

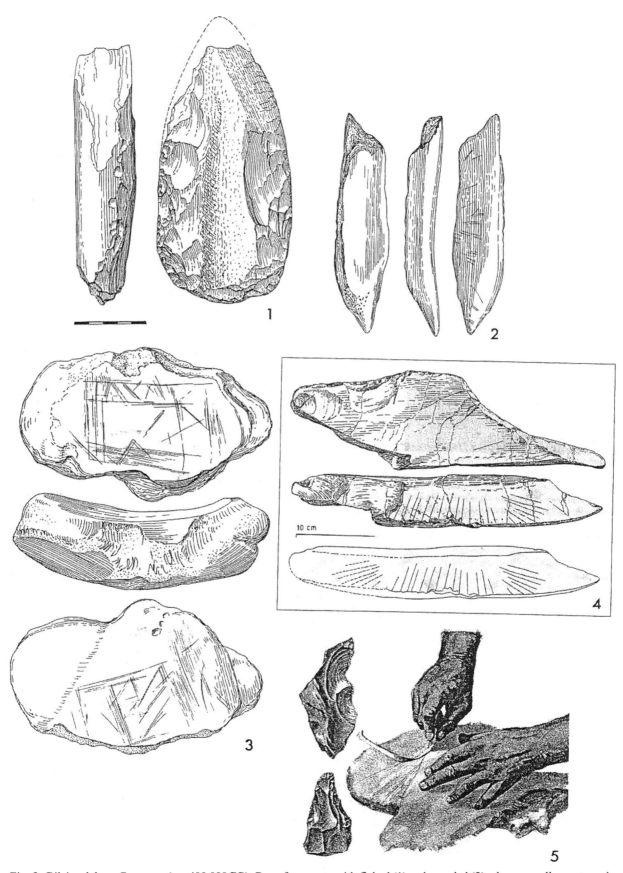

Fig. 2: Bilzingsleben, Germany (ca. 400,000 BP). Bone fragments with flaked (1) and rounded (2) edges as well as cut-marks (3, 4; scale in cm or as indicated; compiled from Mania 1990b: 183, 165; Mania 1990a: 20); 5 reconstructed cutting of leather (from Mania & Dietzel 1980: 95).

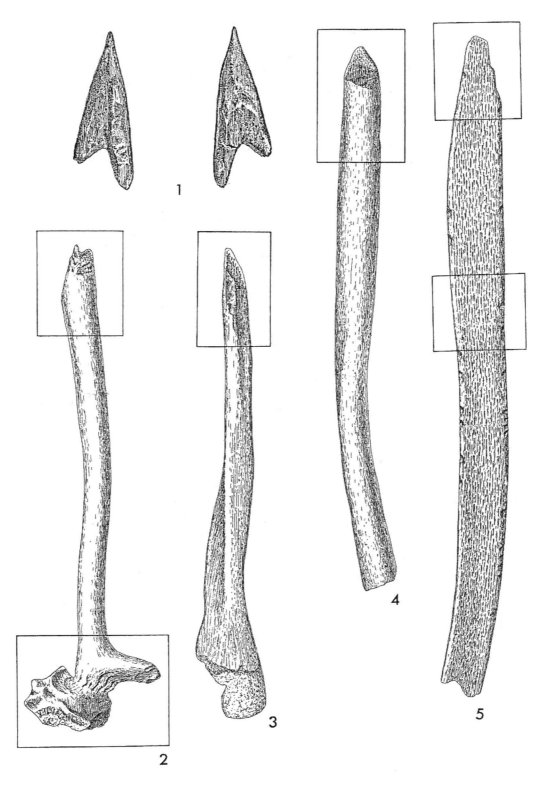

Fig. 3: Salzgitter-Lebenstedt, Germany (Middle Palaeolithic). Bone and antler artefacts (framed areas = sections with specific traces of manufacture and/or use): 1 bone point (original length: 6,3 cm, redrawn by H. Hähnl after Gaudzinski 1998b plate 9), 2 reindeer antler (original length: 55,2 cm), 3 mammoth fibula (original length: 54,4 cm), 4 and 5 mammoth ribs (original length: 63,3 and 72,5 cm; compiled from Gaudzinski 1998b plates 15, 26, 31, 34).

al. 1996; Henke & Rothe 1998), care of weak and sick members of the group (the dawn of empathy?), and a variety of hunting strategies as well as scavenging (Stringer & Gamble 1993: 163). Through hunting, scavenging and food processing, Neandertals necessarily became acquainted with bone material. Abundant fossil and archaeological records attest a continuing refinement of abilities and techniques: Neandertals worked pieces of flint into a variety of utensils such as knives, scrapers, points and blades. Considering all this one would expect at least a certain amount of bone artefacts within the repertoire of finds at Neandertal sites. But there are almost no such items, apart from those mentioned earlier. Poor preservation cannot be blamed because there are several Neandertal sites with excellent preservation conditions for bone material. A lack of skill or manipulative prowess can also be excluded: although the arms and hands of Neandertals were extremely muscular and their grip was powerful, the hands could perform delicate operations (Henke et al. 1996: 30, 101; Stringer & Gamble 1993: 93; Henke & Rothe 1998: 248). A lack of attention on the part of the excavators, ignoring exactly this category of finds, seems to be less probable, although roughly processed pieces of bones such as the ad-hoc implements could have been easily disregarded. There is one small observation which would support such an assumption: during flint-knapping particular tools of stone, antler, bone and teeth were used, the retouchers. The kind of damage which characterise a piece of bone, antler or teeth as a retoucher, is minute: small sections with closely clustered grooves do occur, in German called "Narben-Felder" (Taute 1965: 86ff. and plate 14). Such features are not conspicuous and could in any case be misinterpreted as gnawing marks (Fig. 4). Actually, W. Taute (1965) has collected more than one hundred references of retouchers from a widely scattered literature; many of them were made of bone material and at least twenty of them are considered to be from European Middle Palaeolithic contexts. Consequently, at least during a certain period of their existence, Neandertals or Pre-Neandertals did use such implements. In the light of Tautes' assessment, the idea that the Neandertals were totally ignorant of bone utensils has to be rejected (on which see Chase 1990; Vincent 1988).

Nonetheless, the paucity of bone artefacts remains striking. A solution may be sought in the Neandertals themselves: were they uninventive in the matter of bones or simply not sufficiently curious to recognise the potential of bone as raw material? Were they for some reason simply not stimulated to use bones for utensils? Did they not <u>need</u> such implements? Maybe their demands were met by wooden items which have left no trace in the archaeological record[3]. Or maybe, they used their teeth for certain procedures which in later periods were carried out using bone implements. This "teeth-as-utensils-hypothesis" (Henke et al. 1996: 98) is indicated from particular features such as extreme wear, cracks and grooves at the Neandertals' front teeth as well as from their skull morphology. It points to a major use of the masticatory apparatus for activities which were not primarily connected with eating. "Despite all our enticing new clues, the Neandertals remain a people of mystery" Rick Gore claimed in 1996 (p. 6). The case of bone artefacts quite obviously belongs to these black holes in our knowledge.

However, decorative items associated with Neandertals do occur, although very late and in small numbers. G. Bosinski draws our attention to a pendant, made from a wolf's caudal vertebra; it was found in Bocksteinschmiede in the Lone valley, Germany (Bosinski 1991: 44f.). The Mousterian site of Bacho Kiro has yielded some remarkable items in the form of perforated animal teeth. They are said to be some of the earliest fully credible ornaments of the period (Koslowski 1982). A great deal of attention has been given to the bone finds from La Grotte du Renne at Arcy-sur-Cure, one of the rare Upper Palaeolithic sites in France which yield stone and bone collections both from Mousterian and Châtelperronian levels, the latter dating to ca. 33,000 BP (Movius 1969; Combier 1990). The Mousterians as well as the people of the Châtelperronian were Neandertals (Bosinski 1990: 37ff.; White 1995a: 32). From both periods at Arcy-sur-Cure, major numbers of faunal remains were recovered, wild horse being the dominating species. Although the spectrum of hunted mammals remained the same in both periods, the pattern of exploitation changed (David & Poulain 1990): in the Mousterian an exploitation purely of meat and marrow was evidenced, whereas in the Châtelperronian, bones and teeth in addition to fossil crinoids and molluscs came to be used as utensils and ornaments (Fig. 5; cf. Taborin 1990; Marshack 1988).

A famous cave in Slovenia, Divje Babe, has yielded five bone artefacts: three points, a needle-like implement (Turk & Kavur 1997: 122) and a "flute" (Turk, Dirjec & Kavur 1997: 157ff.). The latter was made from a femur of a cave bear (Fig. 6. 1). This species was also represented in the macro-fauna of the site. The "flute" is dated to about 35,000 BP. The excavators carried out microscopic analysis and experimental bone processing. Unfortunately they could not decide whether the

---

3. It was presumed that Neandertals and Pre-Neandertals lacked elaborate weapons for hunt, but wooden spears of 1,82 to 2,25m length were found in a bog near Schoeningen, Germany, dating to an age of 0.4 million years (Thieme 1996, 1999). This discovery shed a completely new light on ancient hunting tactics. As has already been acknowledged for the Clacton-on-Sea spear-point (Oakley et al. 1977) and the wooden finds from Lehringen (Thieme & Veil 1985), once again the attention of the archaeologists was drawn to the fact that the material culture unearthed to date, reflects only a small part of former activities: many things are not yet known and probably never will be.

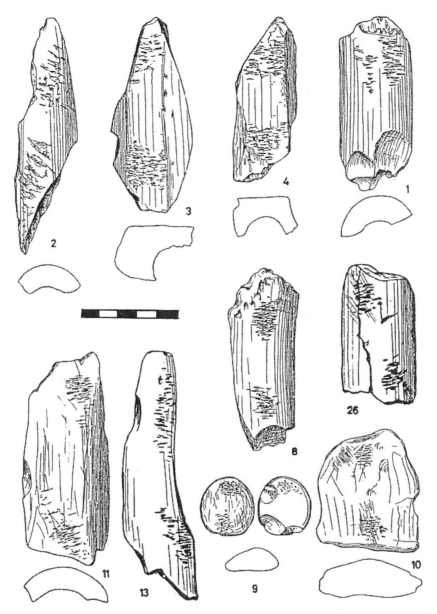

Fig. 4: Retouchers found at Middle Palaeolithic sites (made of bone, except no. 9 and 10; from Taute 1965 plate 14). Scale in cm.

traces were produced by carnivores (although counter-bites were missing) or indicative of intentional drilling carried out by "presumably Neandertal men" (ibid.: 175). However, flutes of a comparable simple technique of manufacture are well known from other Palaeolithic sites which would underline that the artefact from Divje Babe was not unique (Fig. 6).

It is striking that all the ornaments and musical instruments mentioned above are dated to this sensitive phase between 40,000 and 30,000 BP when in Europe modern humans began to thrive and the Neandertals declined. It is a generally accepted view that on various occasions contacts between Neandertals and modern humans must have taken place. An exchange of ideas and goods gains probability (cf. Bar-Yosef & Vandermeersch 2000). The pivotal question remains: did Neandertals or modern humans create these implements? If Neandertals did so, was it a matter of spontaneous artistic creativity or merely an imitation of what they had experienced through such contacts and "cross-cultural" influences? The underlying assumption that artistic expression was rather under-developed if not completely lacking in Neandertals, cannot be settled definitively at this point. It could also be possible that Neandertals put all their creativity into activities that have left no traces such as music-making or wood carving. That a certain sensibility for decoration did exist among Neandertals, has been attested from finds of haematite and manganese found at several Neandertal camps (Bosinski 1991: 45) and also from the recovery of cups with traces of ochre, e.g. at the cave of Cioarei,

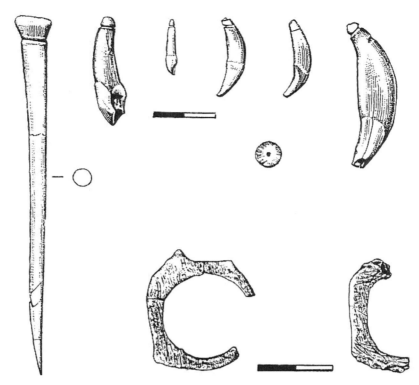

Fig. 5: Arcy-sur-Cure, France (Châtelperronian). Points, ornaments and pendants (from Bosinski 1990: 40f.). Scale in cm.

Romania (Carciumaru & Ulrix-Closset 1996). The excavators there claim "une fois de plus, des capacités intellectuelles et du potentiel spirituel de l'homme de Néandertal." (ibid.: 149). However, it might be suggested that the concept that everything connected with the ability of abstraction or with a desire for decoration and self-expression is solely related to *Homo sapiens sapiens* is far too exclusive.

The Upper Palaeolithic

Simultaneously to the Châtelperronian dominated by Neandertals, the Aurignacian (40,000-28,000 years BP) developed which was closely associated with modern humans (Gambier 1989). An explosion in technology and artistic expression occurred which was part of a world-wide phenomenon (Henke & Rothe 1980, 1998; Bosinski 1990; White 1995b; Bahn 1992, 1998; Oliva 1987). The complexity of cultural expression enlarged dramatically as witnessed also by the increasing range and complexity of tools made of bone and antler. Hitherto, bone artefacts had been rare, but from the Aurignacian onwards we see them in greater quantities and in a variety of forms. Typical for the Aurignacian are, for example, points with split bases and plaques with enigmatic imprints, e.g. from Abri Lartet aux Eyzies (Bosinski 1990) or from Abri Blanchard (White 1995b: 171). Not only from Aurignacian contexts in Europe, but also from Eurasia, a striking enlargement of artistic expression can be evidenced, e.g. from the Sungir-Kostenki culture, dating to $32,700 \pm 700$ BP. In particular, the site of Sungir is famous for its extraordinarily richly equipped burials. Thousands of ivory beads and weapons in form of ivory spears, measuring 2,40 m in length were excavated. Additionally, the site has yielded a variety of antler and ivory artefacts (Bosinski 1990: 46ff.; Binant 1991: 80ff.).

The spectrum of bone and antler artefacts enlarged further with the traditions of the Gravettian, the Solutrean and the Magdalenian: we find incised or perforated teeth, beads, pendants, magnificently carved spearthrowers (Fig. 7. 1-6), *bâtons de commandement*, harpoons, *saigaies*, buttons, discs and many more (Albrecht 1987; Bosinski 1987; Klíma 1987; Clottes 1997). Only the Upper Palaeolithic has so far produced such a variety of figurative objects, using innovative techniques such as carving not only lengthwise but also circumferentially, the technique of *champleve* (scraping away the bone around the figure; Bahn & Vertut 1988: 80) and the creating of free-standing figurines (Müller-Beck et al. 2000). Most spectacular are the figurines of ivory found at the Aurignacian caves of Vogelherd, Geissenklösterle and Hohenstein-Stadel, Germany (Hahn 1987). The latter has yielded one of the most impressive pieces of art: a lion-headed anthropomorphic statuette measuring 28 cm in length (Fig. 7. 7; *Der Löwenmensch* 1994). Equally famous are three statuettes of a bear, a mammoth and a bison,

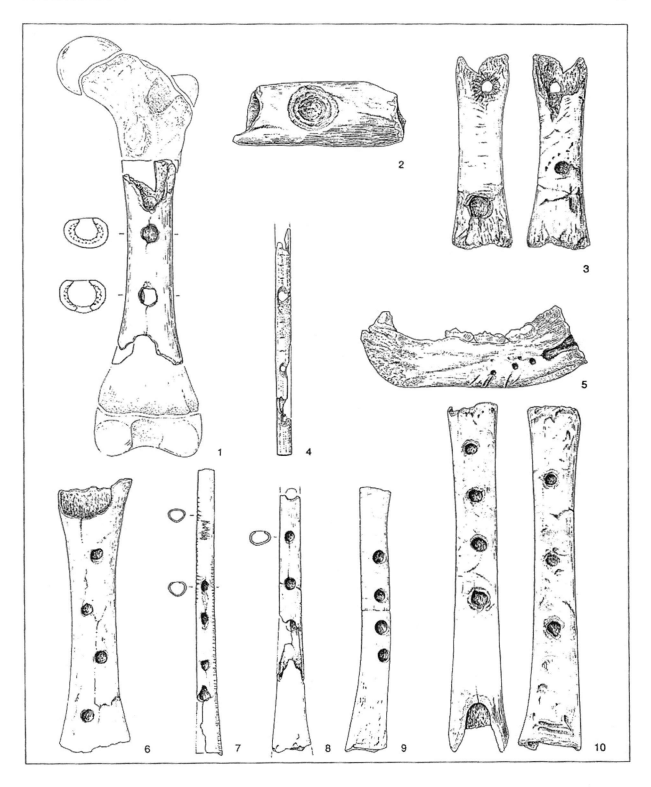

Fig. 6: "Flutes" from Palaeolithic sites: 1 Divje Babe (Slovenia, 35.000 BP), 2 Haua Fteah (Lybia, 60,000- 40,000 BP), 3 Istállóskö (Hungary, ca. 30,000 BP), 4 Geissenklösterle (Germany, ca. 36,000 BP), 5 Potočka zijalka (Slovenia, Aurignacian), 6 Liegloch (Austria, without date), 7, 8 Isturitz (France, Aurignacian), 9 Molodova V (Ukraine, 17,000 BP) 10 Pas du Miroir (France, Magdalenian). Different scales, 1:1 reprint from Turk & Kavur 1997: 180.

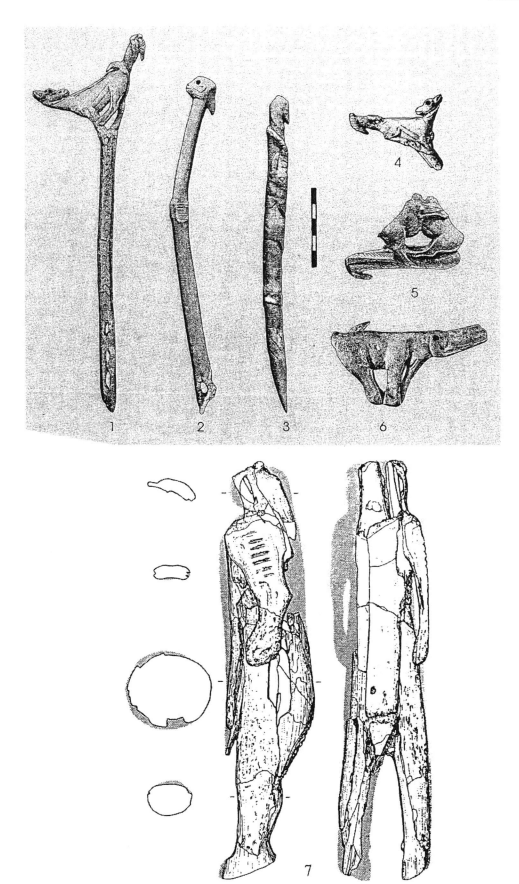

Fig. 7: Figurative art from the Upper Palaeolithic: 1-6 spear-thrower (French sites; from Stodiek & Paulsen 1996: 27; scale in cm), 7 ivory statuette from Hohlenstein (original height 28,1 cm; from Bahn & Vertut 1988: 85).

in addition to an anthropomorphic bas-relief, recovered from the cave of Geißenklösterle, Germany (Hahn 1986; Hahn et al. 1995). Seen in a general pan-European Palaeolithic tradition, these Swabian statuettes have to be interpreted not only as objects of an artistic expression, but rather provide insights into the world of shamanism (Dowson & Porr 1996).

Such extraordinary pieces of art are quite often associated with more simple ornaments and tools that were used for mundane purposes. As an example, the assemblage of Geißenklösterle, Germany may be examined (Tab. 3; Fig. 8). Projectile points predominate; they are produced from bone, antler and ivory and differ in length, diameter and in the basal profile (split, unsplit, plain or obliquely shaped; Fig. 8. 1, 2). Smoothers (Fig. 8. 3), awls (fig. 8. 4) and chisels (fig. 8. 5) are less numerously found. The most outstanding item is a perforated *bâton* (fig. 8. 10), mainly known from the Magdalenian. Furthermore, some incised bird bones (fig. 8. 11) and antler bands (fig. 8. 13) are worth noticing. Retouchers of antler (Fig. 8. 9) and so-called beating-tools or "punches" (Fig. 8. 8) are found which are believed to be associated with the production of flint daggers and adzes. Perforated canines of foxes (Fig. 8. 12) and small ivory beads belong to the category of personal ornaments.

The emergence of artistic expression in the Upper Palaeolithic is also exemplified in the sphere of musical activities. Although evidence for music making is very scarce, one of the connoisseurs of archaeo-musicology, Marcel Otte, draws our attention to three types of instruments: percussion equipment (*bâtons*, scrapers), aerophones (flutes, pipes, whistles and bull-roarers) and stringed instruments in form of the music-bow (Otte 2000: 98). As an example we might focus more closely on the finds from Mezin, a site located in the Ukraine. In terms of western European chronology, it dates to the Gravettian (precise dates are doubtful; see Allsworth-Jones 1998: xix-xxi). Mezin can be characterised as a complex winter camp-site. The faunal material (n = 7763)[4] is dominated by mammoth (45% of the bone material), whereas species such as arctic fox, wolf, wild horse, reindeer and musk ox occur in much lower frequencies (Pidoplichko 1998: 51). Mezin has also yielded many artefacts (Fig. 9): they are made from mammoth bones and ivory, covering mundane as well as ritual activities (Šovkolpjas 1965: 179ff.). Many half-finished pieces and fabrication debris indicate manufacture on-site. The most remarkable finds are some worked and decorated bones of mammoth: mandibles, scapulae and a pelvis were painted with ochre and display flattened, abraded areas which indicate that the bones were used as sounding-boards (Fig. 10. 2, 3); long bone and ivory fragments as well as ribs were shaped as drumsticks and as scrapers (Fig. 10. 1); bracelets of ivory and mollusc shells were used as rattles (Fig. 9, bottom line). According to Šovkolpjas (in Quitta 1957: plate 46) all this was arranged in an orchestral-like round[5]. One can hardly imagine the complexity of shamanistic performance that took place in that dwelling.

Typology	Bone	Teeth	Antler	Ivory
Projectile points	25		6	66
Smoothers	3		1	
Awls	5		1	
Chisels			1	
Retouchers	8		2	
"Flaked" bones	8			
Beating tools			2	
Batons				1
Pendants		2		14
Curved bands			1	1
Incised implements	2			4
Figurative items				4
Unspecific parts of figurative items				3
Total	51	2	14	93

Tab. 3: Geißenklösterle, Germany. Range and number of objects made of bone, teeth, antler and ivory (from Hahn 1988).

The uniqueness of the Mezin finds becomes evident in comparison with another important Ukrainian site, Mehzirich, lying some 300 km south (Pidoplichko 1998: 66ff.). This camp-site is dated to ca. 18,000 BP and is also characterised by mammoth bone dwellings. Mehzirich, too has yielded a rich collection of bone and ivory artefacts. The spectrum includes spear-points, *bâtons de commandement*, hammers, mattocks, scrapers, polishing tools, shovels, points, borers, skewers, needles, arrow-heads, awls, worked ribs and antler, cups and a major assemblage of half-finished items (ibid.: 133ff.). These mundane implements contrast sharply to the ritual equipment: brooches, anthropomorphic statuettes, beads, amulets and pendants, as well as tusks and skulls decorated with ochre. Although the Mehzirich finds represent an impressive collection of artefacts, they lack any hint of the practising of music.

The use of instruments obviously had (and sometimes still has) a spiritual dimension. The context in which this "music" was produced, may have served to

---
4. in addition to non-food animals: n = 329 (Pidoplichko 1998: 51f.).
5. Pidoplichko (1998: 61) more cautiously speaks about a "cult corner" or "shrine" and does not mention the aspect of music-making.

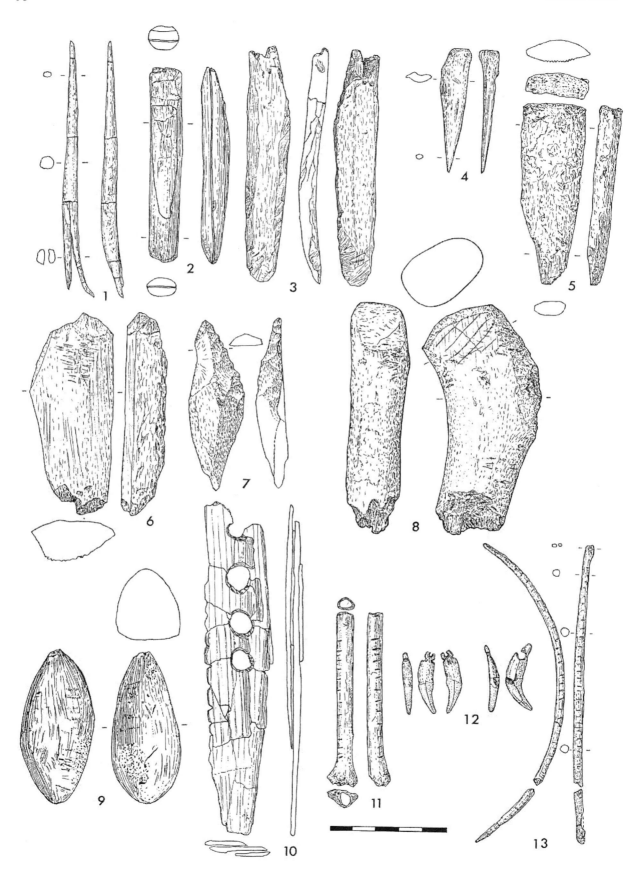

Fig. 8: Geißenklösterle, Germany (Aurignacian). Selection of bone and antler artefacts (compiled from Hahn 1988 plates 35-45). Scale in cm.

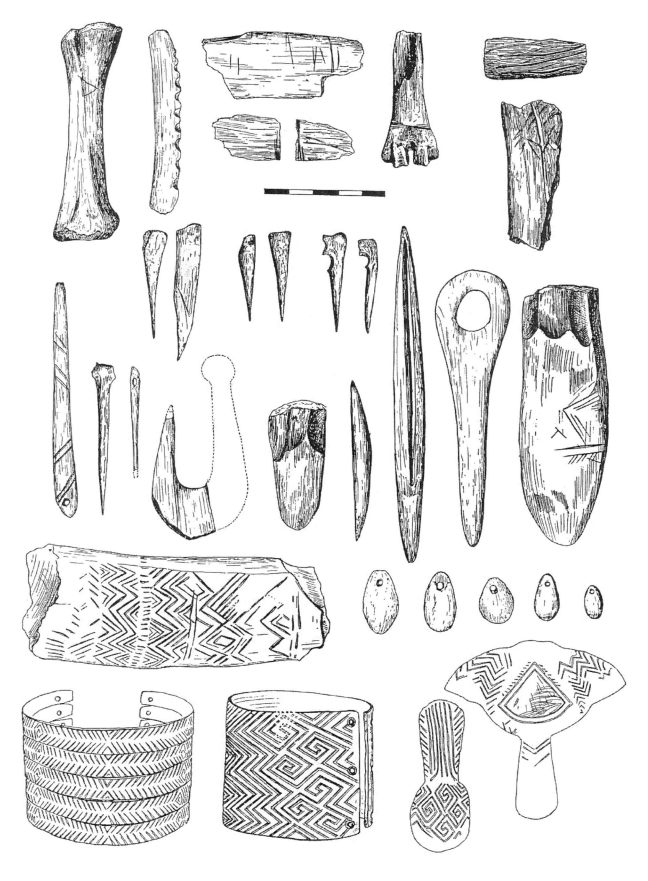

Fig. 9: Mezin, Ukraine ("Gravettian"). Selection of bone, antler and ivory artefacts, including manufacturing debris (compiled from Šovkolpjas 1965). Scale in cm.

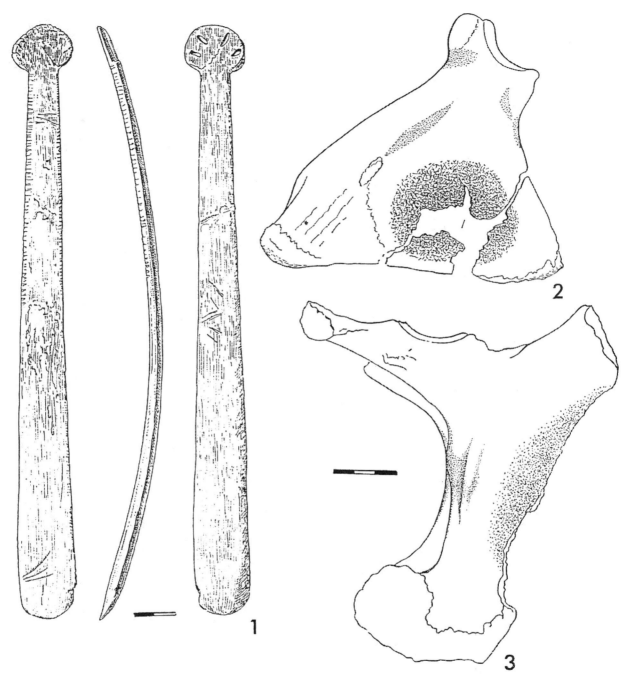

Fig. 10: Mezin, Ukraine ("Gravettian"). A selection of presumed musical instruments: scraper (1) and sounding boards (2, 3; compiled from Šovkolpjas 1965). Scales in cm.

strengthen solidarity within a band of hunters or kinship group or might have been practised at special occasions such as initiations or funerals. It may have been accompanied by singing or by percussion on wooden or stone material (cf. Watson 1996). Music making is inevitably connected with emotion, fantasy and perception (Gibson 1966; Schafer 1977). Emotion is intimately related to communication in a wider sense and hence it should not be excluded from our interpretation and imagination of the past (Tarlow 2000). "Considéré dans ses diverses dimensions, le phénomène musical se présente donc tel qu'un reflet de l'esprit humain à la fois dans sa subtilité, dans son élaboration et dans son adéquation judicieuse aux autres activités spirituelles (religion, language, pensée)," (Otte 2000: 99). Bone artefacts in the form of musical instruments seem to be the most convincing objects in illustrating the relationship between artefact and man. They demonstrate supremely that an artefact resides in a web of meaning that can be understood only with respect to its cultural context.

The Mesolithic

The close of the Palaeolithic period in Atlantic-Baltic Europe coincided with a change in climate whereby the previous arctic conditions vanished. This interval between the last stage of the Palaeolithic, the Magdalenian, and the Neolithic was filled by a number of cultures, collectively called "Epipalaeolithic", "Proto-Neolithic" or "Mesolithic". In Germany alone, thousands of campsites have been discovered and archaeologically researched (e.g. Gramsch 1994; Kind 1999). The economic basis of these Mesolithic cultures was gathering, hunting and fishing, no matter if they were located in Germany, Scandinavia (Clark 1975), Latvia (Zagorska & Zagorska 1989) or elsewhere. This situation is clearly reflected in the spectrum of bone and antler artefacts, in which antler emerges as preferred raw material. Mesolithic technology is characterised by a highly specialised production of barbed spear-heads, harpoons and fish-hooks (Fig. 11 top line and mid right). Furthermore, leatherwork is indicated through pins and awls, limpet scoops were used for the collection and/or processing of shellfish (Griffitts & Bonsall 2001), woodworking is reflected by antler chisels, worked beaver mandibles[6] and perforated antler sleeves which are used for the hafting of axes and adzes, antler mattocks were most probably used for digging (Smith 1989). A compilation of such finds from northern European Mesolithic and Neolithic sites has recently been published by S.B. Dellbrügge (2002). She covers the Maglemose and the Kongemose as well as the Ertebølle culture and describes both the exploitation of raw material and the processing of bone and antler as well as the many of the techniques involved in manufacture. In Mesolithic sites, the frequency of items can be striking: to give one example, in Friesack, Germany, on a site which is contemporaneous with the Manglemose culture in Denmark, more than 700 artefacts were recovered, in addition to 34 perforated teeth and some decorated *bâtons* (Pratsch 1994; Gramsch 1994).

Another source of information on bone tools is provided by burials. Judith M. Grünberg (2000) has recently analysed more than one hundred Mesolithic burials, scattered all over Europe. Another useful and detailed compilation on "Pratiques funéraires..." was given by Christian Jeunesse (1997) and included some critical comments on the highly debatable chronology and possible contemporaneity of Mesolithic graves with early Neolithic developments.

What concerns us here are grave goods in the form of artefacts of bone, antler and teeth. Several categories can be differentiated: ordinary bone and antler tools, ornaments (mostly perforated teeth and shells), and elaborate items of prestigious character such as masterly decorated axes (Clark 1975). All this is evidenced in changing frequencies and quality according to the sex, age and social rank of the interred (Grünberg 2000: 110ff.). Among the common tools, seventeen different types can be distinguished, mainly hunting equipment (hooks, projectiles, harpoons etc.) and items of everyday use such as awls, perforators and double points, but also axes, adzes and hammers. It is remarkable that even some animal burials were equipped with such grave goods: in Skateholm II, Sweden, a dog was buried with a large hammer made of red deer antler (Larsson 1990, 1994) while other carefully buried dogs from Skateholm I were covered with ochre (Larsson 1989).

One of the most important cemeteries of the Mesolithic period is located on an island in lake Onéga, Russia, named Oleneostrovski[7]. Aside from points and harpoons, a large number of decorative items was recovered such as deer and bear canines in association with other grave goods. These finds led to a significant grouping of the buried persons not only according to sex and age, but also to kinship relations (O'Shea & Zvelebil 1984, critically reviewed by Jacobs & Hart 1996).

Another source of objects for personal adornment are sites from which stray beads and pendants have been recovered (Clark 1975). Sites in South Scandinavia are particularly rich in decorated bone and antler artefacts which stand out on account of their geometric, zoomorphic and anthropomorphic designs. According to B. Cunliffe (2001: 135), they reflect a rich visual culture and the development of mechanisms of communication, social selfconsciousness and a specific identity within a community (Fig. 11 mid left; Liversage 1966; Clark 1975; Nash 1998, 2001). The Ryemarksgaard axe is one of the most striking examples (Fig. 11 bottom line). G. Nash assumes the figures and zigzag lines to be "arranged into a multiple-phased narrative that is spatially organised.... witnessing either a scene involving life...or death ... or a journey from a state of consciousness to unconsciousness." (ibid.: 234). Whether one agrees with this or not, Nash's analysis highlights the need for an urgent reconsideration of similar recoveries.

In the Mesolithic, ritual activities or expressions of art beyond burial practices are scarce. One striking example may be mentioned here: at Starr Carr in England, a site noted for its bone and antler industry, some red deer frontlets were excavated with antlers still attached. "They were carefully worked by thinning the antlers but retaining their frontal profile, and by piercing the scull cap as though for attachment to the human heads by means of

---

6. More than 400 were found at Mesolithic peat sites of Northern and Eastern Europe (Zhilin 1998: 26).
7. dating to 7500 BP, including 141 graves with 177 individuals, excavated to date.

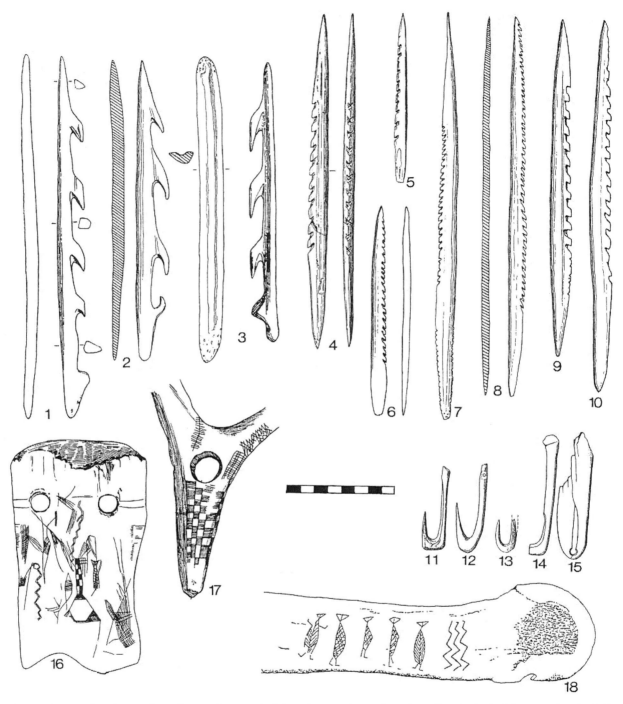

Fig.11: Bone and antler artefacts from early post-glacial settlements in southern Scandinavia: barbed harpoon-heads (1-3), finely toothed spear-heads (4-10), bone fish-hooks (11-15), ornamented red deer antler (16, 17; compiled from Clark 1975: 132, 135, 143, 152), *Bos primigenius* metapodium: the Ryemarksgaard axe (18; Nash 2001: 237). Scale in cm.

cords." (Legge & Rowley-Conwy 1988: 3). Almost identically processed frontlets with intact antlers were found for example at the Mesolithic camp-sites of Duvensee (Lehmann 1991: 39) and Bedburg-Königshoven, Germany (Street 1999: 25). Street (ibid.) proclaims that these frontlets could have been masks or head-gear of schamans, a traditional form of equipment which was already used in the Upper Palaeolithic (Clottes & Lewis-Williams 1997) and which from ethnographic sources, can be traced to the 19th century in Siberia (Tromnau 1985; Braehm 1994: 200; Devlet 1996).

The Neolithic

The Neolithic which in Europe started in the 7th/6th millennium BC (Dickinson 1994; Papathanassopoulos 1996), is an era with dramatic changes: sites are inhabited permanently, hunting and gathering is mostly replaced by livestock management and agriculture, pottery becomes part of the material culture, social structures as well as burial practices change in appearance, to name but a few characteristic aspects. All over Europe bone and antler were now highly esteemed raw materials for the production of tools and ornaments. Such items covered the demands of a broad spectrum of mundane as well as ritual and prestigious activities. The enormous array of Neolithic artefacts published in a widely scattered archaeological and archaeozoological literature cannot be reviewed in an exhaustive manner, not least because major regional and chronological differences have to be taken into consideration (cf. Müller 1964, 1980; Camps-Fabrer 1982, 1985; Schlenker 1996; Sidéra 2001). Instead, I would like to restrict myself to some informative highlights and a few selected results.

For the Neolithic of western Europe, Isabelle Sidéra (2001) has carried out interesting research on the frequency of particular bone artefacts at settlements, enclosures, mines and burials. She emphasises that the context chosen for an investigation of bone artefacts, can be crucial, not only for the interpretation of such finds, but more particularly for the impression one gains of the importance of particular types of artefacts at the location being investigated. It should be emphasised that although in the 5th millennium BC meat supply was mainly based on livestock management, burials yield artefacts almost exclusively made from wild mammal bones: antler arrowheads, beads in form of perforated canines of red deer, bear and boar, incisors of beaver as well as claws of birds of prey and hedgehog mandibles. Such artefacts were of great symbolic value, indicating a particular connection of the Neolithic people to the world of wild beasts. Other extraordinary items took the form of human figurines (Fig. 12). The emergence of such a symbolic system took on vast proportions later on in the Neolithic (cf. Heege & Heege 1989: 50ff.). The procurement of burials of the Linear Pottery culture is also of major interest (Jeunesse 1997: 72ff.): the most frequently found items are metapodial points while other implements such as needles, polishers and *baguettes* occur infrequently; antler is encountered much less frequently than bone (ibid.). Ornaments occur in greater numbers and with a wide range of variability: perforated shells of marine gastropods and bivalves (*Spondylus gaederopus)*, pierced animal teeth such as red deer canines (and others which imitate them), teeth of dogs, foxes and pigs may be mentioned. Outstanding yet often neglected items are the so-called *Gewandknebel,* slightly curved rectangular antler plaquettes, carefully decorated, which resemble antler cheekpieces. About thirty of them have been noted to date. They occur in a restricted geographical area, e.g. in the cemeteries of Schwetzingen, Aitershofen and Ensisheim, Germany. These plaquettes are regularly associated with burials of adult men and evidently are of major chronological, regional and socio-cultural significance (ibid.: 76).

In Central Europe, mainly Swiss sites are particularly prominent in their broad spectrum of objects, in the comprehensive level of inquiry lavished in their finds and in their scholarly publication[8]. In geographical and chronological terms, one has to reckon with a high degree of diversification in implement manufacture. The most common types are points, chisels, axes, adzes and hammers, retouchers, hafts, harpoons and fishhooks, needles, projectiles and arrow-heads, scrapers, knives, spatulae and ornaments such as pendants (e.g. perforated teeth, pierced mandibles of small mammals, ornamented plates) and buttons (Schibler 1980, 1981, 1995, 1998). A typical assemblage of that region has been excavated at Arbon Bleiche 3, a lake-side settlement dating to 3384 to 3370 BC, in the transitional period between the Pfyn and Horgen cultures. Due to excellent preservation conditions, a total of 2872 artefacts, half-finished items and type-specific raw material of bone, tooth and antler was recovered (Fig. 13). The analysis (Deschler-Erb et al. 2002) has revealed that antler comprised only 28% of the material, whereas bone predominated. Long bones of red deer were primarily chosen. Points (n = 697) and chisels (n = 594) prevail whereas other categories comprise only 136 items. Personal ornaments are unexpectedly common (n = 209). Most of the tools and ornaments are scattered over the entire settlement area. Therefore the furnishing of households could be reconstructed almost completely. Particular places of tool manufacturing, however, could not be recognised. Antler beakers belong to the most outstanding objects of the Swiss Neolithic (Fig. 14. 2): in terms of chronology, these small containers are of high significance, although their function still is under discussion (Suter 1981). They are typical for the late 4th and the 3rd millennium BC in the Swiss Cortaillod culture (Schibler 1995: 146). A very puzzling form of artefacts, also evidenced from Swiss lake villages, are red deer skulls with pedicles which were carved to models of human breasts (Clutton-Brock 1991). Nine of them have been found at Yvonand IV on the shore of Lake Neuchâtel. "The process of manufacture was to chip away the thick bone of the pedicle and frontal bone to make the rounded shape of a breast, while the central final breaking-point of the cast antler was kept to present the nipple...." (ibid.: 910). They are interpreted either as ornaments or as items of symbolic value.

Comparably remarkable are finds from six sites of the early Neolithic Körös culture in south-east Europe

---

8. A complete compilation of papers and books published by members of the archaeozoological laboratory/University of Basel can be found in the internet: www.unibas.ch/arch/ArchBiol/biblioaz.htm.

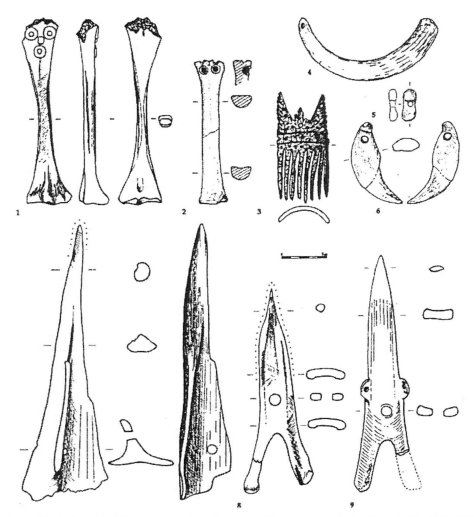

Fig. 12: Rare bone objects from Neolithic grave contexts in western Europe (original plate from Sidéra 2001: 228). Scale in cm.

which were analysed by J. Makkay (1990). More than 600 items were recovered, including points, spatulae, pins, needles, net-needles, burnishers, knives, handles, pendants and chisels, in its entirety a repertoire of more than 30 specific forms. The most striking category are bone spoons (n = 47 plus two half-finished items). They were manufactured from cattle metapodials (mainly *Bos primigenius*), measured about 10-15 cm in length and were carefully shaped. Most of the spoons, in particular their bowls, seemed to have been used rather intensively, nevertheless their function remained obscure (ibid.; more about bone spoons see Nandris 1969).

Linen, the fibre of flax (*Linum usitatissimum*), was the earliest weaving material to be used in the Neolithic, existing before wool yarn. In sharp contrast to wool, flax requires a complicated processing, including retting, drying, breaking, beating, scutching and hackling. Breaking, scutching and hackling are done with special tools – flax hackles – which are made of a group of ribs (Fig. 14. 1).

Such implements are repeatedly found in Swiss lake-sites (Barber 1991: 14; Schlenker 1996: 48; Schibler 1980: 50, 1981: 37ff., 1997: 152f.). J. Schibler (ibid.) has pointed out that in proportion to the importance of flax, flax-hackles occurred in varying frequencies in different cultural contexts. From ca. 3800 BC onwards with a peak during 3200/3100 BC, flax was highly esteemed, for example by the inhabitants of the classical Cortaillod and the Pfyn settlements. In the succeeding Auvernier culture, however, the importance of flax decreased considerably, probably due to the use of other plant fibres or even wool yarn. Excellent conditions of preservation in some Swiss lake-sites even allowed the recovery of wooden flax-working implements (Barber 1991: 14). Flax has to be spun like wool and thus, spindles and spindle-whorls also occurred during this period[9]. They were mostly made from clay, less frequently from wood and stone (Dunning 1992; Rast-Eicher 1997: 322). Spindle-whorls made of bone or antler, however, are found even more scarcely: from Egolswil 4 one spindle whorl of bone is evidenced, while

---

9. The problem with such whorls is the following: they are not yarn-specific, yet the recovery of a spindle whorl need not necessarily indicate flax processing but could also be indicative of the spinning of wool. The pivotal criterion is the length of the fibre: the shorter the fibre, the lighter the spindle whorl has to be (Barber 1991: 43; Leuzinger 2002: 119).

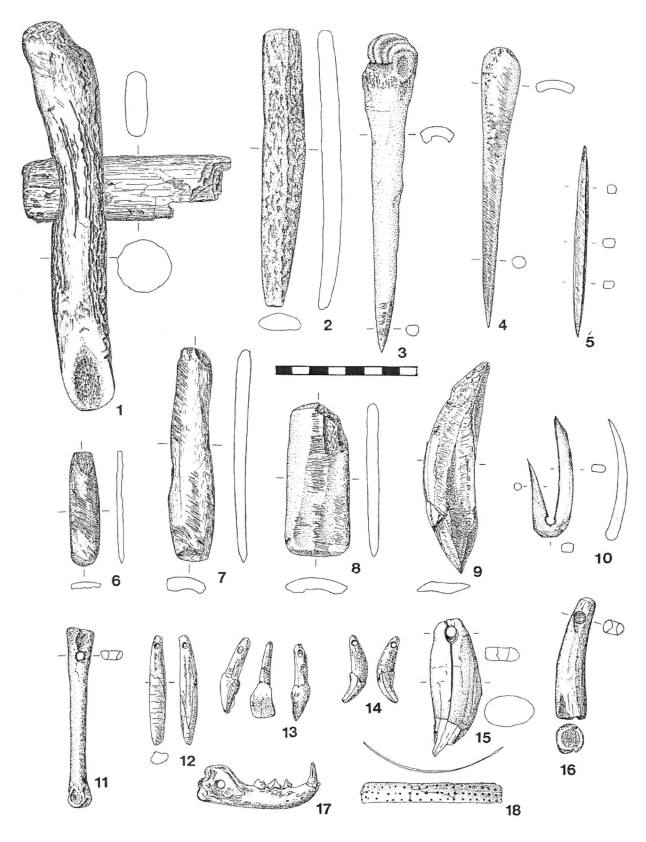

Fig. 13: Arbon Bleiche 3, Switzerland (Neolithic). A selection of bone and antler artefacts: 1 axe, 2 retoucher, 3 massive point with joint, 4 large point, 5 double point, 6 rib chisel, 7, 8 massive chisels, 9 knife made of a boar canine, 10 fish-hook made of a boar canine, 11 pendant made of wolf metapodium, 12-15 teeth pendants (pig, cattle, dog, bear), 16 antler pendant, 17 mandible pendant (wild cat), 18 ornamented plate made of antler compiled from: Deschler-Erb et al. 2002: 342-366). Scale in cm.

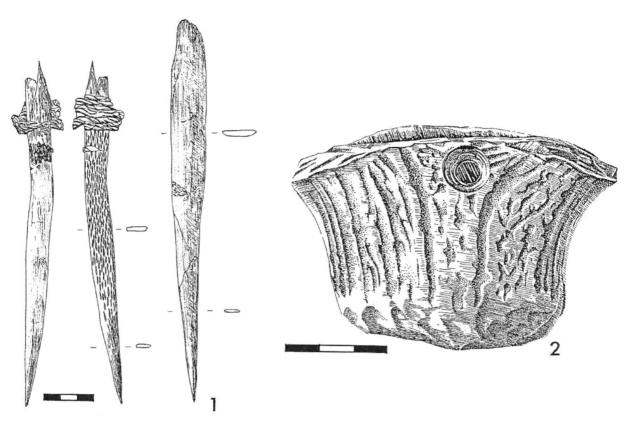

Fig. 14: Arbon Bleiche 3 and Twann, Switzerland (Neolithic). 1 flax-hackles from Arbon Bleiche 3 (Deschler-Erb et al. 2002: 344); 2 antler beaker from Twann (front page of Suter 1981). Scale in cm.

Twann, Feldmeilen-Vorderfeld and Thayngen-Weier have yielded few additional examples made of antler (Suter 1981: 61; Furger & Hartmann 1983: 141; Wininger 1971: 52). Additionally, at Feldmeilen-Vorderfeld, other weaving equipment in the form of wooden weaving swords was recovered (Wininger 1981: 60). Some of the antler combs evidenced at Zürich-Mozartstrasse (Horgen layers) are also assumed to have been used as weaving utensils (Rast-Eicher 1997: 304; Gross et al. 1992: plates 177, 137-139). Two similarly worked combs, made of antler, too, are evidenced from Late Neolithic Servia in Macedonia, Greece (Heurtley 1939: 165, fig. 35e, f). If they were used during the weaving of flax or of wool, has to remain open, since according to recent studies both materials coexisted in that region (Tsachili 1996).

The production of pottery was another major innovation of the Neolithic. Burnishers of bone were used for the shaping and smoothing of pots and vessels; denticulated ribs, mandibles of small mammals or antler stamps were utilised for decorative purposes (Schlenker 1996; cf. Soeffing 1988: 59; Zeeb 1998: plate 75). A Bronze Age ensemble of pottery tools – quite similar to Neolithic items in this category – is illustrated in figure 15.

At various places in Europe, flint mines were exploited. With the help of antler picks and bone shovels, with rakes and wedges, the mine shafts were cut through solid chalk, sometimes many metres deep and branching from simple pits into regular galleries (cf. Böckner 1980; Pelegrin & Richard 1995; Gayck 2000: 44ff.).

In the Late Neolithic, a development was initiated which in many ways should have led to dramatic changes: the domestication of the horse. It most probably happened independently in Spain and south-west France (Uerpmann 1995: 25) as well as in Central Europe (Benecke 1994: fig. 4 and page 129) at the turn of the 4th to the 3rd millennium BC; for south-east Europe, another area with a high potential for autochthonous equid domestication, precise data still are awaited. In areas outside the natural range of the wild horse such as Scandinavia, domesticated animals were introduced from the mid of the 3rd millennium BC onwards (Benecke 1999). However, activities which would demand a certain equipment for these animals – i.e. horse harness – and thus would be reflected in the spectrum of bone and antler artefacts, belong to a much later period (cf. Boroffka 1998).

Copper and Bronze Age

The Bronze Age period in Europe can be seen as a phase in which again major socio-cultural changes occurred, e.g. the establishment of a socially stratified society

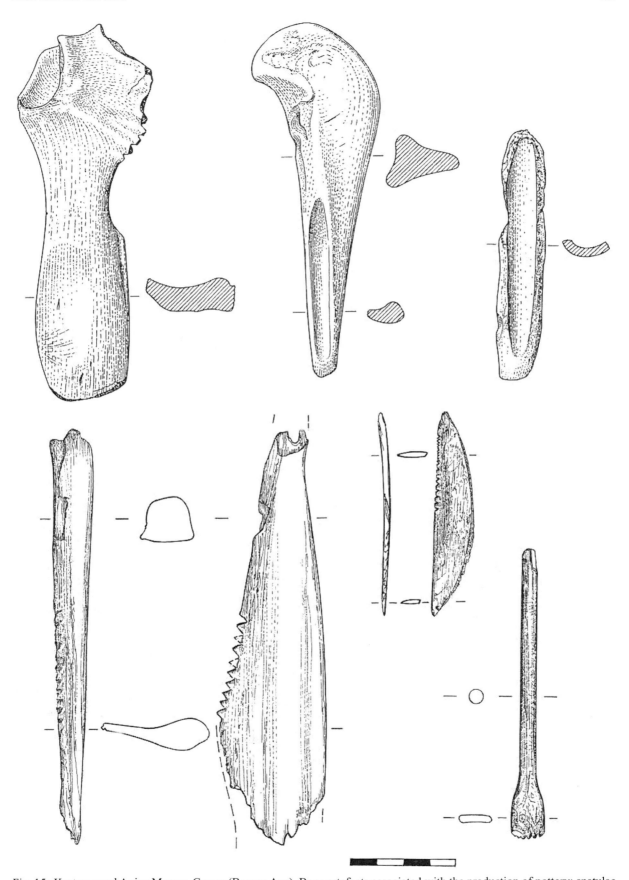

Fig. 15: Kastanas and Agios Mamas, Greece (Bronze Age). Bone artefacts associated with the production of pottery: spatulae, polishers and denticulated items (top line = Kastanas, compiled from Hochstetter 1987 pl. 17; bottom line = Agios Mamas, unpublished results, drawing: H. Hähnl). Scale in cm.

with new forms of leadership and the development of distinct religious systems. Throughout Europe the network of communications and contacts seemed to have been intensified, as evidenced through the transmission of material culture (Coles & Harding 1979; Harding 2000). With the discovery and use of new materials – copper and its alloy bronze – bone and antler implements decrease in importance. Many items of everyday life, but mainly those with prestigious or ritual connotations were then fabricated in metal. The advantages are evident: metal allows for the production of utensils of greater dimensions variable shape, and diverse ornamentation. Furthermore, metal objects have another effect in terms of prestige and monetary value, in that their socio-economic importance is much greater than bone implements. Finally, metal objects can be re-used through re-casting. That the number of bone objects seemed to be reduced – not so much in the Copper Age, but noticeably so in the Bronze Age (cf. Schibler 1997: fig. 125) – may to a lesser extent be the result of a reduced attention being paid to them by archaeologists who – quite understandably – directed their energies more intensively towards metal finds rather than bone (Becker 1993: 134). The fact that from Neolithic contexts it is mainly settlements that have been investigated while for the Bronze Age mainly graves (cf. Fabiš 2000) and hoards, may to a certain degree also have influenced our knowledge of bone artefacts from this period. It is generally accepted, that the provision of bone artefacts differs considerably from one context to the other.

On occasion, bone tools may imitate metal objects in shape and ornament. For example in Lerna IV-V, Greece, elaborate pins were also carved of bone which according to O. Dickinson (1994: 180) reflects "the relative impoverishment of Helladic culture at the time". Another example can be found in the repertoire of bone and antler artefacts, excavated at Kastanas, a hill-site in northern Greece (Hochstetter 1987). Kastanas may act as an example of a site well furnished with bone and antler artefacts in the Bronze and Iron Ages. The spectrum in Kastanas included points, pins, chisels, polishers, handles, drilled astragali, decorated plates, arrowheads, perforated teeth, buttons, small containers and needles (Fig. 16, 17). Fabrication debris such as longitudinally incised metapodials (Fig. 17. 6) and several half-finished items are indicative for a local production of bone needles, of which considerable amount was excavated (see Fig. 17. 3-5, 7-15). Some of them quite obviously imitated bronze models such as the so-called *Warzenhalsnadel,* having a knobbed neck (Fig. 17. 5). This feature turned out to be chronologically significant, since parallel finds could be identified from southeastern European hoards (ibid.: 98). An antler plate decorated with circular ornament (Fig. 16. 17) was found in an Iron Age layer. It is interpreted as possibly part of a quiver of a type well known from the Hallstatt period. Both finds support the idea of far-reaching contacts between Kastanas and sites in the Balkans (ibid.). In contrast, other small finds from Bronze Age layers in Kastanas (but unfortunately no bone implements) suggest influences from the south, from the Mycenaean civilisation.

The impressive remains of the cyclopean walls of Mycenae and the Middle Bronze Age of Greece and the Aegean palaces, are well-known even beyond the sphere of archaeology (Harding 2000). Evidenced from written sources and reflected in the archaeological record, the existence of a hierarchically organised civilisation and a sophisticated administrative system can be confirmed (Halstead 1992, 1994). Due to this evolved social structure, an elite of high-ranking persons was established. Members of this elite were buried with great extravagance in conspicuously constructed graves, equipped with very particular grave goods. Such graves are found in Attica, Messenia, in the Argolid, in Boeotia and on the islands of Keos and on Aegina. Aegina, one of the most famous shaft graves of the Middle Bronze Age, has produced among other prestigious items, about 80 pieces of perforated boars' tusks (Kilian-Dirlmeyer 1997: 35ff.). Such plaques have also been recovered in other Middle to Late Bronze Age sites in mainland Greece and on Aegean islands (ibid.: 49; Dickinson 1994: 204). Unspectacular as they are as single finds, they gain in importance as an ensemble, originally attached to a leather helmet (Fig. 18). These helmets are well known from burials of high-ranking persons. Occasionally, heads carved in ivory and wearing such helmets are found (Demakopoulou 1988: 236; Borchardt 1972). According to C.E. Morris (1990) such boars' tusks helmets are not only attributes of a god of warfare and a reflection of the boar as a source of meat, but above all "a multivocal symbol, whose interlinked spectrum of references included both warfare and hunting, the protection of man and his territory... man's interaction with the natural world" (ibid.: 155). The interpretation of these tusks, however, still lacks one pivotal facet: the identification as to what animal these tusks once belonged – *Sus scrofa* or *Sus domesticus?* This problem has never been investigated in detail (Becker 2000: 174).

To return from the exceptional to the mundane it may be observed that in Central and Northern Europe, a specialisation in food requirement existed as evidenced for example by socketed harpoon-heads as found at Bronze Age sites in southern Germany and Switzerland (Auler 1993). In quantitative terms they form the most important group of utensils used for fishing. Another group of special bone artefacts comprises amulets in the form of carefully ornamented discs of bone and perforated or socketed teeth. The latter most often derive from wild

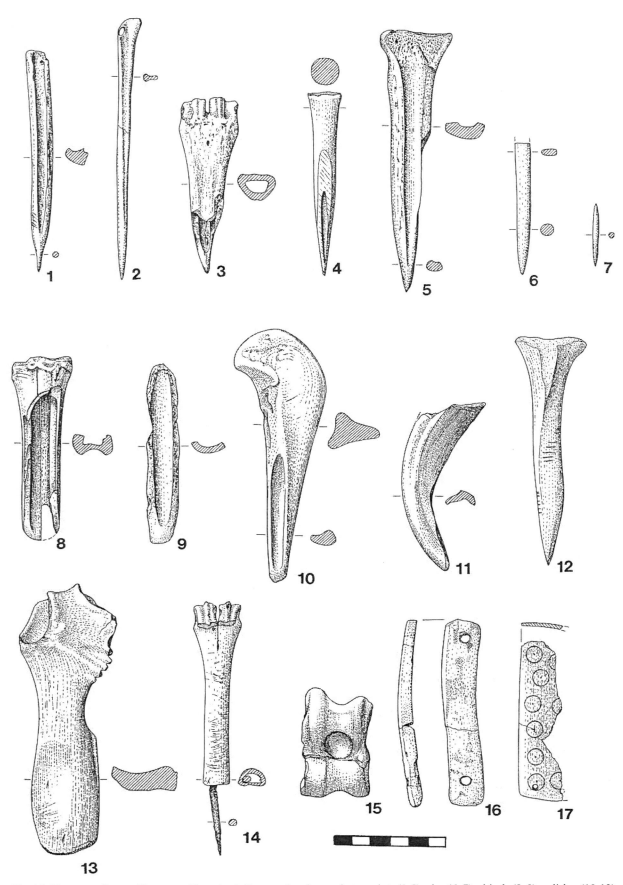

Fig. 16: Kastanas, Greece (Bronze and Iron Age). Bone and antler artefacts: points (1-5), pins (6, 7), chisels (8, 9), polisher (10-13), handle (14), drilled talus (15), antler plates (16, 17; compiled from Hochstetter 1987 plates 16, 17, 23). Scale in cm.

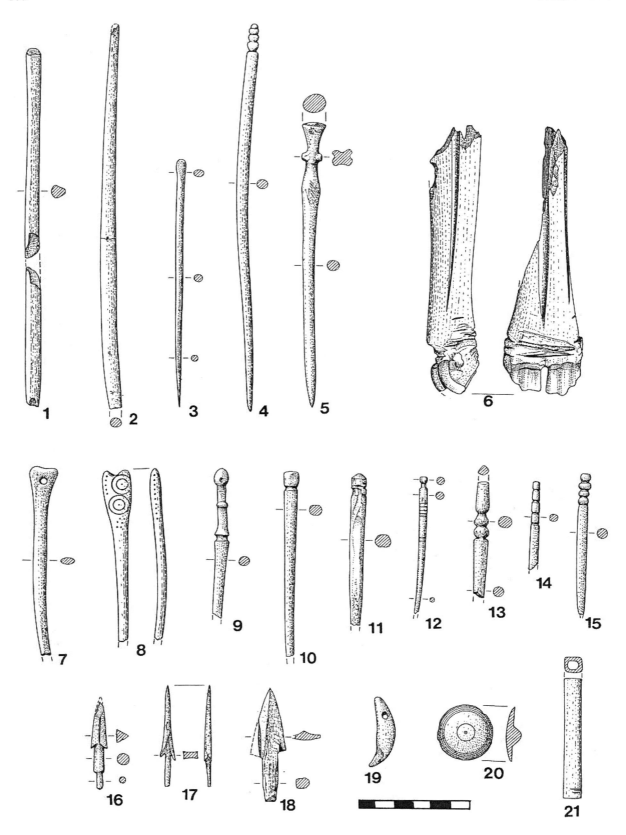

Fig. 17: Kastanas, Greece (Bronze and Iron Age). Bone artefacts: spindles (1, 2), pins (3-5, 7-15), raw material from pin production: metatarsus of a fallow deer (6), arrow-heads (16-18), perforated tooth (19), button (20), bone case (21; compiled from Hochstetter 1987 plates 14, 15, 19). Scale in cm.

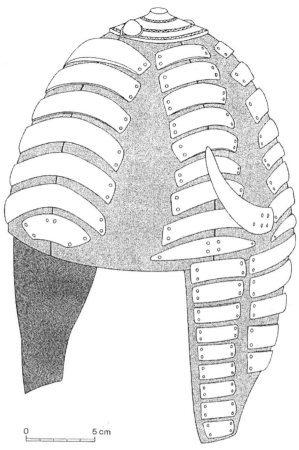

Fig. 18: Reconstruction of boars' tusk helmet from Middle Bronze Age shaft graves (Kilian-Dirlmeyer 1997: 39). Scale in cm.

pig and bear (Biel 1996). Spindle-whorls are very common finds in Bronze Age sites all over Europe, no wonder, since the woolly sheep was distributed over large parts of the continent and the technology of spinning wool was carried out almost everywhere. The fabrication of woollen textiles was different to flax, since it took no elaborate preparation to spin wool; it can virtually be spun off the sheep. Today, women all over the world twist armfuls of loose wool on to a wooden or bone spindle, using a technique known since the Late Neolithic (Barber 1991). The secret of the enormous success of wool lies in the structure of its fibres: they absorb moisture, insulate against heat and cold, resist flame and retain their resilience (Hyde 1988). In later periods, a manifold equipment for the preparation of the fibres and for the weaving process was used such as wool combs, weaving combs (Hodder & Hedges 1977 in K. von Kurzynski 1996: 14), pin beater and weaving tablets. Many of these items are manufactured from bone and antler (Barber 1991). Their invention, however, may origin in much earlier periods. That counts for example for weaving tablets which are said to be characteristic for the Iron Age particularly in Central Europe. However, at Abri Mühltal I near Göttingen, Germany, a perforated bone plaquette was found in a Late Bronze Age context (von Kurzynski 1996: 15); in shape and manufacture, it is almost identical to some Iron Age samples. The close association of this find to some other weaving equipment clearly supports its interpretation and points to an older tradition in the utilisation of such implements.

Another paramount Bronze Age development which was rooted in the Late Neolithic has already been mentioned: the domestication of the wild horse. During the Bronze Age, an increasing number of domestic horses spread into many parts of Europe. Of course, the raising of such an animal had an immense impact on the cultural development. Changes in transport, trade, technology, social life and even burial rites can be evidenced (Azzaroli 1985). But what about bone and antler objects – do they also mirror these new trends? Particularly in Central Europe, parts of harness equipment in the form of antler cheek-pieces and strap distributors are repeatedly found (Piggott 1983: 98ff.; Littauer 1969: 284ff.; Lichardus 1980; Schibler 1998: 275; Bandi 1963). From graves and more seldom from settlements, bar and discoid cheek-pieces as well as tabular forms have been recovered (Fig. 19). The oldest cheek-pieces were ascertained from the Ural-Volga region and the Carpathian Basin at the end of the 3rd millennium BC and at the beginning of the 2nd millennium BC, respectively. From their shaping and distribution, considerable chronological and cultural links emerge (Boroffka 1998).

If we now consider the whole spectrum of bone and antler artefacts, there is a certain continuity from the Neolithic to the Bronze Age as far as the repertoire of the most ordinary utensils such as awls, pins, hafts, hammers or axes is concerned. On the other hand, a radical change in bone crafting is indicated, presenting some very new types of artefacts which are closely linked to technological innovations (Biel 1996: 57f.). A comparable diversification can also be recognised for the Iron Age.

The Iron Age

The introduction of iron technology followed a southeast-northwest route from Greece (at about 1200 BC), the Balkans and Italy to central and northern Europe and finally Scandinavia, which was reached at a much later date at about 500 BC (Hjärthner-Holder & Risberg 2001). From a technical and innovative point of view, the Iron Age is a time of continuity, refinement and further development of known technologies. It is characterised by the use of iron for tools and weapons, by enclosed settlements, the widely introduction of the potter's wheel and of the lathe, the introduction of domestic fowl – to name but a few aspects.

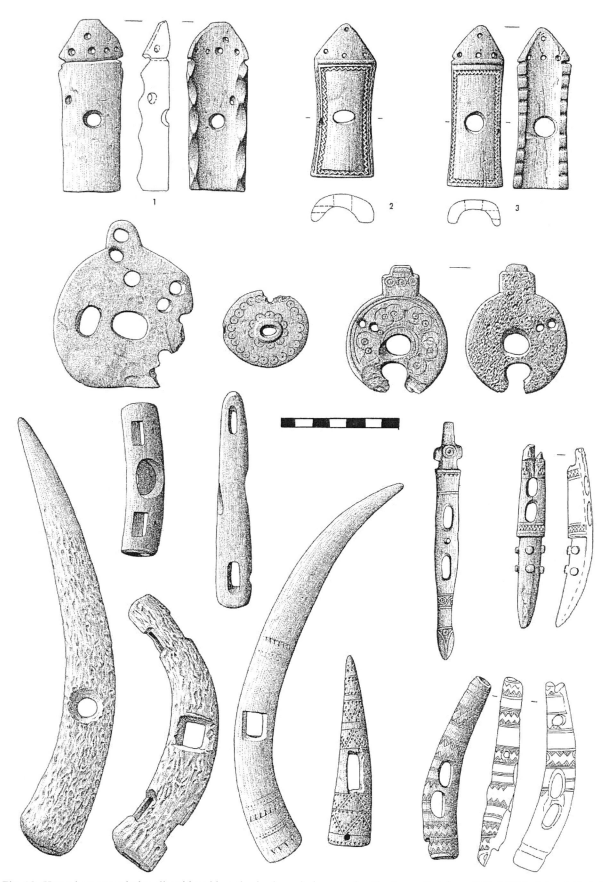

Fig. 19: Horse harness: tabular, discoid and bar cheek-pieces in bone and antler (compiled from Hüttel 1981 plates 1, 3, 6, 8, 12, 14). Scale in cm.

There are many settlements that have provided only small amounts of bone artefacts and others which have yielded larger assemblages (Biel 1996). Iron Age sites in Britain, for example, have yielded more than 25 different types of bone artefacts (Adkins & Adkins 1982: 83f.). This contrasts sharply with the findings documented in a recently published handbook about the Iron Age in Switzerland in which bone and antler artefacts are mentioned only peripherally and in very small numbers (Müller et al. 1999: 270f.). A category of Iron Age artefacts, worth mentioning are amulets made of burrs of red deer antler. Such pieces are quite regularly found in Celtic graves, in combination with perforated teeth, other types of skeletal material and sometimes with molluscs (Pauli 1975). Another category of very special artefacts are the richly decorated handles of bone, which are found in tumuli of the eastern Alpine region during the Hallstatt period (Dobiat 1980: 316). In association with other grave-goods such as spindle-whorls, weights and ornaments, such handles play a major role in answering questions as to gender and social ranking, two of the most widely discussed themes in archaeology to date (Hays-Gilpin & Whitley 1998).

In the Hallstatt period, an important invention which was developed in the South and spread throughout Europe was the lathe. It appeared north of the Alps at about 700 BC, as indicated by finds with characteristic evidence of turning, for example from the Heuneburg, one of the most famous and extensive sites of the 6th century BC (Bittel et al. 1981: 231ff.). An astonishing variety of carved implements made of bone, antler, ivory, amber, coral, jet and sapropelite[10] has been excavated there from which the wealth of this central settlement is demonstrated. In the Heuneburg record, the following artefacts are regularly found (Fig. 20; Sievers 1984): globular pins, a variety of amulets (boars' tusks, astragali, human skull), double-headed points, stamps, handles, ornamental discs, bone skates[11] and ornaments, in addition to spacer-beads and beads. The latter were found in graves near the head and shoulders of the buried persons and most probably formed part of splendid necklaces (Rieckhoff & Biel 2001; Drescher 1984). Some of the spacer-beads and beads evidently were formed on a lathe (Fig. 21).

At the end of the La Tène period, with the emergence of large urban centres, the *oppida,* considerable economic changes occurred. *Oppida* played a regionally dominant role with respect to political, military and religious affairs. At the same time, the emergence of a monetary system allowed the intensification of trade. Excavations in the *oppidum* of Manching, a famous site dating to the end of the La Tène period (mid of 3rd century to 50 BC), has provided a manifold assemblage of bone and antler artefacts (Maier 1961; Jacobi 1974), among which hinges and other parts of furniture form the most outstanding items (Fig. 22). The Iron Age is conventionally considered to end with the Roman invasion, although Iron Age traditions persisted in many respects.

The Roman period

Before Roman times in Europe, rural societies existed essentially with the majority of the population gaining a living from the land. The urban populations were small, and bone crafting was carried out mostly at a household level, practised by certain members of a family with more or less skill. There might have existed some craftsmen who already specialised in bone manufacturing, but it was only with the Roman expansion and the major social, political and economic changes in Europe that a bone working on an industrial scale emerged (Schallmeyer 1996). Professional bone crafting has been repeatedly evidenced in Roman towns and centres in different parts of Europe; specialised craftsmen, labelled e.g. *cornuarius, pectinarius* and *eborarius* (von Petrikovits 1981), lived in industrial quarters and produced a broad spectrum of bone and antler implements (Béal 1984; Boessneck & von den Driesch 1982; Ciugudean 2001; Deschler-Erb 1997, 1998, 2001; MacGregor 1985; Martin-Kilcher 1991; Peters 1998: 252ff.; Schmid 1968 and many others). By contrast, a so-called "do-it-yourself manufacturing" also existed, which was carried out in villas and rural settlements (Deschler-Erb 2001: 31ff.). Additionally, major trade activities, which also included the import and export of bone artefacts from industrial centres to many parts of the Roman empire, can be evidenced. A comprehensive analysis of all these aspects would demand a separate and detailed study. Here we may close the story of bone artefacts and man with a comment of H. Mikler which could characterise most research hitherto in bone manufacturing: "Römische Funde aus Bein erfuhren lange Zeit nicht die gleiche Beachtung wie andere Fundgruppen z.B. aus Keramik, Glas und Metall. Sie fanden zwar Erwähnung in Fundberichten und übergreifenden Publikationen, wurden aber meist nur im globalen Kontext mitbesprochen." (Mikler 1997: 3).

Methodological considerations and concluding remarks

Bone artefacts are, it may be claimed, among those archaeological finds which stimulate most trans-disciplinary thinking. A bone artefact mirrors not only facets

---

10. = fossil material from organic lacustrine sediments
11. In Central Europe, the use of bone skates can even be traced back to the Late Neolithic (Becker 1999: 110).

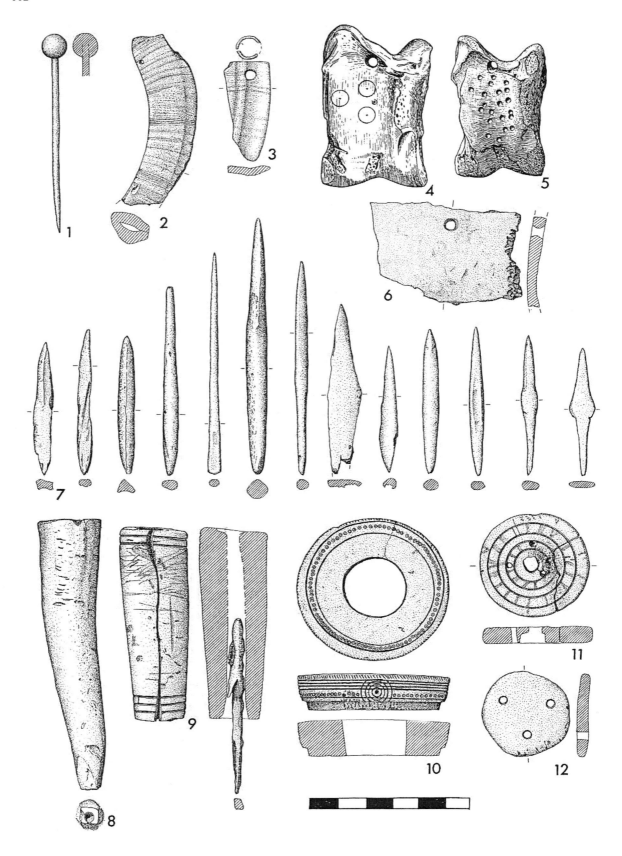

Fig. 20: Heuneburg, Germany (Hallstatt period). Bone and antler artefacts: bulb-headed pin (1), amulets (2, 3 boars' tusks, 4, 5 astragali, 6 human skull), double-headed points (7), stamp (8), handle (9), ornamental discs (10-12; compiled from Sievers 1984 plates 75-77, 79, 117, 118, 120, 121, 124). Scale in cm.

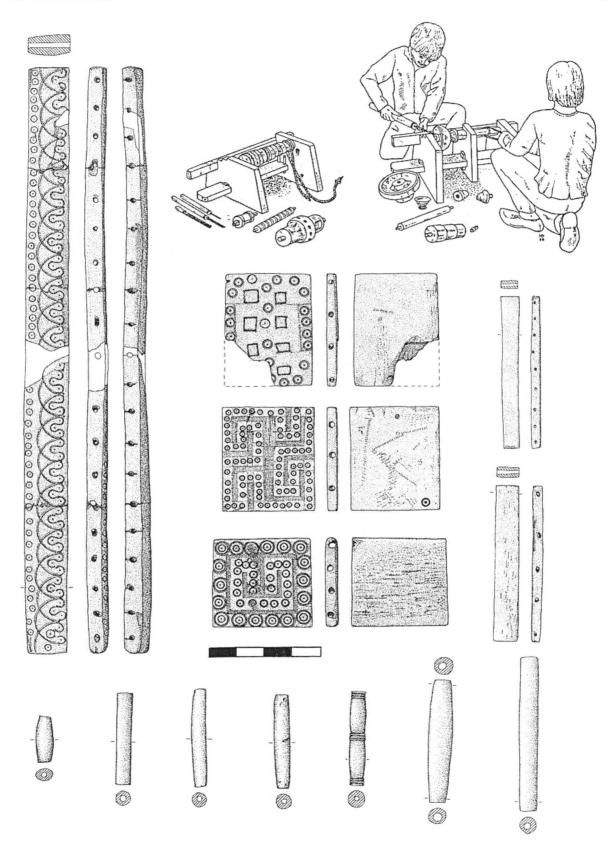

Fig. 21: Heuneburg, Germany (Hallstatt period). Reconstructed lathe and possible lathe-work: spacer-beads and beads (compiled from Sievers 1984 plates 27-31 and Drescher 1984 fig. 24). Scale in cm.

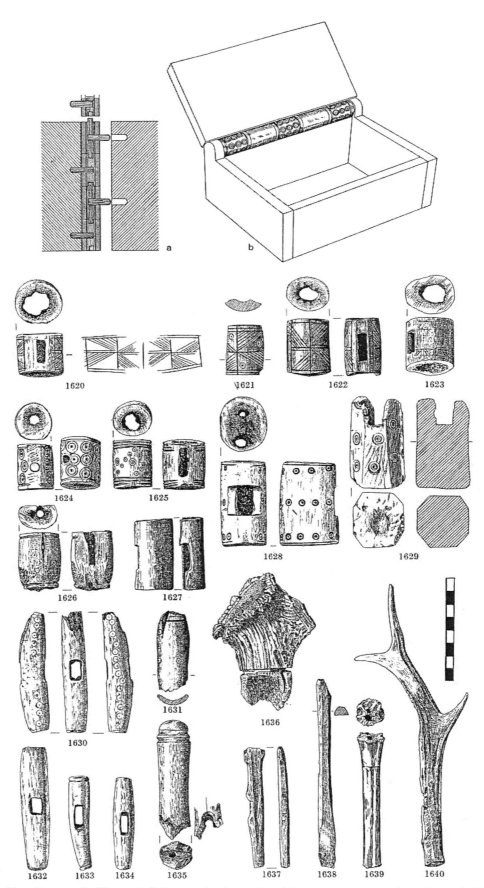

Fig. 22: Manching, Germany (La Tène period). Bone and antler artefacts: hinges, hammers, loop pins and some half-finished items (original plate from Pauli 1975 plate 82); top line: hinges in reconstructed position of use (ibid. fig. 56). Scale in cm.

of manufacturing and use, but also represents the remains of an animal which was hunted or kept in a particular environment, which was butchered, used for its meat, marrow and sinews as well as its bones. Particular skeletal elements were selected by men, providing useful raw material for manufacturing. Consequently, bone artefacts have the capacity to become vehicles for the study of behavioural expression. They mirror human skills and abilities and reflect aspects of cultural processes in which the artefact was embedded. Hence, for a comprehensive evaluation of a bone artefact, one has to collect data from various disciplines and, at the same time, consider the importance of small details.

Up to the present day, a general agreement as to a comprehensive method of documenting and interpreting bone artefacts has not been achieved, despite major publications written in past decades for example by Russian (Semenov 1964), French (Camps-Fabrer 1977; Patou-Mathis 1994), British (MacGregor 1985), Swiss (Schibler 1980, 1981), German (Ulbricht 1984; Müller 1964, 1980, 1982) and Hungarian (Choyke 1982-83, 1984; Choyke & Bartosiewicz 2001) specialists. Even for the measuring and drawing of finished items, comprehensive suggestions have long existed (Laurent 1977; Voruz 1984; Hahn 1992) but have nonetheless been ignored. The interpretation of bone artefacts mostly was (and still is) based on common sense and on the imagination of the excavator. A step-by-step analysis which considers every facet of possible information, has been carried out only rarely.

To improve the investigation of bone artefacts worldwide, under the auspices of the International Council of Archaeozoology (ICAZ) a "Worked Bone Research Group" has been established since 1997. In the U.S.A. similarly forward-looking approaches exist, mirrored at the 64th Annual Meeting of the Society for American Archaeology in Chicago in 1999, entitled "Technology of Skeletal Materials: Considerations of Production, Method and Scale" (see contributions in Choyke & Bartosiewicz 2001). In current papers about bone tools (cf. ibid. and especially LeMoine 2001; Schibler, in print) a step-by-step-analysis is followed taking the following aspects into consideration:

1. Investigation of osteological data;
2. Microscopic analyses of traces of manufacture and use;
3. Measuring and drawing of finished items according to international standards;
4. Recognition of different procedures occurring during fabrication by means of a detailed analysis of half-finished pieces and type-specific raw material;
5. Consideration of the context between subsistence strategy, exploitation or resources and bone crafting (*chaîne opératoire*);
6. Combination of all data within the stratigraphical context and habitation structures;
7. Comparison with ethnographic studies, written and pictorial evidence;
8. Consideration of results from experimental archaeology.

Of course, such analysis also has its specific problems. If one examines microscopic wear, for example the question arises as to whether the traces recognised were primarily produced during the manufacturing process or through the utilisation of bone utensils, namely through the contact with particular materials such as stone, pottery, soil, wood or skin. Nowadays, an increasing number of researchers are attempting to experiment with materials similar to those found at archaeological sites. Inquiries can be made to replicate wear patterns observed on particular artefacts. This type of work will help to identify traces of manufacturing and use as well as the kind of activity in which specific artefacts were involved (see the various issues of "Experimentelle Archäologie in Deutschland", e.g. Stodiek 1996 and Lobisser 1996; cf. also Hein & Hahn 1998; Stodiek & Paulsen 1996; d'Errico & Giacobini 1985). Others study the fabrication and utilisation of artefacts by groups of people who still employ a relatively primitive technology (Soeffing 1988; Steinbring 1966; Burford 1978; Leskov & Müller-Beck 1993). Another fruitful study area is provided by written sources or depictions of situations in which the function of particular bone artefacts is illustrated (Lehmkuhl & Müller 1995; Peters 1998; Bohnsack 1985: 37ff.; Fittà 1998; Otte 2000). Such attempts have existed in literature for 30 years (e.g. Redman 1973; Becker 1993) but only in recent years have they found an increasing resonance in archaeology.

Another crucial part of these analyses are considerations of subsistence strategy, exploitation of resources and bone crafting which may be combined in order to design a *chaîne opératoire*. These operational chains meticulously study the manufacturing process from the requisition and selection of raw materials to finished products and wast material (Lemonnier 1992). The combination and evaluation of all data available finally leads ultimately to the functional interpretation of a bone artefact, and beyond that, to a recognition of its meaning. In most cases, the archaeologist may not be in a position to recover the meaning that particular artefacts held for specific individuals, but it may nonetheless be possible to deduce their agreed significance within a given society. In any case, interpretation by a modern observer raises the issue of whether this accords with the actual meaning that a given artefact had for its user.

Nonetheless, a step-by-step analysis may prompt further questions, for there are always divergent opinions about the range of possible functions for an artefacts, about the relevance of ethnographic comparisons (Lee & DeVore 1968) and about the results of experimental archaeology. Therefore, when studying newly excavated assemblages of bone artefacts, it is vital to adopt a multidisciplinary approach which also takes into account diachronic comparisons. With this aim in view, it is essential to study bone artefacts not only from the published data, from photographs and drawings, but also in nature. Such opportunities need to be created through the establishment of representative collections of bone artefacts, where scientists may carry out comparative studies on the material itself. Perhaps the State Collection of Anthropology and Palaeoanatomy at Munich would be the adequate location for such an undertaking which would be a magnificent contribution to further fruitful research on bone artefacts.

Acknowledgements

This paper was based on a talk, initiated by Joris Peters and Gisela Grupe on the occasion of the 100th anniversary of the State Collection of Anthropology and Palaeoanatomy in Munich; both of them I owe many thanks for their invitation to join this conference. I am extremely grateful to Arthur MacGregor for his useful comments on the first draft of the paper; he also most kindly improved my english text. I also extend thanks to my colleagues, in particular Carola Metzner-Nebelsick and Elke Kaiser who have given me valuable archaeological information on a variety of topics. H. Hähnl redrew some bone artefacts, displayed in figures 1, 3 and 15.

References

Adkins L. & Adkins R.A., 1982.
   The Handbook of British Archaeology. London: MacMillan.

Albrecht G., 1987.
   Kunstobjekte aus dem Magdalénien. In: Müller-Beck H. & Albrecht G. (eds.). *Die Anfänge der Kunst vor 30000 Jahren*: 43-51. Stuttgart: Theiss.

Allsworth-Jones P., 1998.
   Introduction: Mammoth bone dwellings in Eastern Europe: a reassessment. In: Pidoplichko I. G. (ed.). *Upper Palaeolithic Dwellings of Mammoth Bones in the Ukraine* (translated by P. Allensworth-Jones): i-xxxvi. British Archaeological Reports International Series 712. Oxford: Archaeopress.

Auler J., 1993.
   Zur Nachbildung und Funktion bronzezeitlicher Tüllenharpunen. Ein Beitrag zur experimentellen Archäologie. *Archäologisches Korrespondenzblatt* 23: 197-206.

Azzaroli A., 1985.
   An Early History of Horsemanship. Leiden: Brill & Backhuys.

Bahn P. G., 1992.
   Ancient Art. In: Jones S., Martin R. & Pilbeam D. (eds.). *The Cambridge Encyclopedia of Human Evolution*: 361-364. Cambridge: Cambridge University Press.

Bahn P. G., 1998.
   The Cambridge Illustrated History of Prehistoric Art. Cambridge: Cambridge University Press.

Bahn P. G. & Vertut J., 1988.
   Images of the Ice Age. New York: Facts on File.

Bar-Yosef O. & Vandermeersch B., 2000.
   Koexistenz von Neandertaler und modernem *Homo sapiens*. *Spektrum der Wissenschaften* 3: 48-55.

Bandi G., 1963.
   Középső Bronzkori Lószerszám-szíjelosztó csontlemezek Kérdése a Kárpát-Medencében (Summary: Die Frage der Riementeiler des mittelbronzezeitlichen Pferdegeschirrs im Karpatenbecken). *Archaeologiai Értesítő* 90: 46-60.

Barber E. J. W., 1991.
   Prehistoric Textiles. Princeton, New Jersey: Princeton University Press.

Béal C.-J., 1984.
   Les objets de tabletterie antique du musée archéologique de Nîmes. Nîmes: Cahiers des musées et monuments de Nîmes 2.

Becker C., 1993.
   Zur Aufdeckung von Kausalitäten zwischen Ernährungsgepflogenheiten und Knochenverarbeitung. In: Friesinger H., Daim F., Kanelutti E. & Cichocki O. (eds.). *Bioarchäologie und Frühgeschichtsforschung*. Archaeologia Austriaca, Monographien 2: 133-157. Wien: Institut für Ur- und Frühgeschichte der Universität Wien.

Becker C., 1999.
   Neue Tierknochenanalysen zum Spätneolithikum des Mittelelbe-Saale-Gebietes. *Bericht der Römisch-Germanischen Kommission* 80: 91-121.

Becker C., 2000.
   Tierknochenfunde – Zeugnisse ritueller Aktivitäten. *Altorientalische Forschungen* 27: 167-183.

Benecke N., 1994.
   Zur Domestikation des Pferdes in Mittel- und Osteuropa. Einige neue archäozoologische Befunde. In: Hänsel B. & Zimmer S. (eds.). *Die Indogermanen und das Pferd*. Festschrift für B. Schlerath. Archaeolingua 4: 123-144. Budapest: Archaeolingua Alapítvány.

Benecke N., 1999.
   Pferdeknochenfunde aus Siedlungen der Bernburger Kultur – ein Beitrag zur Diskussion um die Anfänge der Pferdehaltung in Mitteleuropa. In: Kokabi M. & May E. (eds.). *Beiträge zur Archäozoologie und Prähistorischen Anthropologie II*: 107-120. Konstanz: Gesellschaft für Archäozoologie und Prähistorische Anthropologie e.V.

Biel J., 1996.
   Bronze- und Eisenzeit. In: Kokabi M. et al.: 57-70.

Binant P., 1991.
   Les sépultures du Paléolithique. Collection Archéologie Aujourd'hui. Paris: Editions Errance.

Bittel K., Kimmig W. & Schiek S. (eds.), 1981.
   Die Kelten in Baden-Württemberg. Stuttgart: Konrad Theiss.

Böckner G., 1980.
Geweihgezähe neolithischer Silexabbauanlagen am Beispiel Loewenburg – Neumühlefeld III – ein Beitrag zur Methodik. In: Deutsches Bergbaumuseum Bochum – Ausstellungskatalog *5000 Jahre Feuersteinbergbau*: 48-66. Bochum: Deutsches Bergbaumuseum.

Boessneck J. & von den Driesch A., 1982.
Tierknochenabfall aus einer spätrömischen Werkstatt in Pergamon. *Archäologischer Anzeiger* **1982**: 563-754.

Boetzkes M., Schweitzer I. & Vespermann J., 1999.
Eiszeit. Das grosse Abenteuer der Naturbeherrschung. Begleitbuch zur gleichnamigen Ausstellung. Hildesheim, Stuttgart: Roemer- und Pelizaeus-Museum, Jan Thorbecke.

Bohnsack A., 1985.
Spinnen und Weben. Hamburg: Rohwolt.

Borchardt J.,1972.
Homerische Helme. Mainz: Philipp von Zabern.

Boroffka N., 1998.
Bronze- und früheisenzeitliche Geweihtrensenknebel aus Rumänien und ihre Beziehungen. *Eurasia Antiqua* **4**: 81-135.

Bosinski G., 1987.
Die Kunst des Magdalénien im Rheinland. In: Müller-Beck H. & Albrecht G. (eds.). *Die Anfänge der Kunst vor 30000 Jahren*: 52-59. Stuttgart: Theiss.

Bosinski G., 1990.
Homo sapiens. L'histoire des chasseurs du Paléolithique supérieur en Europe (40 000 – 10 000 avant J.-C.). Paris: Editions Errance.

Bosinski G., 1991.
Der Neandertaler, seine Zeit, sein Schicksal. *Archäologie im Ruhrgebiet* **1**, Jahrgang 1991 (1993): 25-48.

Bosinski G. & Henke W., 1993.
Der Neandertaler, seine Zeit / sein Schicksal. Gelsenkirchen: Edition Archaea.

Braehm H., 1994.
Die magische Welt der Schamanen und Höhlenmaler. Köln: Du Mont.

Brain C. K., 1976.
Bone weathering and the problem of bone pseudo-tools. *South African Journal of Science* March 1976: 97-99.

Breuil A., 1932.
Le feu et l'industrie de pierre et de l'os dans le gisement du *Sinanthropus* à Chou-Kou-Tien. *L'Anthropologie* **42**: 1-17.

Burford I. R., 1978.
Nunamiut Ethnoarchaeology. New York: Academic Press.

Camps-Fabrer H. (ed.), 1977.
Méthodologie appliquée à l'industrie de l'os préhistorique. Deuxième colloque international sur l'industrie de l'os dans la Préhistorie, Abbaye de Sénanque (Vaucluse) 9-12 juin 1976. Paris: Edition du Centre National de la Recherche Scientifique.

Camps-Fabrer H. (ed.), 1982.
L'industrie en os et bois de cervidés durant le Néolithique et l'Age des Métaux. Deuxième réunion du groupe de travail n° 3 sur l'industrie de l'os préhistorique, Saint-Germain-en-Laye 1980. Paris: Editions du Centre National de la Recherche Scientifique.

Camps-Fabrer H. (ed.), 1985.
L'industrie en os et bois de cervidés durant le Néolithique et l'Age des Métaux. Troisième réunion du groupe de travail n° 3 sur l'industrie de l'os préhistorique, Aix-en-Provence, 26, 27, 28 Octobre 1983. Paris: Editions du Centre National de la Recherche Scientifique.

Carciumaru M. & Ulrix-Closset M., 1996.
Paleoenvironment et adaptation culturelle des Néandertaliens de la grotte Cioarei à Borosteni (Roumanie). In: Otte M.: 143-160.

Chase P. G., 1990.
Tool-making Tools and Middle Palaeolithic Behaviour. *Current Anthropology* **31/4**: 443-447.

Choyke A. M., 1982-1983.
An analysis of bone, antler and tooth tools from Bronze Age Hungary. *Mitteilungen des Archäologischen Institutes der Ungarischen Akademie der Wissenschaften* **12-13**: 13-38, plates 1-5 and tables 1-12.

Choyke A., 1984.
Faunal information offered by worked bone assemblages. *Acta Archaeologica Hungarica* **36**: 53-58.

Choyke A. M. & Bartosiewicz L. (eds.), 2001.
Crafting Bone: Skeletal Technologies through Time and Space. British Archaeological Reports International Series 937. Oxford: Archaeopress.

Ciugudean D., 2001.
Workshops and manufacturing techniques at Apulum (AD 2nd-3rd century). In: Choyke A. & Bartosiewicz L.: 61-72.

Clark G., 1975.
The earlier Stone Age Settlement of Scandinavia. Cambridge: Cambridge University Press.

Clottes J., 1997.
Niaux, Die altsteinzeitlichen Bilderhöhlen in der Ariège. Sigmaringen: Jan Thorbecke.

Clottes J. & Lewis-Williams D., 1997.
Schamanen. Trance und Magie in der Höhlenkunst der Steinzeit. Sigmaringen: Jan Thorbecke.

Clutton-Brock J., 1991.
Representation of the female breast in bone carvings from a Neolithic lake village in Switzerland. *Antiquity* **65**: 908-910.

Coles J. M. & Harding A. F., 1979.
The Bronze Age in Europe. London: Methuen.

Combier J., 1990.
De la fin du Moustérien au Paléolithique Supérieur – les données de la région Rhodanienne. In: Farizy C.: 267-277.

Cunliffe B., 2001.
Facing the Ocean. The Atlantic and its Peoples 8000 BC – AD 1500. Oxford: University Press.

David F. & Poulain Th., 1990.
La faune de grands mammifères des niveaux XI et XC de la Grotte du Renne à Arcy-sur-Cure (Yonne). Ètude préliminaire. In: Farizy C.: 319-323.

Dellbrügge S. B., 2002.
Steinzeitliche Knochen- und Geweihfunde im nördlichen Schleswig-Holstein. Universitätsforschungen zur Prähistorischen Archäologie 83. Bonn: Habelt.

Demakopoulou K. (ed.), 1988.
Das mykenische Hellas, Heimat der Helden Homers. Ausstellung im Rahmen von "Berlin Kulturstadt Europas 1988". Berlin: Dietrich Reimer.

Deschler-Erb S., 1997.
Bone, antler, tooth and ivory: raw materials for Roman artifacts. *Anthropozoologica* **25-26**: 73-78.

Deschler-Erb S., 1998.
Römische Beinartefakte aus Augusta Raurica. Rohmaterial, Technologie, Typologie und Chronologie. *Forschungen in Augst* **27/1-2**. Augst: Römermuseum Augst.

Deschler-Erb S., 2001.
Do-it-yourself manufacturing of bone and antler in two villas in Roman Switzerland. In: Choyke A. & Bartosiewicz L.: 31-40.

Deschler-Erb S., Marti-Grädel E. & Schibler J., 2002.
Die Knochen-, Zahn- und Geweihartefakte. In: de Capitani A., Deschler-Erb S., Leuzinger U., Marti-Grädel E. & Schibler J. (eds.). *Die jungsteinzeitliche Seeufersiedlung Arbon Bleiche 3. Funde.* Archäologie im Thurgau 11: 277-366. Arbon: Departement für Erziehung und Kultur des Kantons Thurgau.

Devlet E., 1996.
Rock art and the material culture of Sibiria and Central Asian shamanism. In: Price N. S. (ed.). *The Archaeology of Shamanism*: 43-55. London & New York: Routledge.

Dickinson O., 1994.
The Aegean Bronze Age. Cambridge World Archaeology. Cambridge: Cambridge University Press.

Dobiat C., 1980.
Das hallstattzeitliche Gräberfeld von Kleinklein und seine Keramik. *Schild von Steier – Beiträge zur Steirischen Vor- und Frühgeschichte und Münzkunde Beiheft* 1. Graz: Landesmuseum Joanneum.

Dowson T. A. & Porr M., 1996.
Special objects – special creatures: Shamanistic imaginary and the Aurignacian art of south-west Germany. In: Price N. S. (ed.). *The Archaeology of Shamanism*: 165-177. London & New York: Routledge.

Drescher H., 1984.
Bemerkungen zur Metallverarbeitung auf der Heuneburg und zu einigen besonderen Fundstücken. In: Sievers S. (ed). *Die Kleinfunde der Heuneburg.* Römisch-Germanische Forschungen **42**: 95-136.

Dunning C., 1992.
Le filage. *Helvetia archaeologica* **90**: 43-50.

d'Errico F. & Giacobini G., 1985.
Approche méthodologique de l'analyse de l'outillage osseux. Un exemple d'étude. *L'Anthropologie* **89/4**: 457-472

d'Errico F. & Giacobini G., 1988.
L'apport des études de surface pour l'interpretation des modifications sur l'os au Paléolithique Moyen. In: Binford L. & Rigaud J.-P. (eds.). *L'Homme de Néandertal 4. La Technique*: 39-47. Paris: Etudes et Recherches Archéologiques de l'Université de Liège n° 31.

Fabiš, M., 2000.
Die Knochenindustrie von dem frühbronzezeitlichen Gräberfeld in Jelšovce. In: Bátora J. (ed.). *Das Gräberfeld von Jelšovce/Slowakei. Ein Beitrag zur Frühbronzezeit im nordwestlichen Karpatenbecken 2.* Prähistorische Archäologie in Südosteuropa **16/2**: 601-611. Kiel: Oetker/Voges.

Farizy C. (ed.)., 1990.
Paléolithique moyen récent et Paléolithique supérieur ancien en Europe. Actes du colloque international de Nemours 9-11 mai 1988. Mémoires du Musée de Préhistoire d'Ile de France 3. Némours: Association pour la Promotion de la Recherche Archéologique en Ile de France.

Fittà. M., 1998.
Spiele und Spielzeug in der Antike. Unterhaltung und Vergnügen im Altertum. Stuttgart: Theiss.

Furger A. R. & Hartmann F., 1983.
Vor 5000 Jahren ..... so lebten unsere Vorfahren in der Jungsteinzeit. Bern: Haupt.

Gabunia L., Jöris O., Justus A., Lordkipanidze D., Muskhelishvili A., Nioradze M., Swisher III. C.C. & Vekuna A., 1999.
Neue Hominidenfunde des altpaläolithischen Fundplatzes Dmanisi (Georgien, Kaukasus) im Kontext aktueller Grabungsergebnisse. *Archäologisches Korrespondenzblatt* **29/4**: 451-488.

Gambier D., 1989.
Fossil Hominids from the early Upper Palaeolithic (Aurignacian) of France. In: Mellars P. & Stringer P. (eds.). *The Human Revolution. Behavioural and Biological Perspectives on the Origins of Modern Humans*: 194-211. Edinburgh: Edinburgh University Press.

Gaudzinski S., 1998a.
Vorbericht über die taphonomischen Arbeiten zu Knochengeräten und zum faunistischen Material der mittelpaläolithischen Freilandfundstelle Salzgitter-Lebenstedt. *Archäologisches Korrespondenzblatt* **28**: 323-337.

Gaudzinski S., 1998b.
Knochen und Knochengeräte der mittelpaläolithischen Fundstelle Salzgitter-Lebenstedt. *Jahrbuch des Römisch-Germanischen Zentralmuseums Mainz* **45**: 163-220.

Gaudzinski S., 1999.
Ein mittelpaläolithisches Rentierjägerlager bei Salzgitter-Lebenstedt. In: Boetzkes M. et al.: 165-175.

Gayck S., 2000.
Urgeschichtlicher Silexbau in Europa. Beiträge zur Ur- und Frühgeschichte Mitteleuropas 15. Weissbach: Beier & Beran.

Gibson J. J., 1966.
The Senses Considered as Perceptual Systems. Boston: Houghton Mifflin Company.

Gore R., 1996.
Neandertals. *National Geographic* **189/1**: 1-35.

Gramsch B., 1994.
Mesolithische Jäger in Brandenburg. *Archäologie in Deutschland* 3: 6-9.

Griffitts J. & Bonsall C., 2001.
Experimental determination of the function of antler and bone 'bevel-ended tools' from prehistoric shell middens in western Scottland. In: Choyke A. & Bartosiewicz L.: 207-220.

Gross E., Bleuer E., Hardmeyer B., Rast-Eicher A., Ritzmann Ch., Ruckstuhl B., Ruoff U. & Schibler J. (eds.), 1992.
Zürich "Mozartstrasse". Neolithische und bronzezeitliche Ufersiedlungen 2. Berichte der Züricher Denkmalpflege Monographien 17. Zürich: Züricher Denkmalpflege.

Grünberg J. M., 2000.
Mesolithische Bestattungen in Europa. Ein Beitrag zur vergleichenden Gräberkunde. Teil I: Auswertung, Teil II: Katalog. Internationale Archäologie 40. Rahden/Westfalen: Marie Leidorf.

Hahn J., 1986.
Kraft und Aggression. Die Botschaft der Eiszeitkunst im Aurignacien Süddeutschlands. Archaeologica Venatoria 7. Tübingen: Archaeologica Venatoria, Institut für Urgeschichte der Universität Tübingen.

Hahn J., 1987.
Die ältesten figürlichen Darstellungen im Aurignacien. In: Müller-Beck H. & Albrecht G. (eds.). *Die Anfänge der Kunst vor 30000 Jahren*: 25-33. Stuttgart: Theiss.

Hahn J., 1988.
Die Geißenklösterle-Höhle im Achtal bei Blaubeuren I. Forschungen und Berichte zur Vor- und Frühgeschichte in Baden-Württemberg 26. Stuttgart: Theiss.

Hahn J., 1992.
Zeichnen von Stein- und Knochenartefakten. Archaeologica Venatoria 13. Tübingen: Archaeologica Venatoria, Institut für Urgeschichte der Universität Tübingen.

Hahn J., Menu M., Taborin Y., Walter Ph. & Widemann F. (eds.), 1995.
Le travail et l'usage de l'ivoire au Paléolithique Supérieur. Actes de la Table Ronde Ravello, 29-31 Mai 1992. Rom: Instituto Poligrafica e Zecca dello Stato.

Halstead P., 1992.
The Mycenaean Palatial Economy. Making the most of the gaps in the Evidence. *Proceedings of the Cambridge Philological Society* 38: 57-86.

Halstead P., 1994.
The North-South Divide: Regional Paths to Complexity in Prehistoric Greece. In: Mathers C. & Stoddart S. (eds.). *Development and Decline in the Mediterranean Bronze Age*. Sheffield Archaeological Monographs 8: 195-219. Sheffield: J. R. Collins Publications.

Harding A. F., 2000.
European Societies in the Bronze Age. Cambridge World Archaeology. Cambridge: Cambridge University Press.

Hays-Gilpin K. & Whitley D. S. (eds.), 1998.
Reader in Gender Archaeology. London, New York: Routledge.

Heege E. & Heege A., 1989.
Die Häuser der Toten. Jungsteinzeitliche Kollektivgräber im Landkreis Northeim. Wegweiser zur Vor- und Frühgeschichte Niedersachsens 16. Hildesheim: August Lax.

Hein W. & Hahn J., 1998.
Experimentelle Nachbildung von Knochenflöten aus dem Aurignacien der Geissenklösterle-Höhle. Experimentelle Archäologie in Deutschland, Bilanz 1997. *Archäologische Mitteilungen Nordwestdeutschland* Beiheft 19: 65-73.

Henke W., Kieser N. & Schnaubelt W., 1996.
Die Neandertalerin, Botschafterin der Vorzeit. Gelsenkirchen/Schwelm: Edition Archaea.

Henke W. & Rothe H., 1980.
Der Ursprung des Menschen. Stuttgart, New York: Gustav Springer.

Henke W. & Rothe H., 1998.
Stammesgeschichte des Menschen. Berlin: Gustav Springer.

Heurtley W. A., 1939.
Prehistoric Macedonia. Cambridge: University Press.

Hjärthner-Holder E. & Risberg C., 2001.
The Innovation of Iron. From Bronze to Iron Age Societies in Sweden and Greece. In: Werbart B. (ed.). *Cultural Interactions in Europe and the Eastern Mediterranean during the Bronze Age (3000-500 BC)*. British Archaeological Reports International Series 985: 29-42. Oxford: Archaeopress.

Hochstetter A., 1987.
Kastanas. Ausgrabungen in einem Siedlungshügel der Bronze- und Eisenzeit Makedoniens 1975-1979. Die Kleinfunde. Prähistorische Archäologie in Südosteuropa 6. Berlin: Spiess.

Hüttel H.-G., 1981.
Bronzezeitliche Trensen in Mittel- und Osteuropa. Prähistorische Bronzefunde 16, 2. München: C. H. Beck.

Hyde N., 1988.
Fabric of History – Wool. *National Geographic* 173/5: 552-591.

Jacobi G., 1974.
Werkzeuge und Gerät aus dem Oppidum von Manching. Die Ausgrabungen in Manching 5. Wiesbaden: Franz Steiner.

Jacobs K. & Hart P., 1996.
Des riches, des pauvres, et de la chasse au statut social en Karélie Mésolithique. In: Otte M.: 403-416.

Jeannerod M., 1997.
The Cognitive Neuroscience of Action. Cambridge: Blackwell Publisher.

Jeunesse C., 1997.
Pratiques funéraires au Néolithique ancien. Sépultures et nécropoles des sociétés danubiennes (5500-4900 av. J.C.). Paris: Errance.

Jünger K., 2002.
Herbei, herbei, was Löffel sei.... Ausstellung der Wilhelm Wagenfeld Stiftung 3. März – 30. Juni 2002. Bremen: Anabas.

Kilian-Dirlmeyer I., 1997.
Das mittelbronzezeitliche Schachtgrab von Ägina. Kataloge vor- und frühgeschichtlicher Altertümer 27 = Alt-Ägina IV, 3. Mainz: Zabern.

Kind C.-J., 1999.
Waldjäger der Nacheiszeit. *Archäologie in Deutschland* **4**: 14-17.

Klíma B., 1987.
Die Kunst des Gravettien. In: Müller-Beck H. & Albrecht G. (eds.). *Die Anfänge der Kunst vor 30000 Jahren*: 34-52. Stuttgart: Theiss.

Kokabi M., Schlenker B. & Wahl J. (eds.), 1996.
"Knochenarbeit" – Artefakte aus tierischen Rohstoffen im Wandel der Zeit. Begleitheft zur Ausstellung im Saalburg-Museum. Saalburg-Schriften 4. Bad Homburg v.d.H.: Saalburgmuseum.

Koslowski J., 1982.
Excavation in the Bacho Kiro Cave, Bulgaria: final report. Warszawa: Polish Scientific Publishers.

Kuckenburg M., 2001.
Als der Mensch zum Schöpfer wurde. Stuttgart: Klett-Cotta.

von Kurzynski K., 1996.
"...und ihre Hosen nennen sie bracas". Textilfunde und Textiltechnologie der Hallstatt- und Latènezeit und ihr Kontext. Internationale Archäologie 22. Espelkamp: Marie Leidorf.

Larsson L., 1989.
Ethnicity and traditions in Mesolithic mortuary practises of southern Scandinavia. In: Shennan S. J. (ed.). *The Mesolithic in Europe*. Actes of the Symposium at Edinburgh 1985: 366-378. Edinburgh: Donald.

Larsson L., 1990.
Dogs in Fraction – Symbols in Action. In: Vermeersch P. M. & Van Paar P. (eds.). *Contributions to the Mesolithic in Europe*: 153-160. Leuwen: University Press.

Larsson L., 1994.
Pratique mortuaires et sépultures de chiens dans les sociétés mésolithiques de Scandinavie mériodonale. *L'Anthropologie* **98**: 562-565.

Laurent P., 1977.
Le dessin de l'industrie osseuse préhistorique. In: Camps-Fabrer H.: 27-47.

Legge A. J. & Rowly-Conwy P. A., 1988.
Starr Carr Revisited. London: Centre for Extra-Mural Studies Birbeck College, University of London.

Lee R. B. & DeVore I., 1968.
Man the Hunter. Chicago: Aldine-Atherton.

Lehmann Th., 1991.
Göttinger Typentafeln zur Ur- und Frühgeschichte Mitteleuropas: Mesolithikum. Oerlinghausen: Archäologisches Landesmuseum.

Lehmkuhl U. & Müller H.-H., 1995.
Werkzeug – Spielzeug – Waffen. *Archäologie in Deutschland* **1**: 22-25.

LeMoine G., 2001.
Skeletal technology in context: an optimistic overview. In: Choyke A. M. & Bartosiewicz L.: 1-7.

Lemonnier P., 1992.
Elements for an Anthropology of Technology. Ann Arbor: University of Michigan Anthropological Papers 88.

Leskov A. M. & Müller-Beck H., 1993.
Arktische Waljäger vor 3000 Jahren. Unbekannte sibirische Kunst. Mainz, München: von Hase & Köhler.

Leuzinger U., 2002.
Textilherstellung. In: de Capitani A., Deschler-Erb S., Leuzinger U., Martin-Grädel E. & Schibler J., (eds.), 2002. *Die jungsteinzeitliche Seeufersiedlung Arbon Bleiche 3. Funde*. Archäologie im Thurgau 11: 115-134. Arbon: Departement für Erziehung und Kultur des Kantons Thurgau.

Lichardus J., 1980.
Zur Funktion der Geweihspitzen vom Typ Osdorf. *Germania* **58**, 1980: 1-24.

Littauer M. A., 1969.
Bits and pieces. *Antiquity* **33**: 289-300.

Liversage D., 1966.
Ornamented Mesolithic Artefacts from Denmark. *Acta Archaeologica* (København) **37**: 221-237.

Lobisser W., 1996.
Versuche zum Bau von dreilagigen Beinkämmen. Experimentelle Archäologie in Deutschland Bilanz 1996. *Archäologische Mitteilungen Nordwestdeutschland* Beiheft **18**: 67-85.

Der Löwenmensch. Tier und Mensch in der Kunst der Eiszeit. Begleitheft zur Ausstellung im Ulmer Museum 11. September – 13. November 1994.
Sigmaringen: Jan Thorbecke.

Lyman R. L., 1984.
Broken bones, bone expediency tools, and bone pseudotools: lessons from the blast zone around Mount St. Helens, Washington. *American Antiquity* **49/2**: 315-333.

MacGregor A., 1985.
Bone, Antler, Ivory & Horn. The Technology of Skeletal Materials Since the Roman Period. London, Sydney: Croom Helm.

Maier R. A., 1961.
Zu keltischen Würfelfunden aus dem Oppidum von Manching. *Germania* **39**: 354-360.

Makkay J., 1990.
Knochen-, Geweih- und Eberzahngegenstände der frühneolithischen Körös-Kultur. *Communicationes Archaeologicae Hungariae* 1990: 23-58.

Mania D., 1990a.
Urmenschen in Thüringen. Bilzingsleben – vor 350.000 Jahren. *Archäologie in Deutschland* **3**: 12-31.

Mania D., 1990b.
Auf den Spuren des Urmenschen. Die Funde aus der Steinrinne von Bilzingsleben. Berlin: Deutscher Verlag der Wissenschaften.

Mania D., 1998.
Die ersten Menschen in Europa. Archäologie in Deutschland (Sonderheft). Stuttgart: Theiss.

Mania, D. & Dietzel, A., 1980.
Begegnung mit dem Urmenschen. Die Funde von Bilzingsleben. Leipzig, Jena, Berlin: Urania.

Marshack A., 1988.
The Neanderthals and the human capacity for symbolic thought: cognitive and problem-solving aspects of Mousterian symbol. In: Otte M. (ed.). *L'Homme de Néandertal 5, La Pensée*: 57-91. Paris: Etudes et Recherches Archéologiques de l'Université de Liège n° 32.

Martin-Kilcher S., 1991.
Geräte und Geräteteile aus Knochen und Hirschhorn aus dem Vicus Vitudurum-Oberwinterthur. *Beiträge zum römischen Oberwinterthur-Vitudurum* **5**: 1-61.

Mikler H., 1997.
Die römischen Funde aus Bein im Landesmuseum Mainz. Monographies *instrumentum* 1. Montagnac: éditions monique mergoil.

Mithen S., 1998.
The Prehistory of the Mind. Phoenix: Orion House.

Morris C. E., 1990.
In pursuit of the White Tusk Boar. In: Hägg R. & Nordquist G. C. (eds.). *Celebration of Death and Divinity in the Bronze Age Argolid*. Skrifter Udgivna av Svenska Instituteit i Athen 4, 40: 149-156.

Movius H., 1969.
The Châtelperronian in French archaeology: the evidence of Arcy-sur-Cure. *Antiquity* **43**: 111-123.

Müller F., Kaenel G. & Lüscher G. (eds.), 1999.
Die Schweiz vom Paläolithikum bis zum frühen Mittelalter IV. Eisenzeit. Basel: Verlag Schweizerische Gesellschaft für Ur- und Frühgeschichte.

Müller H.-H., 1964.
Bandkeramische Knochengeräte – zoologisch betrachtet. *Varia Archaeologica* **3**: 25-38, plate 1.

Müller H.-H., 1980.
Neolithische Gerbereiwerkzeuge aus Knochen. *Alt-Thüringen* **17**: 7-18.

Müller H.-H., 1982.
Knochengeräte aus dem mittleren Neolithikum – zoologisch betrachtet. *Zeitschrift für Archäologie* **16**: 149-156.

Müller-Beck H., Conard N. C. & Schürle W. (eds.), 2000.
Eiszeitkunst. Stuttgart: Theiss.

Nandris, J., 1969.
Bos primigenius and the Bone Spoon. *Bulletin of the Institute of Archaeology* **9**: 63-82.

Nash G., 1998.
Exchange, Status and Mobility: Mesolithic Portable Art of Southern Scandinavia. British Archaeological Reports International Series 710. Oxford: Archaeopress.

Nash G., 2001.
Altered states of conciousness and the afterlife: a reappraisal of a decorated bone piece from Ryemarksgaard, Central Zealand, Denmark. In: Choyke A. M. & Bartosiewicz L.: 231-240.

Oakley K. P., Andrews P., Keeley L. H. & Clark J. D., 1977.
A Reappraisal of the Clacton Spearpoint. *Proceedings of the Prehistoric Society* **43**: 13-30.

Oliva M., 1987.
Aurignacien na Morave. Kromeriz: Studie muzea Kromeriszka.

Olsen S., 1989.
Distinguishing natural and cultural damage on archaeological antlers. *Journal of Archaeological Science* **16**: 125-135.

Orschiedt J., Auffermann B. & Weninger G.-C., 1999.
Familientreffen Deutsche Neanderthaler 1856-1999. Ausstellungskatalog Neandertal Museum. Mettmann: Neandertal Museum.

O'Shea J. M. & Zvelebil M., 1984.
Oleneostrovski mogilnik: Reconstructing the social and economic organization of prehistoric foragers in northern Russia. *Journal of Anthropological Archaeology* **3**: 1-40.

Otte M. (ed.), 1988.
L'Homme de Neandertal. Actes du colloque international de Liège 4-7 decembre 1986, volumes 1 – 8. Paris: Etudes et Recherches Archéologiques de l'Université de Liège n° 28 – 35.

Otte M. (ed.), 1996.
Nature et Culture. Actes du colloque international de Liège 13-17 décembre 1993, volume I, II. Liège: Etudes et Recherches Archéologiques de l'Université de Liège n° 68.

Otte M., 2000.
Regards sur la musique paléolithique. In: Hickmann E. & Eichmann R. (eds.). *Studien zur Musikarchäologie I*. Orient-Archäologie 6: 97-102. Rahden/Westfalen: Marie Leidorf.

Papathanassopoulos G. A. (ed.), 1996.
Neolithic Culture in Greece: 137-139. Athens: N. P. Goulandris Foundation.

Patou-Mathis M., 1994.
Outillage peu élaboré en os et en bois de cervidés IV. Artefacts 9, 6e Table Ronde Taphonomie/Bone Modification Paris 1991. Bruxelles: Treignes.

Paul A., 1998.
Von Affen und Menschen. Darmstadt: Wissenschaftliche Buchgesellschaft.

Pauli L., 1975.
Keltischer Volksglauben. Amulette und Sonderbestattungen am Dürrnberg bei Hallein und im eisenzeitlichen Mitteleuropa. Münchner Beiträge zur Vor- und Frühgeschichte 28. München: C.H. Beck.

Pelegrin J. & Richard A., 1995.
Les Mines de Silex au Néolithique en Europe: Avancées récentes. Documents Préhistoriques 7. Nancy: Comité des Travaux Historiques et Scientifiques.

Peters J., 1998.
Römische Tierhaltung und Tierzucht. Eine Synthese aus archäozoologischer Untersuchungen und schriftlich-bildlicher Überlieferung. Passauer Universitätsschriften zur Archäologie 5. Rahden/Westfalen: Marie Leidorf.

von Petrikovits H., 1981.
Die Spezialisierung des römischen Handwerks. In: Jankuhn H., Janssen W., Schmidt-Wiegand R. & Tiefenbach H. (eds.). *Das Handwerk in vor- und frühgeschichtlicher Zeit. Teil 1.* Historische und rechtshistorische Beiträge und Untersuchungen zur Frühgeschichte der Gilde: 63-132. Göttingen: Vandenhoeck & Ruprecht.

Pidoplichko I. G., 1998.
Upper Palaeolithic Dwellings of Mammoth Bones in the Ukraine (translated by P. Allensworth-Jones). British Archaeological Reports International Series 712. Oxford: Archaeopress.

Piggott S., 1983.
The earliest wheeled transport. London: Thames & Hudson.

Pratsch S., 1994.
Die Geweihartefakte des mesolithisch-neolithischen Fundplatzes von Friesack 4, Kr. Havelland. *Veröffentlichungen des Museums für Ur- und Frühgeschichte Potsdam* **28**: 7-98.

Quitta H., 1957.
Neue Hüttengrundrisse aus dem ukrainischen Jungpaläolithikum. *Ausgrabungen und Funde* **2**: 312-322.

Rast-Eicher A., 1997.
Die Textilien. In: Schibler J., Hüster-Plogmann H., Jacomet S., Brombacher C., Gross-Klee E. & Rast-Eicher A. (eds.). *Ökonomie und Ökologie neolithischer und bronzezeitlicher Ufersiedlungen am Zürichsee.* Monographien der Kantonsarchäologie Zürich 20: 300-328. Zürich, Egg: Verlag Fotorotar AG.

Redman C. L., 1973.
Early Village Technology. A View through the Microscope. *Paléorient* **1**: 249-261.

Rieckhoff S. & Biel J., 2001.
Die Kelten in Deutschland. Stuttgart: Theiss.

Schafer R. M., 1977.
The Tuning of the World. Toronto: McClelland & Stewart.

Schallmeyer E., 1996.
Die Verarbeitung von Knochen in römischer Zeit. In: M. Kokabi et al.: 71-82.

Schibler J., 1980.
Osteologische Untersuchungen der cortaillodzeitlichen Knochenartefakte. Die neolithischen Ufersiedlungen von Twann 8. Bern: Haupt.

Schibler J., 1981.
Typologische Untersuchungen der cortaillodzeitlichen Knochenartefakte. Die neolithischen Ufersiedlungen von Twann 17. Bern: Haupt.

Schibler J., 1995.
Geweih, Knochen. In: Stöckli E., Niffeler U. & Gross-Klee E. (eds.). *Die Schweiz vom Paläolithikum bis zum frühen Mittelalter II. Neolithikum*: 142-151. Basel: Verlag Schweizerische Gesellschaft für Ur- und Frühgeschichte.

Schibler J., 1997.
Knochen- und Geweihartefakte. In: Schibler J., Hüster-Plogmann H., Jacomet S., Brombacher C., Gross-Klee E. & Rast-Eicher A. (eds.). *Ökonomie und Ökologie neolithischer und bronzezeitlicher Ufersiedlungen am Zürichsee.* Monographien der Kantonsarchäologie Zürich 20: 122-219. Zürich & Egg: Fotorotar AG.

Schibler J., 1998.
Knochen- und Geweihartefakte. In: Hochuli S., Niffler U. & Rychner V. (eds.). *Die Schweiz vom Paläolithikum bis zur frühen Mittelalter III. Bronzezeit*: 274-278. Basel: Verlag Schweizerische Gesellschaft für Ur- und Frühgeschichte.

Schibler J. (ed.), in print 2002/3.
Bone, Antler and Teeth. Raw materials for tools from archaeological contexts. Proceedings of the 3rd meeting of the (ICAZ) Worked Bone Research Group, Basel/Augst 4.-8. September 2001. Archäologie International. Rahden/Westfalen: Marie Leidorf.

Schlenker B., 1996.
Knochen- und Geweihgeräte in der Jungsteinzeit. In: Kokabi et al.: 41-56.

Schmid E., 1968.
Beindrechsler, Hornschnitzer und Leimsieder im römischen Augst. Provincialia. Festschrift Rudolf Laur-Belart: 185-197. Basel, Stuttgart: Schwabe & Co.

Semenov A. S., 1964.
Prehistoric technology. London: Cory, Adams & Mackay.

Sidéra I., 2001.
Domestic and funerary bone, antler and tooth objects in the Neolithic of western Europe: a comparison. In: Choyke A. M. & Bartosiewicz L.: 221-229.

Sievers S., 1984.
Die Kleinfunde der Heuneburg. *Heuneburgstudien 5 = Römisch-Germanische Forschungen* 42. Mainz: Zabern.

Singer R., Wymer J., Gladfelter B. G. & Wolff R. G., 1973.
Excavation of the Clactonian Industry at the Golf Course, Clacton-on-Sea, Essex. *Proceedings of the Prehistoric Society* **39**: 6-74.

Smith C., 1989.
British Antler Mattocks. In: Bonsall C. (ed.). *The Mesolithic in Europe.* Papers presented at the Third International Symposium, Edinburgh 1985: 272-283. Edinburgh: John Donald Publications.

Soeffing H., 1988.
Die Töpferei bei den For im Jebel Marra – ein bedrohtes Handwerk. In: Vossen R., (ed.). *Töpfereiforschung zwischen Archäologie und Entwicklungspolitik.* Töpferei und Keramikforschung 1. Bonn: Rudolf Habelt.

Šovkolpjas I., 1965.
Mezinskaja stojanke. Kiew: Naukova Dumka.

Steinbring J., 1966.
The manufacture and use of bone defleshing tools. *American Antiquity* **31/4**: 575-581.

Stodiek U., 1996.
Die Speerschleuder – Erfindung einer verbesserten Jagdtechnik in der späten Altsteinzeit. Experimentelle Archäologie in Deutschland. *Archäologische Mitteilungen aus Nordwestdeutschland* **13**: 73-77.

Stodiek U. & Paulsen H., 1996.
"Mit dem Pfeil, dem Bogen..." Technik der steinzeitlichen Jagd. Archäologische Mitteilungen aus Nordwestdeutschland Beiheft 16. Oldenburg: Issensee.

Street M., 1999.
Haushunde und Schamanen. *Archäologie in Deutschland* **4**: 24-25.

Stringer C. & Gamble C., 1993.
In Search of the Neanderthals. London: Thames and Hudson.

Suter P. J., 1981.
Die Hirschgeweihartefakte der Cortaillod-Schichten. Die neolithischen Ufersiedlungen von Twann 15. Bern: Haupt.

Taborin Y., 1990.
Les prémices de la parure. In: Farizy C.: 335-344.

Tarlow S., 2000.
Emotion in archaology. *Current Anthropology* **41/5**: 713-746.

Taute W., 1965.
Retoucheure aus Knochen, Zahnbein und Stein vom Mittelpaläolithikum bis zum Neolithikum. *Fundberichte aus Schwaben* N.F. **17**: 76-102, table 1 and plates 14-22, 32.

Thieme H., 1996.
Altpaläolithische Wurfspeere aus Schöningen, Niedersachsen – ein Vorbericht. *Archäologisches Korrespondenzblatt* **26**: 377-393.

Thieme H., 1999.
Jagd auf Wildpferde vor 400.000 Jahren. In: Boetzkes M. et al.: 122-136.

Thieme H. & Veil S., 1985.
Neue Untersuchungen zum eemzeitlichen Elefanten-Jagdplatz Lehringen, Ldkr. Verden. *Die Kunde* N.F. **36**: 11-58.

Trinkaus E., 1992.
Evolution of human manipulation. In: Jones S., Martin R. & Pilbeam D. (eds.). *The Cambridge Encyclopedia of Human Evolution*: 346-349. Cambridge: Cambridge University Press.

Trinkaus E. & Shipman P., 1993.
The Neandertals. Changing the Image of Mankind. London: Jonathan Cape.

Tromnau G., 1985.
Steinzeitliche Tanzmasken. In: Ausstellungskatalog des Niederrheinischen Museums der Stadt Duisburg. *Rentierjäger und Rentierzüchter Sibiriens, früher und heute*: 90-96. Duisburg: Dezernat für Kultur und Bildung.

Tsachili I., 1996.
Weaving – Basketry. In: Papathanassopoulos G. A.: 137-139.

Turk I., Bastiani G., Culiberg M., Dirjec J., Kavur B., Krystufek B., Ku T.-L., Kunej D., Nelson D. E., Omerel-Terlep M. & Sercelj A. (eds.), 1997.
Mousterian "Bone Flute" and other finds from Divje Babe I Cave Site in Slovenia. Ljubljana: Znanstvenoraziskovalni SAZU.

Turk I., Dirjec J. & Kavur B., 1997.
Description and explanation of the origin of the suspected bone flute. In: Turk I. et al.: 157-178.

Turk I. & Kavur B., 1997.
Survey and description of Palaeolithic tools, fireplaces and hearths. In: Turk I. et al.: 119-156.

Uerpmann H.- P., 1995.
Domestikation of the Horse – When, Where and Why? In: Bodson L. (ed.). *Le cheval et les autres équidés: aspects de l'histoire de leur insertion dans les activités humaines.* Colloque d'histoire des connaissances zoologiques 6: 15-29. Liège: Université de Liège.

Ulbricht I., 1984.
Die Verarbeitung von Knochen, Geweih und Horn im mittelalterlichen Schleswig. Ausgrabungen in Schleswig. Berichte und Studien 3. Neumünster: Karl Wachholtz.

Valoch K., 1982.
Die Beingeräte von Předmostí in Mähren (Tschechoslowakei). *Anthropologie* **20**: 57-69.

Veil S., 1999.
Kultur vor dem modernen Menschen. Fragen zu den archäologischen Spuren aus der Zeit des Neandertalers. In: Boetzkes M. et al.: 137-164.

Vincent A., 1988.
L'os comme artefact au Paléolithique Moyen: principes d'étude et premiers résultats. In: Binford L. & Rigaud J.-P. (eds.). *L'Homme de Néandertal 4. La Technique*: 185-196. Paris: Etudes et Recherches Archéologiques de l'Université de Liège n° 31.

Vincent A., 1993.
L'outillage osseux au Paléolithique moyen: une nouvelle approche. Thèse de doctorat, Université Paris.

Voruz J.-L., 1984.
Outillages osseux et dynamisme industriel dans le Néolithique jurassien. Cahiers d'Archéologie Romande 29. Lausanne: Bibliothèque historique vaudoise.

Watson A., 1996.
The sounds of transformation: Acoustics, monuments and ritual in the British Neolithic. In: Price N. S. (ed.). *The Archaeology of Shamanism*: 178-192. London & New York: Routledge.

White R., 1995a.
Ivory personal ornaments of Aurignacian age: technological, social and symbolic perspectives. In: Hahn J., Menu M., Taborin Y., Walter Ph. & Widemann F. (eds.). *Le Travail et l"usage de l'ivoire au Paléolithiques Supérieur.* Actes du Table Ronde à Ravello, 29-31 Mai 1992: 29-62. Roma: Instituto Poligrafico a Zecca dello Stato.

White R., 1995b.
Bildhaftes Denken in der Eiszeit. In: Streit B. (ed.). *Evolution des Menschen*: 166-173. Heidelberg: Spektrum Akademischer Verlag.

Wilson F. R., 2000.
Die Hand – Geniestreich der Evolution. Stuttgart: Klett-Cotta.

Wininger J., 1971.
Das Fundmaterial von Thayngen-Weier im Rahmen der Pfyner Kultur. Monographien zur Ur- und Frühgeschichte der Schweiz 18. Basel: Birkhäuser.

Wininger J., 1981.
Feldmeilen-Vorderfeld. Der Übergang von der Pfyner zur Horgener Kultur. Antiqua 8. Frauenfeld: Huber.

Zagorska I. & Zagorska F., 1989.
The Bone and Antler Inventory from Zvejnieki II, Latvian SSR. In: Bonsall C. (ed.). *The Mesolithic in Europe.* Papers presented at the Third International Symposium, Edinburgh 1985: 414-423. Edinburgh: John Donald Publications.

Zeeb A., 1998.
Die Goldberg-Gruppe im frühen Jungneolithikum Südwestdeutschlands. Universitätsforschungen zur prähistorischen Archäologie 48. Bonn: Rudolf Habelt.

Zhilin M. G., 1998.
Artifacts made of animals' teeth and jaws in the Mesolithic of Eastern Europe. In: Pearce M. & Tosi M. (eds.). *Papers from the European Association of Archaeologists, Third Annual Meeting at Ravenna 1997*, volume 1: Pre- and Protohistory. British Archaeological Reports International Series 717: 26-28. Oxford: Archaeopress.

Zilhão J. & d'Errico F., 2000.
Die unterschätzten Neandertaler. In: Wong K. (ed.). Der Streit um die Neandertaler. *Spektrum der Wissenschaft* **3**: 68-69.

Hiding in Plain Sight:
The value of Museum Collections in the study of the origins of animal domestication

Melinda A. Zeder, Smithsonian Institution, Washington DC

Abstract

More than 50 years ago interdisciplinary teams of scientists used combined perspectives from the biological, physical, and human sciences to study the initial domestication of plants and animals in the Zagros Mountains of present-day Iran and Iraq. Their research succeeded in drawing the general outlines of plant and animal domestication in Southwest Asia. This early work also generated the first systematic collections of archaeological animal remains that document the transition from hunting to herding. Recent advances in analytical methods, plus enhanced understanding of the process of animal domestication in other parts of the Fertile Crescent region, argue for a return to these long dormant collections. This paper presents the results of a reconsideration of sheep and goat domestication in the eastern arc of the Fertile Crescent applying new techniques for identifying and dating initial animal domestication to the entire corpus of caprine remains generated by these early investigations. Using these techniques clear evidence of the management of morphologically wild goats is first detected within the natural habitat of wild goats and directly dated to 8900 uncalibrated B.P. Morphological change usually taken as a leading edge marker of domestication seems instead a delayed artifact of this process that may only be evident once domestic herds move out of this heartland region of their initial domestication. The new evidence for goat domestication in the Zagros has parallels in emerging scenarios for sheep and pig domestication in other parts of the Fertile Crescent. Drawing from this work it is proposed that a full understanding of livestock domestication can only be gained with a change methodological approaches for detecting initial domestication, a change in regional focus, and a new openness to sharing data and access to collections by international researchers working on the domestication and dispersal of domestic animals in the Old World.

Es ist mehr als 50 Jahre her, seit interdisziplinär arbeitende Teams von Wissenschaftlern sich eine Natur- und Geisteswissenschaften verknüpfende Sichtweise zu eigen machten, um der frühen Domestikation von Tieren bzw. Kultivierung von Pflanzen im Zagrosgebirge, d.h. in Gebieten des heutigen Iraq und Iran, auf die Spur zu kommen. Durch ihre Untersuchungen gelang es, die Grundzüge dieser Entwicklung in Südwestasien nachzuzeichnen. Diese frühen Untersuchungen führten gleichzeitig zum Aufbau der ersten systematischen Sammlungen von Knochenmaterialien archäologischer Provenienz, die den Übergang vom Jagen zur Herdenhaltung dokumentierten. Jüngste Weiterentwicklungen analytischer Methoden und ein verbessertes Verständnis zum Ablauf der Domestikation in anderen Teilen des Fruchtbaren Halbmondes sind der Grund für das erneute Zurückgreifen auf diese seit langer Zeit ruhenden Sammlungen. In diesem Beitrag wird die Domestikation von Schaf und Ziege im östlichen Teil des Fruchtbaren Halbmondes einer neuerlichen Bewertung unterzogen, in dem für den gesamten Corpus der Ovicaprinidenreste aus diesen alten Untersuchungen neue Techniken zur Identifizierung und Datierung zur Anwendung gebracht werden. Mithilfe dieser Techniken gelang der erste eindeutige Nachweis eines beginnenden Managements von morphologisch wilden Ziegen innerhalb ihres natürlichen Verbreitungsgebietes durch den Menschen bereits um 8900 B.P. Demnach scheinen die bisher als wichtige Eckpfeiler für das Erkennen von Domestikationsvorgängen geltenden, morphologischen Veränderungen vielmehr eine verspätete Sonderentwicklung darzustellen, die erst wirklich sichtbar wird, wenn Herden domestizierter Tiere das Gebiet ihrer Erstdomestikation verlassen. Dieser neue Nachweis für die Domestikation von Ziegen im Zagrosgebirge findet Parallelen in ähnlichen strukturierten Szenarien zur Domestikation von Schafen und Schweinen in anderen Teilen des Fruchtbaren Halbmondes. Aus der vorliegenden Arbeit erwächst die Erkenntnis, dass ein wirkliches Verständnis um die Vorgänge der Domestikation nur erreicht werden kann, wenn sich sowohl der methodische Ansatz als auch der Fokus auf bestimmte Regionen ändern und die Bereitschaft wächst, Daten auszutauschen und Sammlungen zugänglich zu machen für solche Wissenschaftler, die sich mit Fragen der Domestikation und der Verbreitung von Haustieren in der Alten Welt auseinandersetzen.

Key words: domestication, goats, Zagros Mountains, Fertile Crescent
Domestikation, Ziegen, Zagrosgebirge, Fruchtbarer Halbmond

Introduction

More than fifty years ago Robert Braidwood began a revolutionary program of archaeological investigation in Southwest Asia aimed at tracing the origins of plant and animal domestication and the development of food producing economies. To accomplish this goal he brought together a team of archaeologists, biologists, geomorphologists, and physical scientists with the skills and training needed to understand the complex biological, climatological, and cultural forces at work in this major transition in human history. The newly developed technology of radiocarbon dating was a special feature of this visionary work, as were the insights of zoologists like Charles Reed and botanists like Hans Helbaek in unraveling the evolutionary processes that resulted in the transformation of wild plant and animal species into domesticated crops and livestock. Braidwood's work focused on the eastern arc of the Fertile Crescent in the highlands and foothills of the Zagros Mountains of present day Iraq and Iran (Figure 1). He was joined here by other pioneer archaeologists like Ralph and Rose Solecki, Phillip Smith, Frank Hole and, Kent Flannery, who shared his interest in tracing the transition from hunting and gathering to food production. The combined efforts of the early researchers working in the Zagros was the first comprehensive outline of the transition to farming and herding in the Fertile Crescent – an outline that still carries currency in the study of Near Eastern agricultural origins (Braidwood and Howe 1960; Braidwood et al. 1983; Hole et al. 1969; Mortensen 1972; Smith 1976; Solecki, R.S. 1963; Solecki, R.L 1981).

Nearly two decades of work in this region was brought to a close by political upheaval, first in Iraq in the 1960s and then in Iran in the late 1970s. The spotlight of interest in agricultural origins shifted to the western arc of the Fertile Crescent of present day Israel, Jordan and western Syria, and a bit later to its center in northeastern Syria and southeastern Anatolia. Collections from the early work in the eastern Fertile Crescent were scattered among museums and research laboratories in the United States and Europe, where they have remained largely untouched for several decades.

There have been a lot of advances in the study of Near Eastern agricultural origins in the intervening years. Work across the Fertile Crescent has confirmed that the process of plant and animal domestication was not isolated to only one portion of this vast region, but seems to have been widely dispersed throughout, although with a considerable amount of regional variation in timing, target plant and animal species, and outcomes. Methods of archaeobiological analysis have been developed and refined over the years, resulting in more systematic and replicable approaches to identifying initial domestication in the archaeological record. New analytical techniques are having a profound impact. In particular small volume Atomic Mass Spectrometry (AMS) radiocarbon dating techniques developed over the last decade now allow for a much more precise temporal placement of this transition. Earlier radiocarbon techniques could only date large volumes of carbonized material found in loose association with the remains of early domesticates. Standard deviations of the dates derived from these samples were often as large as a hundred years or more. AMS radiocarbon dating makes it possible to directly date small fragments of the remains of domestic plants and animals, often with only a 50 to 60 year margin of error. Advances in the study of ancient DNA in archaeological plant and animal remains promises great breakthroughs in tracing the genetic ancestry of early domesticates. The analysis of chemical isotope signatures from human and animal bones is being used to monitor in shifts in climate, as well as changes in human and animal diet that accompany the transition to agriculture.

The time is then ripe to return to the valuable collections generated by the initial investigations in the eastern Fertile Crescent to update and refine the story they have to tell about agricultural origins in this pivotal region. This paper presents ongoing research that applies the new analytical approaches and techniques in a systematic and comprehensive way to the entire corpus of curated sheep and goat skeletal collections from the eastern Fertile Crescent. The ultimate goal of this research is to provide a more complete and coherent picture of the course of animal domestication in the Zagros, and to assess the bearing of this regional scenario on the process of animal domestication throughout the Fertile Crescent as a whole.

Reconsideration of the evidence for sheep and goat domestication in the Zagros

Identifying Initial Domestication in the Archaeological Record

Early Research

One of the primary challenges facing researchers who study the origins of domestication is how to identify domesticates in the archaeological record. Earlier analysts used a variety of approaches to distinguish the remains of domesticated sheep and goats from their wild progenitors. Flannery (Hole et al. 1969) pointed to changes in horn morphology, zoogeographic evidence, and slaughter profiles at the site of Ali Kosh, to argue that initial goat domestication took place about 9500 years ago in the lowlands of Iran, well outside the natural habitat for wild goats. Hesse (1978, 1984) developed a novel technique that combined data on the size of animals and age at death

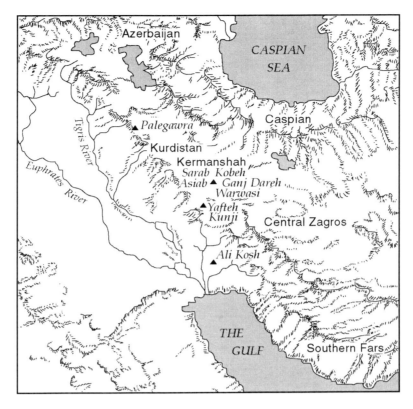

Fig. 1: Map of the Zagros Region of Iraq and Iran showing the location of archaeological sites mentioned in the text.

to reconstruct slaughter strategies for goats at the highland Iranian site of Ganj Dareh. Using this technique he detected a distinctive culling pattern in which males were slaughtered at young ages and females allowed to survive through their prime reproductive years – a practice that mirrors management strategies used today for domestic herd animals. Thus even though he found no evidence for morphological change in the goats from Ganj Dareh, Hesse argued that this management strategy marked them as domestic herd animals. Ganj Dareh is located well within the heartland habitat of wild goats, and given dating techniques of the time appeared roughly contemporary with basal levels at Ali Kosh, leaving open the question of the initial context of goat domestication in the Zagros. Somewhat paradoxically, Bökönyi maintained that an emphasis on mature male goats at near-by, and possibly contemporary, Asiab was an indicator of an ongoing process of domestication. Uerpmann (1979), Meadow (Bar Yosef & Meadow 1995), Helmer (1992), and Legge (1996) have subsequently pointed to an apparent reduction in the size of goats from Ganj Dareh to argue that these animals had in fact undergone morphological change and were, therefore, irrefutably domesticated. They maintain that the larger Asiab goats, on the other hand, were wild, hunted animals.

There was also disagreement over the timing and locale of initial sheep domestication in the Zagros. Perkins based his argument for initial sheep domestication in northwestern Iraq at 11,000 years ago on zoogeographic and demographic data (Perkins 1964). Flannery on the other hand, maintained that the presence of a hornless female sheep cranium in basal levels at Ali Kosh was evidence that sheep were first domesticated in lowland Iran at about the same time as goats, more than a millennium later than Perkins' claim for sheep domestication in Iraqi Kurdistan.

Thus while some researchers looked to morphological changes as key indicators of domestication, others relied more on demographic profiling to mark this transition, while still others relied more heavily on zoogeographic evidence. Moreover, there was little agreement on the standards for proof needed to demonstrate domestication using any of these methods. To make matters worse, analyses of assemblages with bearing on the question of sheep and goat domestication in the eastern Fertile Crescent were unevenly published. The Ali Kosh assemblage was fully documented in Hole et al. (1969), and Hesse published his analysis of the Ganj Dareh material in full in his dissertation (Hesse 1978). Bökönyi gave a relatively complete account of his study of animal assemblages from Asiab and other sites in the Kermanshah valley (Bökönyi 1977). However, other relevant assemblages were published only as brief notes (Perkins 1964), or are available only as unpublished manuscripts. Many assemblages were never studied in any systematic fashion. Clearly the data from Zagros sites bore evidence to an early transition to herding, especially for goats. But questions of when, where, and how the transition took place were largely unanswered. Without answers to these questions the overarching question of why domestication occurred was entirely out of reach.

Current Research

My ongoing study of animal domestication in the eastern Fertile Crescent seeks to develop new standards of proof for the identification of caprine domestication and to apply these standards in a consistent fashion to all curated assemblages from the Zagros that bracket the transition to sheep and goat herding (Zeder 1999, 2001; Zeder & Hesse 2000). Of special concern is sorting out which of the two primary methods used to identify animal domestication today, size reduction or demographic profiling, is most responsive to detecting the earliest phases of the transition from hunting to herding. To clarify the temporal framework of this transition, this research also includes an extensive program of AMS radiocarbon dating of the bones of animals from sites on either side of the domestic divide. It also includes a comprehensive study of modern skeletal collections of wild and domestic sheep and goats from Iran and Iraq in an effort to establish empirical baseline data for the study of size variation and age profiles in archaeological samples.

Analysis of 40 modern goat post-crania housed in the collections of the Field Museum of Natural History (FMNH) showed that sex is the single most important characteristic affecting size in goats (Zeder 1999, 2001; Zeder & Hesse 2000). All bones and all dimensions measured show the same strong bimodal pattern with almost no overlap between male and female animals (Figure 2). All bones also show a striking similarity in the range of variation in the size of male and female animals. Even the unfused elements of young males older than one year of age are absolutely larger than the fully fused bones of adult female animals. In addition to sex, regional variation also has a significant impact on the size of modern goats from the Zagros (Figure 3). There is a marked and regular decrease in the size of both male and female animals moving southward along the spine of the Zagros, probably resulting from a combination of increasing aridity and decreasing quality of forage. In contrast, domestic status seems to have little influence on the size of these modern animals. Domestic female goats from Iraq and Iran are virtually indistinguishable from wild females on the basis of long bone breadth measurements (Figures 2, 3). Domestic males fall within the small end of the wild male size range, but are still within this range and are uniformly larger than both domestic and wild females. These results call into question the conclusion that domesticated goats are distinguishable from wild goats on the basis of size. They also highlight the importance of factors like sex and regional variation when considering size patterning in the archaeological record.

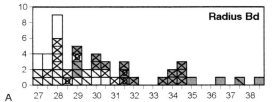
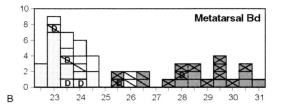
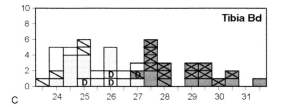
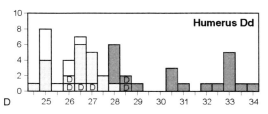

Fig. 2: Size distribution of six long bones of modern FMNH goats. Females shown in light grey bars, males in dark grey bars, fused bones in solid bars, fusing bones noted by a single hatch mark, unfused bones by a double hatch. Female animals younger than 1 year of age shown in white, unshaded bars and males younger than 1 year of age in stippled bars. Domestic specimens indicated by a "D".

Archaeological assemblages analyzed in this research are curated by the Smithsonian Institution, the Field Museum of Natural History, the Royal Copenhagen Museum, and the University Museum of the University of Pennsylvania. They include the remains of sheep and goats recovered from Paleolithic sites in the highland Zagros regions of Iran and Iraq that date back to the Middle Paleolithic (ca. 50,000 years ago), as well as from Upper Paleolithic cave sites occupied from 35,000 to 15,000 years ago (Shanidar, Kobeh, Kunji, Bisetun, Tamtama, Warwasi, Palegawra, and Yafteh Cave). Essentially all analysts accept these as wild, hunted animals, on the basis of both their size and demographic data. Also included is Epipaleolithic material from the site of Asiab, which, as we have seen, was considered domestic by Bökönyi, while most other researchers interpret the goats from the site to be wild.

Candidate domestic assemblages include the remains from Ganj Dareh, widely accepted as domestic on the basis of both slaughter profiles and size, and Ali Kosh, where changes in horn morphology and slaughter profiles indi-

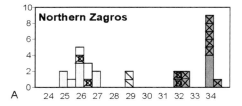
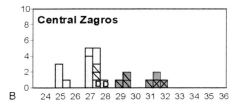
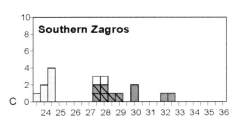

Fig. 3: Metacarpal distal breadth (Bd) measurements for modern FMNH goat specimens by region. Females shown in light grey bars, males in dark grey bars, fused bones in solid bars, fusing bones noted by a single hatch mark, unfused bones by a double hatch. Female animals younger than 1 year of age shown in white, unshaded bars and males younger than 1 year of age in stippled bars. Domestic specimens indicated by a "D".

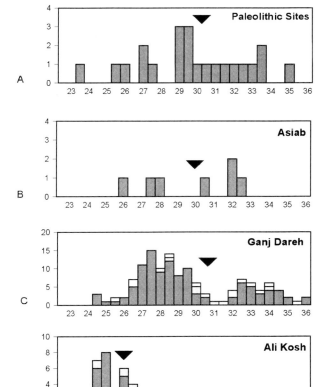

Fig. 4: Comparison of the second phalanx greatest length (GL) for specimens from archaeological assemblages. Fused bones represented by dark grey bars, fusing bones in light grey shaded bars, and unfused bones by white, unshaded bars. Means are indicated by a black triangle. Paleolithic Sites represented include Kobeh, Kunji, Palegawra, and Yafteh Caves.

cate domestic status. Other assemblages thought to lie on the other side of the transition between hunting and herding are the Iranian sites of Tepe Guran, and Sarab, and Jarmo in Iraq.

A comparison of the size of the archaeological remains from the Zagros sites finds no indication of size reduction in goats thought to span the transition from hunting to herding (Figure 4). Contrary to earlier research that claimed reduction in the size of Ganj Dareh goats when compared to a Paleolithic wild benchmark (Uerpmann 1979), there is no change in the size of goats from this site when compared to goats from Paleolithic or Epipaleolithic sites in the same the upland zone as Ganj Dareh. The goats from lowland Ali Kosh are significantly smaller than the upland goats. However, the degree of size reduction noted here is almost indistinguishable from that seen in modern wild goats from highland central Iran when compared to wild populations from more arid southern regions (Figure 3). Thus it is difficult to say whether the difference in the size of goats from the ancient sites of Ganj Dareh and Ali Kosh is an artifact of domestication or of environmental variation between upland and lowland habitats.

All bones measured at both Ganj Dareh and Ali Kosh show a strong bimodality that closely mirrors the sexual dimorphism seen in modern goats. Based on this modern analog data, it is likely that the smaller animals are females, while the larger animals are male. The regularity of this pattern across all bones and dimensions, plus statistical similarities between the archaeological assemblage and the modern material, makes it possible to use these systematic size differences to divide the samples from these sites into male and female sub-populations. Sex-specific age profiles can then be constructed by computing the relative proportions of unfused, fusing, and fused elements for a range of elements where age of fusion is known (see Zeder 1991:91).

The resulting sex-specific age profiles for the goats from Ganj Dareh show an unmistakable domestic signature (Figure 5a). Males are slaughtered young – with few males surviving beyond about 2 years of age, and there is a clear delay in the slaughter of female breeding stock. The goats from Ali Kosh show a similar pattern (Figure 5c), with some indication of a shift toward prolonged survivorship of young males and adult females. In contrast, there is a clear pattern of prolonged survivorship of both male and female goats in the Asiab assemblage (Figure 5b), a pat-

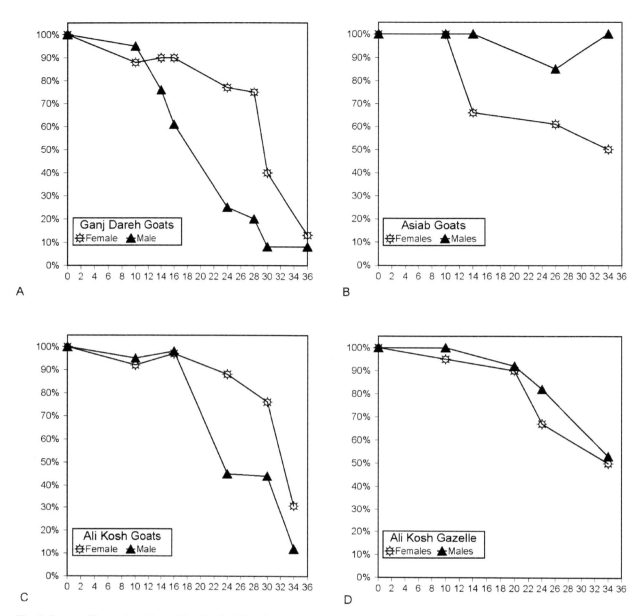

Fig. 5: Sex-specific survivorship profiles. (a) Ganj Dareh goats, (b) Asiab goats, (c) Ali Kosh goats, (d) Ali Kosh gazelle (from a study performed by Anastasia Polous, 2001).

tern more consistent with a hunting strategy focusing on prime age adults. A similar pattern is seen when this technique is applied to gazelle from Ali Kosh (Figure 5d), an animal widely accepted as never domesticated.

It seems then that the earliest indication of domestic status in these samples can be found in demographic data, not in morphological changes in the shape or size of bones. Demographic data clearly show the Ganj Dareh goats were managed in the same fashion that domestic herds are managed today. And yet the Ganj Dareh goats show no evidence of morphological change as a result of their management. There were no clear changes in horn morphology in the Ganj Dareh goats similar to those reported by Flannery at Ali Kosh (Hesse 1978:193-214). Moreover, the apparent reduction in size of the Ganj Dareh goats noted by earlier researchers, seems an artifact of a shift in the demographics of the adult members of a managed herd. Specifically, the bone assemblage generated by the hunting of a wild population is more likely to be dominated by the hardy bones of large adult males selected by hunters interested in maximizing meat yield. In contrast, the fused and more commonly preserved (and measurable) bones of animals harvested from a managed herd will be dominated by smaller adult females allowed to survive through their reproductive years. The more friable bones of males harvested at young ages are not as likely to be preserved in as great a quantity, or if they are may not be as easily identified or measured as the fused bones of adult animals. Thus when a herded population is compared to a hunted one this demographic difference will make it seem as if the herded population is composed of smaller animals, lending weight to the erroneous conclusion that size reduction has occurred as a result of domestication. The natural size difference between hunted and managed populations has also been inadvertently exaggerated by common analytical practices. For example, the restriction of metric analysis to the fused bones of adult animals and the normalization and homogenization of metric data across the skeleton exacerbates the natural taphonomic bias for female specimens in an assemblage from a managed herd. The practice of comparing specimens from widely different regions may mistake clinal variation related to environmental variables for size differences resulting from domestication (see Zeder 2001:76-77). This does not mean that domestication has no impact on size. It likely does. However, detecting these changes requires greater attention to factors like sex, age, and regional variation on size. It also calls for a change in the analytical methods used to measure this change. However, as is the case for changes in horn morphology, any reduction in the size of goats in domestic herds seems to be a delayed artifact of domestication, and is not then as useful as demographic profiling in detecting the leading edge of domestication.

Since the demographic profile of goats from Ganj Dareh and Ali Kosh both display distinctive domestic signatures, the question of the temporal relationship between these two sites becomes particularly important. A very different developmental scenario is needed to account for the initial domestication of goats within their natural habitat (where Ganj Dareh is located), than if goat domestication took place in more marginal areas outside this natural habitat (where Ali Kosh is located). Earlier attempts at radiocarbon dating of these two sites were ambiguous (Hole 1987), and left plenty of room to argue for either developmental scenario.

New AMS dates resolve this long-standing uncertainty (Table 1). Twelve AMS dates of goat bone collagen from Ganj Dareh indicate a very short occupation span of no greater than 200 years at about 8900 uncalibrated B.P. (Zeder & Hesse 2000). In contrast, dates of carbonized bone put initial aceramic levels at Ali Kosh at about 8500 uncalibrated B.P, continuing to about 8000 uncalibrated B.P. Dates of bone collagen from Asiab place the occupation of this site at about 9600 uncalibrated B.P., about 700 years earlier than Ganj Dareh. This means that the management of morphologically wild goats began first in their natural habitat, probably sometime between the occupation of Asiab and that of Ganj Dareh (between 9600 and 8900 uncalibrated B.P.). The Ali Kosh goats, then, represent the movement of managed herds out of this natural habitat zone, perhaps as much as 500 years after the occupation of Ganj Dareh, and into more marginal regions. Moving into the arid lowland plain was a dramatic break with the wild goat populations and their natural habitat, a break that likely required some adjustment in management strategies. The more conservative harvest profile seen at Ali Kosh, which allows for somewhat more prolonged survivorship of young male and adult females, may be a response to the loss of easy access to wild animals for restocking managed herds. Genetic isolation from wild populations and the impact of new selective pressures imposed by tighter human control of breeding are signaled by the changes in horn morphology evident in later phases of occupation at Ali Kosh, and, possibly, by the smaller size of these animals.

The case for sheep domestication in the Zagros remains less clear that that for goats. Perkins's early claims for sheep domestication at Iraqi sites dating to 11,000 years ago cannot be verified because the samples from these sites cannot be located, nor was his analysis published in enough detail to judge the strength of his argument for domestication. Flannery's identification of the Ali Kosh hornless female cranium as a domestic sheep looses some of its weight when it is noted that more than 20% of the modern wild female sheep in the FMNH collections were hornless. The sample of sheep remains from both Ganj Dareh and aceramic levels at Ali Kosh are too small to yield reliable age profiles, but these small samples are dominated by fused bones of adult animals in strong contrast to the goats from these sites.

Ceramic Neolithic sites from all elevations in the Iranian Zagros show an increase in sheep that seems to accompany the appearance of ceramic technology at about 8000 uncalibrated B.P. (Table 2). At aceramic Ganj Dareh, located at an elevation of 1350 m, only about 7% of the identifiable caprine bones belonged to sheep. They jump to 18% of caprine bones in Ceramic Neolithic levels at nearby Tepe Sarab. At lowland Ali Kosh, sheep contribute only 6% of the identifiable caprine bones in aceramic levels, but increase to 10% in ceramic age deposits at the site dating to after about 8100 B.P. An even more marked increase is seen as Tepe Guran, located at about 950 m elevation, where sheep increase from only 7% in basal levels and to 33% in upper post 8000 B.P. levels at the site.

It is tempting to read the increase in sheep utilization at all elevations in the Iranian Zagros as signaling the introduction of sheep domesticated outside the region, perhaps accompanying the introduction of ceramic technology. However, this picture is complicated somewhat by a concomitant, and even more dramatic, increase in the exploitation of hunted gazelle that accompanies both sheep and ceramics in the Iranian Zagros (Table 2). At lowland Ali Kosh, gazelle jump from 24% of the combined caprine and gazelle remains from aceramic levels to 44% in ceramic levels. At mid-elevation Guran gazelles increase from 11% in basal levels to remarkable

	Beta Analytic Number	Level	Depth (cm)	C14 B.P.#	C14 Cal B.P.!	2 Sigma Cal B.P.@	1 Sigma Cal B.P.@@
Palegawra	B-159546		10-20	5130+/-50	5910	5980-5970 5950-5740	5920-5890 5800-5770
	B-159543		20-40	12,510+/-90	15,100	15,530-14,150	15,480-14,240
	B-159545		60-80	8790+/-70	9860	10,150-9560	10,100-10,090 9920-9690
	B-159542		80-100	11,210+/-110	13,160	13,760-13,700 13,470-12,900	13,420-13,000
	B-159544		80-100	10,170+/-70	11,910	12,340-11,550 11,490-11,430	12,290-12,220 12,140-11,650
Asiab	B-159555		30-45	9480+/-80	10,710	11,110-10530	11,050-10,960 10,770-10,590
	B-159554		45-60	9370+/-60	10,570	10,720-10,420	10,670-10,520
	B-159552		75-90	7790+/-60	8580	8660-8420	8610-8460
Ganj Dareh	B-108239	B	165-180	8930 +/-60	9940	10,005-9870	9975-9905
	B-108238	A	180-200	8780 +/-50	9850	9910-9585	9880-9805 9780-9660
	B-108240	B	220-240	8780 +/-50	9850	9910-9585	9880-9805 9780-9660
	B-108241	B	240-260	8720 +/-50	9650	9875-9525	9845-9725 9695-9565
	B-108242	B	280-300	8940 +/-50	9945	10000-9890	9975-9915
	B-108244	D	430-460	8840 +/-50	9890	9945-9820 9765-9665	9915-9860
	B-108243	C	460-480	8920 +/-50	9935	9990-9875	9960-9905
	B-108246	E	580-585	8870 +- 50	9905	9960-9845 9725-9695	9935-9875
	B-108245	D	580-600	8940 +/-50	9945	10,000-9890	9975-9915
	B-108247	E	665-675	8830 +/-50	9880	9940-9805 9780-9660	9910-9850
	B-108248	E	700-710	8900 +/-50	9920	9980-9865	9950-9895
	B-108249	E	765-768/70	8840 +/-50	9890	9945-9820 9765-9665	9915-9860
Ali Kosh	B-137020*	Mohammed Jaffar	50-60	7100 +/- 70	7940	8020-7775	7970-7845
	B-118719*	Mohammed Jaffar	70-80	8130 +/- 70	8995	9245-8940	9185-9110 9090-8970
	B-118720*	Mohammed Jaffar	130-140	8130 +/- 70	9000	9360-8715	9215-8965
	B-118721*	Ali Kosh	180-200	8720 +/- 100	9650	9935-9480	9875-9525
	B-118722*	Ali Kosh	210-230	8110 +/- 80	8985	9245-8750	9040-8960
	B-137021*	Ali Kosh	250-270	8450 +/- 70	9485	9555-9270	9530-9425
	B-118723*	Ali Kosh	280-300	8490 +/- 90	9465	9565-9350	9505-9415
	B-118724*	Ali Kosh	380-400	8340 +/- 100	9375	9485-9000	9440-9220
	B-108256	Bus Mordeh	540-560	8000 +/-50	8945	8985-8620	8965-8705
	B-122721*	Bus Mordeh	630-650	8540 +/- 90	9482	9650-9385	9530-9445
	B-137024*	Bus Mordeh	680-710	8410 +/- 50	9465	9520-9380 9370-9300	9490-9425

Tab. 1a: AMS Dates on Bones from Zagros Sites.

	Beta Analytic Number	Level	Depth (cm)	C14 B.P.#	C14 Cal B.P.!	2 Sigma Cal B.P.@	1 Sigma Cal B.P.@@
Guran	B-147111	D		7630+/-60	8400	8530-8350	8430-8380
	B-147112	F		7260+/-40	8030	8160-7970	8140-8010
	B-147113	H		7950+/-40	8770	9000-8630	8980-8660
	B-147114	K		7080+/-60	7930	7990-7780	7960-7840
	B-147115	L		7940+/-40	8760	9000-8620	8980-8820 8800-8650
	B-147116	N		3690+/-40	4060	4150-3900	4090-3970
	B-147117	P		7890+/-40	8640	8970-8910 8870-8830 8790-8590	8740-8610
	B-147118	Q		8070+/-40	9010	9040-8980	9020-9000
	B-147119	R		8000+/-50	8990	9020-8650	9000-8770
	B-147122	T		8170+/-40	9100	9260-9020	9140-9030
	B-147120	U		8060+/-40	9010	9030-8980 8820-8800	9020-9000
	B-147121	V		7820+/-50	8600	8710-8450	8630-8550
Sarab	B-159547	1A		7470+/-70	8330	8400-8160	8360-8190
	B-159548	3		7950+/-60	8770	9010-8600	8990-8640
	B-159550	4		8070+/-60	9010	9120-8770	9030-8990
	B-159549	5		7800+/-60	8580	8710-8430	8620-8510

\# – Uncalibrated conventional ^{14}C age of specimens in ^{14}C B.P. (+/- 1δ).
! – Intercept between the conventional ^{14}C age and the dendrocalibrated calendar time scale, in calendar yr. B.P. (Pretoria calibration procedure program, Beta Analytic)
@ – 2δ dendrocalibrated age range for specimens, in calendar yr. B.P.
@ @ – 1δ dendrocalibrated age range for specimens, in calendar yr. B.P.
* – Date based on carbonized bone, all other dates based on collagen.

Tab. 1b: AMS Dates on Bones from Zagros Sites.

42% in ceramic levels. The same pattern is even seen in high elevation areas, not commonly associated with this steppic species. While there are no gazelle in sample of remains analyzed from Ganj Dareh, they comprise 25% of the analyzed sample from Tepe Sarab.

Thus it is unclear whether the increase in sheep in the Ceramic Neolithic of the Iranian Zagros represents evidence a much delayed local domestication of sheep, the introduction of sheep domesticated outside the region, or part of a broadening of the animal resource base to include an increased emphasis on hunting medium sized ungulates that includes both gazelle and wild sheep. Regrettably, even though somewhat larger than in earlier periods, samples of sheep remains from Ceramic Neolithic age sites in the Zagros are still too small to yield reliable age profiles that would allow us to determine if the sheep from these sites were managed. What is clear, is that the Ceramic Neolithic in the Iranian Zagros saw a marked change from Aceramic Neolithic's almost exclusive focus on domestic goats, to a mixed animal exploitation strategy based on both domestic livestock and hunted game.

Animal domestication elsewhere in the Fertile Crescent

The reconsideration of the curated animal bone collections from the Zagros provides unequivocal evidence

Phase	Site	Date	Elevation	% Sheep vs Goat	Total NISP Sheep & Goat	% Gazelle vs Caprines	Total NISP Caprines & Gazelle
Aceramic	Ali Kosh	8500-8300 B.P.	200 m	6%	1299	24%	2398
	Guran	8100-8000 B.P.	950 m	15%	87	11%	144
	Ganj Dareh	8900-8800 B.P.	1350 m	7%	3859	0%	7941
Ceramic	Ali Kosh	8100-7100 B.P.	200 m	10%	156	44%	401
	Guran	7800-7000 B.P.	950 m	33%	101	42%	339
	Sarab	8000-7400 B.P	1300 m	18%	739	25%	1561

Tab. 2: Representation of Sheep and Gazelle in assemblages from the Iranian Zagros based on the reanalysis of caprine and gazelle bones analyzed in this study.

for the management of morphologically wild goats in their natural habitat, directly dated to 8900 uncalibrated B.P. Morphological markers of domestication are delayed by as much as 600 years, and are most clearly marked by changes in horn morphology that take place over the course of the occupation at Ali Kosh, located well outside the natural habitat of wild goats. It is difficult to compare this scenario to data elsewhere in the Near East. Metric and long bone fusion data are not usually presented in a way that allows detailed evaluation of sex and age profiles. Inconsistent reporting of radiometric data obscures temporal patterning. However, there are indications that the pattern emerging for the eastern Fertile Crescent may not be unique, but may instead be repeated with sheep, pigs, and perhaps independently once again with goats in different parts of the broad arc of the Fertile Crescent.

The regional focus of initial sheep domestication now seems more likely to lie to the west of the Zagros in the central Fertile Crescent. Evidence from Hallan Çemi (Rosenberg et al. 1995), Çayönü (Lawrence 1982, Hongo & Meadow 1998, 2000), and most recently the exciting finds at Nevali Çori (Peters et al. 1999) point to the evolution of increasingly intensive hunting into management and domestication of sheep over a period stretching from about 10,000 to about 8500 uncalibrated B.P. Demographic data from central Anatolia at Aşikli Höyük suggests management of morphologically wild sheep at about 9000 years ago (Vigne & Buitenhuis 1999; Vigne et al. 2000). Similarly age and sex profiles of sheep at the site of Shillourokambos on Cyprus have been interpreted as indicating the management of morphological wild sheep (introduced to the island from somewhere in the northern Levant) at about 9400-9000 uncalibrated B.P. Domestic sheep do not seem to arrive in the southern Levant, in western arc of the Fertile Crescent, until about 8500-8000 B.P. (Horwitz & Ducos 1998), about the same time that the representation of sheep increases in assemblages in the Zagros.

A parallel, and roughly contemporary pattern is emerging for pig in oak/pistachio forests of southeastern Anatolia at the apex of the Fertile Crescent. Mounting evidence points to the evolution of increasingly selective hunting of wild boar evolving into swine management over the period of time bracketed by the occupation of Hallan Çemi (Redding & Rosenberg 1998) and the many levels at Çayönü (Ervynck et al. 2001). Clear morphological markers of domestication are not seen until 500 to 1000 years after the first signs of swine management. The radiation of pigs throughout the arc of the Fertile Crescent follows a similar, though somewhat delayed, trajectory as sheep. Domestic pigs appear in the eastern arc of the Fertile Crescent about 7000 B.P. (Hole et al. 1969), and in the western arc at about 7500 B.P. (Horwitz et al. 1999).

A case might also be made for an independent center of goat domestication in the far western arm of the Fertile Crescent in the Jordan Valley. Sometime between 9000-8500 uncalibrated B.P a long-standing emphasis on gazelle is abruptly replaced by a focus on morphologically wild goats (Horwitz et al. 1999). There are no sex-specific demographic profiles for goats from this region. In fact, even basic caprine harvest patterns are not widely available for assemblages from the Southern Levant. However, Horwitz believes that age data from some of these sites may point to local domestication of indigenous wild goats (Horwitz et al. 1999:69). Morphological markers of domestication are not seen in goat remains from the southern Levant until about 8300-8000 uncalibrated B.P.

Genetic data suggest three independent domestication events of goats, at least two of which were likely to have occurred in the Fertile Crescent (Luikart et al. 2001). So an independent domestication of goats in the southern Levant is certainly possible. However, it is also possible that the abrupt increase of goats in the Southern Levant at about 8600 uncalibrated B.P. marks the introduction of managed but morphologically wild goats from somewhere else. Data from Cyprus indicate that morphologically wild goats were also being herded along with sheep at about 9400 to 9000 uncalibrated B.P. The first appearance of goats in the assemblage from Abu

Hureyra in the northern Levant (most securely dated to about 8600 B.P.) is accompanied by demographic data that suggest a similar culling strategy to that detected at Ganj Dareh (Legge 1996; Legge & Rowley-Conwy 2000). Goats dominate in the assemblage from the site after about 8300 uncalibrated B.P., reversing a many millennia emphasis on hunted gazelle. Clearly better reporting of metric and age data, as well as more precise radiocarbon dating of material from across the entire region are needed (perhaps coupled with studies of ancient DNA), to locate the center or centers of goat domestication in the Fertile Crescent, and to trace the direction and the pace of the movement of domestic goats throughout the region.

Conclusions / Zusammenfassung

Thus a new and exciting story for animal domestication is beginning to emerge in the Fertile Crescent. While the specific location and the timing of the process varies, the domestication of all three cornerstone domesticates (sheep, goat, and pigs) appears to have been the outgrowth of process in which intensive exploitation of wild progenitor species evolved into the management of morphologically wild animals within their natural habitat. These developments are coupled with an increasingly sedentary focus on the utilization of cereals, pulses, and nut resources that rebounded across the Fertile Crescent after the end of the last Ice Age, a process that resulted in the initial domestication of plants here as well. Traditional morphological markers of animal domestication are only evident well after this new technology was firmly established, and probably only after it began to move out of this heartland area into new and more challenging environments for man and beast alike.

A constellation of biological, climatic, and cultural factors likely guided this transition, which was played out in highly localized ways in different parts of this broad region depending on the specific mix of topography, biota, and people. Telling this story requires some fundamental changes in the conceptual approach to the problem of animal domestication and in the research methods used to study it.

There has also been a tendency to conflate the process of domestication with morphological change. Recent research indicates, however, that morphological changes should be more properly viewed as artifacts of animal domestication, rather than hallmarks of the process itself. A shift in regional focus is also needed. Research has tended to focus on the foothills and steppe areas where morphological change in animals is first seen. It now seems that we need to look deeper into the mountain arc that serves as home to most progenitor species to catch the initial phases of the domestication process.

Analytical techniques need to be adjusted accordingly. The powerful osteometric methods developed here in Munich were originally directed toward detecting morphological changes thought to be leading edge markers of the ongoing process of domestication. With some modification these same techniques can be used to monitor the subtle demographic shifts in harvested animal that now seem more central to the initial phases of the domestication process. A comprehensive program of direct AMS dating of specimens is the only way to securely fix the chronological sequence of the course of animal domestication and agricultural origins in the Fertile Crescent. Standard and consistent reporting of dates (especially making it clear if dates reported are calibrated or uncalibrated, B.C. or B.P.) is essential. A coordinated effort by archaeologists, archaeozoologists, molecular biologists, and biochemists is needed to maximize what is learned through the application of exciting new tools of DNA and isotope analysis, while minimizing the loss of collections in the application of destructive analytical techniques.

Recovery of faunal remains from ongoing excavations needs to be done in such a way as to assure that large enough samples are collected to effectively use the new tools developed for unraveling the story of animal domestication. Moreover, as this study has shown, there is a treasure trove of curated collections of animal remains stored in museums and private offices around the world, many from areas where research access is currently difficult. Much work needs to be done to locate and consolidate these collections. Policies of proper curation and open access need to be encouraged. For it is these collections, recovered during the initial pulse of interdisciplinary effort to study agricultural origins and tucked away in dusty storerooms and offices for years and years, that may well represent our best hope for gaining new perspectives on this profound transitional period in the history of humankind.

Damit wird für den Fruchtbaren Halbmond eine neue und überaus interessante Seite in der Geschichte der Tierdomestikation aufgeblättert. Während Ort und Zeit dieses Prozesses veränderlich sind, scheint sein Ursprung bei den drei wichtigen Haustieren Schaf, Ziege und Schwein in einer Phase intensiver Nutzung ihrer wilden Vorfahren zu liegen, die zu einem Management morphologisch noch wilder Tiere innerhalb ihres natürlichen Verbreitungsgebietes führte. Diese Entwicklungen sind an eine zunehmende Sesshaftigkeit gekoppelt, einhergehend mit der Nutzung von Getreide, Hülsenfrüchten und Nüssen, welche sich in der Nacheiszeit über das Gebiet des Fruchtbaren Halbmondes ausbreiteten – ein Prozess, der schließlich in die Kultivierung von Wildpflanzen einmündete. Die traditionell relevanten morphologischen Kennzeichen der Tierdomestikation werden erst sichtbar, nachdem diese neue Technologie fest etabliert war. Vermutlich erst danach begann sie sich aus dem Kernland heraus in neue und für Mensch und Tier gleichermaßen herausfordernde Regionen auszudehnen.

Diese Ausbreitung wurde von einer ähnliche Konstellation biologischer, klimatischer und kultureller Faktoren getragen, was sich in diesem großen Gebiet regional dann sehr unterschiedlich ausprägte, je nach Topographie, Umwelt und Bevölkerungsstruktur. Will man diese Geschichte nachzeichnen, so sind fundamentale Änderungen im methodischen Ansatz und den Untersuchungsmethoden erforderlich, um die mit der Domestikation verknüpfte Problematik in den Griff zu bekommen.

Es gab die Tendenz, den Prozess der Domestikation allein auf morphologische Veränderungen zurückzuführen. Aktuelle Forschungen haben jedoch gezeigt, dass morphologische Veränderungen vielmehr eine Folge der Domestikation sind, als ein Meilenstein der Entwicklung selbst. Eine Verschiebung weg von der isolierten Betrachtung bestimmter Regionen ist ebenfalls notwendig. Bisher konzentrierten sich die Untersuchungen auf die randlichen Hügelzonen und die Steppengebiete, wo man zuerst Gestaltsveränderungen bei den Tieren feststellen konnte. Nun hat es den Anschein, dass wir, um die frühesten Phasen des Domestikationsprozesses fassen zu können, stärker in die Gebirgsregionen selbst hineinblicken müssen, das Heimatgebiet der meisten Vorgängerarten unserer Haustiere.

Unsere Analyseverfahren müssen dem angepasst werden. Die überaus wichtigen osteometrischen Methoden, die hier in München entwickelt wurden, waren ursprünglich darauf ausgerichtet, gestaltliche Veränderungen aufzuspüren, die man für Eckpfeiler eines fortschreitenden Domestikationsprozesses ansah. Mit nur geringfügigen Modifikationen können diese Techniken dazu verwendet werden, die nur sehr vagen demographischen Veränderungen bei vom Menschen gehaltenen Tierpopulationen aufzuzeigen, die nun mehr in das Zentrum der eigentlichen Domestikationsvorgänge gerückt sind. Ein entsprechendes Programm zur direkten AMS-Datierung spezieller Stücke ist der einzige Weg, eine chronologische Abfolge im Ablauf der Domestikation von Tieren und des frühen Ackerbaus festzulegen. Eine Einheitlichkeit und das regelmäßige Veröffentlichen von Daten (vor allem der Hinweis auf kalibriertes oder unkalibriertes Material, B.C. oder B.P.) sind hier wesentlich. Ein wohlkoordiniertes Zusammengehen von Archäologen, Archäozoologen, Molekularbiologen und Biochemikern wäre vonnöten, um das zu optimieren, was bisher durch die Anwendung von DNA- und Isotopen-Analysen an Erkenntnissen gewonnen wurde, während man sich gleichzeitig bemühen sollte, den Verlust von Material durch zerstörerische Analysetechniken zu minimieren.

Die Bergung von Knochenmaterialien aus laufenden Ausgrabungen sollte in einer Weise durchgeführt werden, die sicherstellt, dass ausreichend umfangreiche Materialien für die Anwendung neuer Untersuchungsmethoden zur Aufdeckung der Domestikationsprozesse zur Verfügung stehen. Darüber hinaus existiert – wie auch diese Studie gezeigt hat – ein großes und wichtiges Potential durch sorgsam verwahrte Tierknochensammlungen in Museen und an privaten Stellen in aller Welt. Manche dieser Sammlungen stammen aus Gebieten, die heute nur schwer zugänglich sind. Man müsste viel Arbeit investieren, um diesen Fundus jeweils zu lokalisieren und die Kollektionen zusammenzuführen. Vereinbarungen zu sinnvoller Verwaltung und einem offenen Zugang müssten verstärkt getroffen werden. Denn es sind diese Sammlungen, die während einer Phase starker Impulse für interdisziplinäre Zusammenarbeit im Hinblick auf das Erkennen der Wurzeln von Ackerbau und Viehzucht entdeckt wurden, welche nun Jahr um Jahr in staubigen Räumen und Büros vor sich hinschlummern und auf die wir unsere ganze Hoffnung legen, um neue Erkenntnisse zu dieser überaus wichtigen Phase der Menschheitsgeschichte zu erlangen.

References

Bar-Yosef O. & Meadow R., 1995.
    The Origins of Agriculture in the Near East. In: Price D. & Gebauer A.-B. (eds.). *Last Hunters, First Farmers: New Perspectives on the Transition to Agriculture*: 39-94. School of American Research Advanced Seminar Series, Santa Fe: SAR Press.

Bökönyi S., 1977.
    Animal Remains from the Kermanshah Valley, Iran. BAR Supplementary Series, No. 34. Oxford: BAR.

Braidwood R. J. & Howe B., 1960.
    Prehistoric Investigations in Iraqi Kurdistan. The Oriental Institute of the University of Chicago Studies in Ancient Oriental Civilization, No. 31. Chicago: University of Chicago Press.

Braidwood L., Braidwood R.J., Howe B., Reed C.A. & Watson P.J., 1983.
Prehistoric Archaeology Among the Zagros Flanks. Oriental Institute Publications, No. 105. Chicago: The Oriental Institute.

Ervynck A., Dobney K., Hongo H. & Meadow R., 2001.
Born Free? New Evidence for the Status of *Sus scrofa* at Neolithic Çayönü. *Paléorient* **27**: 47-73.

Helmer D., 1992.
La Domestication des Animaux par les Hommes Préhistoriques. Paris: Masson.

Hesse B., 1978.
Evidence for Husbandry from the Early Neolithic Sites of Ganj Dareh in Western Iran. PhD Dissertation, Columbia University, Ann Arbor: University Microfilms.

Hesse B., 1984.
These are Our Goats: The Origins of Herding in West Central Iran. In: Clutton-Brock J. & Grigson C. (eds). *Animals and Archaeology: 3. Early Herders and their Flocks*: 243-264. BAR International Series 202. Oxford: BAR.

Hole F.A., 1987.
Chronologies in the Iranian Neolithic. In: Aurenche O., Evin J. & Hours F. (eds.). *Chronologies in the Near East*: 353-379. BAR International Series. Oxford: BAR.

Hole F.A., Flannery K.V. & Neely J.A., 1969.
Prehistory and Human Ecology on the Deh Luran Plain. Memoirs of the Museum of Anthropology, No. 1. Ann Arbor: The University of Michigan Press.

Hongo H. & Meadow R., 1998.
Pig Exploitation at Neolithic Çayönü Tepesi (Southeastern Anatolia). In: Nelson S. (ed.). *Ancestors for the Pigs: Pigs in Prehistory*: 77-98. Masca Research Papers in Science and Archaeology, Vol 15. Philadelphia: Museum Applied Science Center in Archaeology, University of Pennsylvania Museum of Archaeology and Anthropology.

Hongo H & Meadow R., 2000.
Faunal Remains from Prepottery Neolithic Levels at Çayönü, Southeastern Turkey, A Preliminary Report Focusing on Pigs (*Sus* sp.). In: Mashkour M., Choyke A.M., Buitenhuis H. & Poplin F. *Archaeozoology of the Near East, IVA*:121-140. ARC Publication No. 32. Groningen:ARC.

Horwitz L.K. & Ducos P., 1998.
An Investigation into the Origins of Domestic Sheep in the Southern Levant. In: Buitenhuis H., Bartosiewicz L., and Choyke A.M. (eds). *Archaeozoology of the Near East, III*: 80-95. ARC Publications No. 18. Groningen: ARC.

Horwitz L.K., Tchernov E., Ducos P., Becker C., von den Driesch A., Martin L. & Garrard A. 1999.
Animal Domestication in the Southern Levant. *Paléorient* **25**: 63-80.

Lawrence B., 1982.
The Principal Food Animals at Çayönü. In: Braidwood L.S. & Braidwood R.J. (eds). *Prehistoric Village Archaeology in South-Eastern Turkey*: 175-199. BAR International Series No. 138. Oxford:BAR.

Legge A., 1996.
The Beginning of Caprine Domestication in Southwest Asia. In: Harris, D.R.(ed.) *The Origins and Spread of Agriculture and Pastoralism in Eurasia*: 238-263. Washington:Smithsonian Institution Press.

Legge A.J. & Rowley-Conwy P.A., 2000.
The Exploitation of Animals. In: Moore A.M.T., Hillman G.C. & Legge A.J. (eds). *Village on the Euphrates: From Foraging to Farming at Abu* Hureyra: 423-474. Oxford: Oxford University Press.

Luikart G., Ludovic G., Excoffier L., Vigne J.-D., Bouvet J. & Taberlet P., 2001.
Multiple Maternal Origins and Weak Phylogeographic Structure in Domestic Goats. *Proceedings of the National Academy of Sciences* **98**: 5927-5932.

Mortensen P., 1972.
Seasonal Camps and Early Villages in the Zagros. In: Ucko P., Tringham R. & Dimbleby G.W. (eds.). *Man, Settlement, and Urbanism*: 293-297. London: Duckworth.

Perkins D., 1964.
Prehistoric Fauna from Shanidar, Iraq. *Science* **144**:1565-66.

Peters J., Helmer D., von den Driesch A. & Segui S., 1999.
Animal Husbandry in the Northern Levant. *Paléorient* **25**: 27-48.

Redding R.W. & Rosenberg M., 1998.
Ancestral Pigs: A New (Guinea) Model for Pig Domestication in the Middle East. In: Nelson S. (ed.). *Ancestors for the Pigs: Pigs in Prehistory*: 65-76. MASCA Research Papers in Science and Archaeology, Vol 15. Philadelphia: Museum Applied Science Center in Archaeology, University of Pennsylvania Museum of Archaeology and Anthropology.

Rosenberg M., Nesbitt R.M.A., Redding R.W. & Strasser T.F., 1995.
Hallan Çemi Tepesi: Some Preliminary Observations Concerning Early Neolithic Subsistence Behaviors in Eastern Anatolia. *Anatolica* **21**: 1-12.

Smith P.E.L., 1976.
Reflections on Four Seasons of Excavation at Tappeh Ganj Dareh. In: Bagherzadeh F.(ed.). *Proceedings of the IVth Annual Symposium on Archaeological Research in Iran, 3rd-8th November 1975*:11-22., Tehran: Iranian Center for Archaeological Research.

Solecki R.L., 1981.
An Early Village at Zawi Chemi Shanidar. *Biblioteca Mesopotamica*, Vol. 13. Uneda Publications.

Solecki R.S., 1963.
Prehistory in Shanidar Valley, Northern Iraq. *Science* **139**: 179-193.

Uerpmann, H.-P., 1979.
Probleme der Neolithisierung des Mittelmeeraumes. Beihefte zum Tübinger Atlas des Vordern Orients, Reihe A, Nr. 27. Wiesbaden: Dr Ludwig Reichert.

Vigne J.-D. & Buitenhuis H., 1999.
Les Premiers Pas de la Dometication Animale à l'Ouest de l'Euphrate: Chypre et l'Anatolie Centrale. *Paléorient* **25**: 49-62.

Vigne J.-D., Carrére I., Saliége J.-F., Person A., Bocherens H., Guilaine J. & Briois F., 2000.

> Predomestic Cattle, Sheep, Goat, and Pig During the Late 9th and the 8th Millennium cal. BC on Cyprus: Preliminary results of Shillourokambos (Parekklisha, Limassol). In Mashkour M., Choyke A.M., Buitenhuis H. & Poplin F. (eds.). *Archaeozoology of the Near East, IVA*: 83-106. ARC Publication No. 32. Groningen: ARC.

Zeder M.A., 1991.

> *Feeding Cities: Specialized Animal Economy in the Ancient Near East*. Washington, D.C.:Smithsonian Institution Press.

Zeder M.A., 1999.

> Animal Domestication in the Zagros: A Review of Past and Current Research. *Paléorient* 25: 11-25.

Zeder M.A., 2001.

> A Metrical Analysis of a Collection of Modern Goats (*Capra hircus aegagrus* and *Capra hircus hircus*) from Iran and Iraq: Implications for the Study of Caprine Domestication. *Journal of Archaeological Science* 28: 61-79.

Zeder M.A. & Hesse B., 2000.

> The Initial Domestication of Goats (*Capra hircus*) in the Zagros Mountains 10,000 Years Ago. *Science* 287: 2254-2257

PART IV:

DECYPHERING ANCIENT BONE

Ancient bones and teeth on the microstructural level

Simon Hillson, Daniel Antoine, Institute of Archaeology,
University College London, 31-34 Gordon Square, London, WC1H 0PY, United Kingdom

Abstract / Zusammenfassung

Studies of the microstructure of bone, and the dental tissues enamel, dentine and cement have considerable potential for archaeology. Careful selection of methods of specimen preparation, and techniques of microscopy, is needed so that the effects of diagenetic change (and sample preparation artefacts) can be clearly seen and distinguished from the structures of tissue development and turnover which form the focus of histological studies. In addition, it is important to choose techniques which are capable of showing these structures unambiguously, and to check with other methods where there is uncertainty. For these reasons, back scattered electron imaging in the scanning electron microscope is particularly useful for archaeological bone and cement, and is now a standard facility on many instruments. It is possible to use it to investigate bone diagenesis, the development and turnover of bone with increasing age and disease, and the finely layered development of cement. Ground and polished "thin" sections of teeth, examined in a standard transmitted light microscope, are still the most practical way to study the development of enamel, which has a very useful 24 hourly growth rhythm preserved within its structure, but great care is needed when these structures are interpreted. More recent developments such as confocal light microscopy, which has a very thin plane of focus, can be used to check. Used carefully, however, enamel development has great potential as a way of calibrating the growth sequence, and providing high resolution estimates of age at death.

Untersuchungen des mikrostrukturellen Aufbaues von Knochen und der Zahngewebe Schmelz, Dentin und Zement haben einen hohen Stellenwert in der Archäologie. Eine sorgfältige Auswahl der geeigneten Methoden zur Probenpräparation und Mikroskopie ist erforderlich, um Dekompositions- und Präparationsartefakte von den auf der Entwicklung und des Umbaues der Gewebe resultierenden Strukturen unterscheiden zu können, welche Ziel der histologischen Inspektion sind. Darüberhinaus ist es in bezug auf die angewandten Techniken wesentlich, dass diese auch in der Lage sind, diese Strukturen eindeutig darzustellen, und im Zweifelsfalle mit alternativen Techniken zu prüfen. Aus den genannten Gründen ist insbesondere die Bildgebung mit Rückstreuelektronen in der Rasterelektronenmikroskopie für die Inspektion von Knochen und Zahnzement geeignet, heute eine Standardeinrichtung der meisten Geräte. Der Einsatz dieser Technik ist für die Beurteilung von Knochendekomposition, der Knochenentwicklung und seines Umbaues mit zunehmendem Alter oder im Krankheitsfalle, sowie der fein zonierten Entwicklung des Zahnzementes geeignet. Der nach wie vor praktischste Weg zur Beobachtung der Entwicklung des Zahnschmelzes, welcher in seinem mikrostrukturellem Aufbau einen 24-Stunden-Rhythmus konserviert, besteht in der Inspektion von polierten "Dünnschnitten" der Zähne, betrachtet im Standard-Durchlichtmikroskop. Diese Strukturen sind jedoch mit größter Vorsicht zu interpretieren. Zur Kontrolle eignen sich neue methodische Zugänge wie die konfokale Lichtmikroskopie, welche die Beurteilung sehr feiner Fokussierungsebenen zulässt. Bei sorgfältiger Anwendung der jeweiligen Methode ist die Zahnschmelzentwicklung sehr gut zur Kalibirierung von Wachstumssequenzen geeignet und liefert eine Sterbealterschätzung mit hoher Präzision.

Key words: bone microstructure, tooth microstructure, development, turnover, microscopical techniques
Knochenmikrostruktur, Zahnmikrostruktur, Entwicklung, Umbau, mikroskopische Techniken

Introduction

Bones and teeth, although they are both mineralised tissues and are all that remains of the body on most archaeological sites, are not at all similar in their biology or in their microstructure. Table 1 shows the chemical composition and main characteristics of fresh bone, which makes up the skeleton and fresh cement, dentine and enamel which are the tissues of the teeth. The differences between them dictate how they can be examined, how well-preserved they are in archaeology, and the range of information that can be determined from the microstructure. One major distinction between the dental tissues and bone is that enamel, dentine and cement to do not turn over in life. Most tissues of the body are constantly being created and destroyed, so that they are replaced many times during the life of an individual. Bone is replaced in this way (see below), but the dental tissues, once they are formed in childhood, remain as they were apart from the losses caused by tooth wear and dental disease. This gives rise to a number of possibilities, including the opportunity to compare the chemical composition of tissues formed during childhood, with those representing the last few years of adult life before death. It also means that the dental tissues record the whole sequence of their development in their microstructure. Table 1 shows that enamel is completely different to all the other tissues in that it is almost entirely mineral, contains no living cells and is, in effect, dead even as a fresh tissue. Enamel coats the surfaces of the teeth over 80 or more years, in an abrasive, wet and sometimes acid environment so it is perhaps not surprising that it is nearly always the best preserved of tissues in most archaeological sites. Dentine is next most heavily mineralised, although it falls far short of enamel. It does not actually contain cells in life, because the odontoblasts, which are the cells of dentine, line the pulp chamber inside the tooth and send narrow processes through fine tubules which run right through the tissue. Each odontoblast remains alive throughout the life of the tooth and they are amongst the longest lived cells in the body. Dentine is best preserved deep inside the tooth underneath the solid cap of enamel, which protects it and it has become recognised that this is the best place to take samples for biochemical analysis. Cement and bone have a similar composition, and are somewhat less heavily mineralised than dentine. They are correspondingly less well-preserved on many archaeological sites.

This paper gives three examples which illustrate the contrasting methods and possibilities.

Back scattered electron imaging of bone

A variety of microscopy and specimen preparation techniques can be used for archaeological specimens (Table 2). One of the best for bone and cement is scanning electron microscopy, using back scattered electron mode (SEM-BSE) (Boyde 1972; Boyde 1984; Boyde et al. 1986; Boyde et al. 1990; Boyde & Jones 1984; Reid & Boyde 1987). The first requirement for this method is the preparation of a very flat, cut and polished (or milled) surface, which exposes the tissue to be examined. It is usually necessary to embed the bone in a hard resin such as polymethylmethacrylate (PMMA), to fill the spaces in the tissue so that the edges of any voids do not become rounded on polishing. Usable surfaces can be prepared by cutting the embedded block with a rotating diamond edged blade, followed by careful polishing on abrasive paper, although the best technique is micromilling (the machine for this is not a common piece of laboratory equipment, so many researchers still use simple polishing). The flat surface is coated with carbon and placed in the specimen chamber of the electron microscope so that it is perpendicular to the electron beam. In the most common configuration, a circular solid state back scattered electron detector is fitted directly above the specimen, also perpendicular to the beam. The image produced contains information about the atomic number of elements present in the specimen, their density (together called compositional contrasts), and any uneven-ness of the surface (topographic contrasts). The topographic effects are minimised by the specimen preparation method, and by the way the microscope is set up and operated. SEM-BSE images of bone therefore show mostly the composition of the tissue with brighter areas being more heavily mineralised and darker, less heavily mineralised, although a number of complex factors are involved. The back scattered electrons that form the image are generated in a spherical "interaction volume" in the specimen underneath the point at which the electron beam strikes it. The size of this interaction volume depends upon the operation of the microscope, but it is usually about 1 micrometer (1 µm) in diameter which thus represents the resolution of the image. The magnification is usually set at 100 times so that each pixel on the display represents 1 µm on the specimen. In addition, the image represents the composition of only the top 1 µm of the specimen surface. This is a major advantage over light microscopy of thin sections (see below) which can rarely be made thinner than 60 µm and where the plane of focus is also relatively thick so that the image is produced by many structures inside the tissue superimposed over one another. Another advantage is that relatively little tissue is lost during the sectioning process – usually 250 µm or so. After microscopy has been carried out, it is possible to remove the methylmethacrylate resin in acetone and the specimen can be reconstructed to something approaching its original state.

The basic unit of bone structure seen in these images is the *osteon* (Williams & Bannister 1995). This is a cigar or sausage shaped element, 1 – 2 µm long and around 100 µm in diameter. A canal, the *Haversian canal*, runs along the centre and in life contains the vascular and nervous supply. In section (Figure 1) the osteons are seen as rounded structures of varying brightness, which reflects their relative mineralisation. Around each osteon is a much brighter, and therefore more heavily mineralised, *reversal line* which follows an irregular, or scalloped course. Inside this, it is often possible to make out a series of concentric rings which represent the *laminae*, or layers in which the bone was laid down within each osteon. Dark spaces scattered through the laminae represent the *lacunae* of the osteocytes, or bone cells, which are filled with the embedding resin. The osteons intersect one another and represent the process of bone turnover. The first stage of osteon formation occurs when a group of osteoclasts, or bone removing cells, cuts a tunnel through an area of pre-existing tissue. This marks out the shape of the osteon and creates the more

	Enamel	**Dentine**	**Cement**	**Bone**
Inorganic component	>96%	72%	70%	70%
Collagen	-	18%	21%	21%
Non-collagenous proteins	0.5%	2%	1%	1%
Cells in living tissue	None	Odontoblast processes	Cementocytes (sometimes vascular tissue)	Osteocytes, osteoblasts, osteoclasts, vascular tissue, nervous tissue
Tissue turnover	None	None	Limited	Complete turnover (30-100% per year in children, 5-15% in adults)

Table 1

heavily mineralised reversal line around its edge. Osteoblasts (bone forming cells) follow behind the "cutting cone" of the osteoclasts and lay down successive layers of bone matrix. Once formed, these matrix layers mineralise progressively, so that the more recently formed osteons show as darker (less heavily mineralised) areas in the image, and the older osteons as brighter areas (Figure 1). Back scattered images make it possible to follow the different stages of bone formation. Most bones in the skeleton are formed through endochondral ossification, in which the initial growth is in cartilage. This is heavily mineralised and, in specimens which include it, is strikingly brighter than the rest. The gaps left by the cells of cartilage are lined by so-called primary osteons. Then, as bone turnover proceeds, the whole structure is progressively replaced by larger secondary osteons which clearly cut into the preceding structure. The so-called flat bones of the cranial vault form through intramembranous ossification. Here, the first bone formed is woven bone, which does not have an osteon structure and has a more irregular arrangement of large osteocyte lacunae. Once again, it shows as more heavily mineralised than later-formed components of the tissue. Gaps in the woven bone are lined by primary osteons and then, as the whole structure turns over, secondary osteons cut through. Any one specimen represents the accumulation of these changes and provides a record of the development of the bone. Images of this type can therefore be used for investigations of growth, and in the understanding of pathological lesions, but the most frequent use has been in the estimation of age in adult humans. Most methods of this type use ordinary light microscopy of thin sections, usually taken from the thick cortical bone of the middle part

Microscopy	Material	Preparation
Polarizing microscope	Intact tissue, usually impregnated with acrylic or epoxy resin	Parallel-sided polished section 60-100μm thick
Ordinary transmitted light microscope or polarizing microscope	Small piece (2-3mm) of intact tissue, impregnated with acrylic to make compact pellet	Glass bladed microtome sectioning (followed by staining in some cases)
Ordinary transmitted light microscope (or phase contrast, fluorescence etc.)	Demineralised (or partly demineralised) tissue, usually by buffered acid solution	Microtome sectioning, followed by staining
Confocal light microscopy	Intact tissue, sometimes impregnated with resin	Cut and polished surface
Confocal laser scanning microscopy	Surface	Clean with acetone
Scanning electron microscope, Everhart-Thornley detector	Surface, or fractured surface, or epoxy resin replica	Clean and sputter coat (unless using SEM at very low kV, or ESEM)
Scanning electron microscope, backscattered electron detector	Intact tissue, impregnated with acrylic resin	Cut and polished surface (micromilled), coated with carbon (unless ESEM)

Table 2

of a large long bone shaft (Robling & Stout 2000; Ubelaker 1986). Other methods use the rib, and this has clear advantages in terms of availability of material, because there are many ribs in each skeleton and they are not one of the "key" bones for anthropologists whereas, for example, femora are. The most commonly used schemes involve counting the number of osteons, or fragments of osteons, in a measured area (or areas) of the bone section. In older individuals, there is a gradual accumulation of these structures with progressive remodelling. This has been tested, for example, in the skeletons excavated from the crypt of Christ Church, Spitalfields in

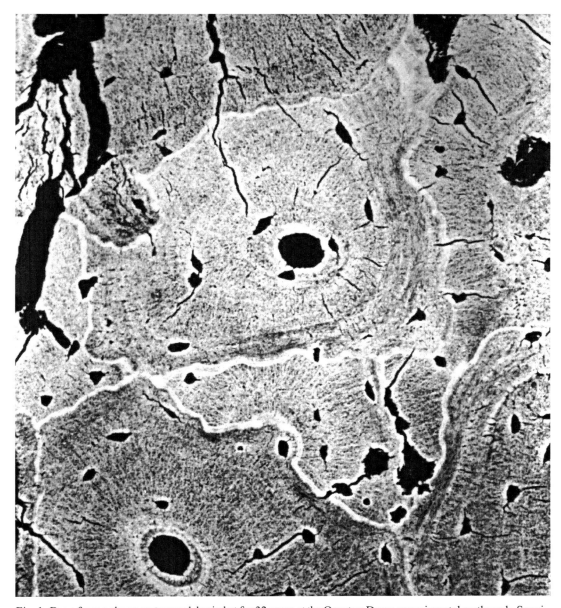

Fig. 1: Bone from a sheep metacarpal, buried at for 32 years at the Overton Down experimental earthwork. Specimen was embedded in polymethymethacrylate, cut and polished with fine abrasive paper, then examined using a Hitachi S570 scanning electron microscope operated in back scattered electron mode. Field width 2.25 mm. The outlines of several secondary osteons can be seen, with a brightly marked reversal line and the outlines of laminae. The round dark voids at the centre of the osteons are the Haversian canals, and smaller lens shaped voids are the osteocyte lacunae. Several of the osteons intersect other, earlier formed (more mineralised and therefore brighter) osteons. The specimen shows prominent cracking, one of the most common diagenetic changes, together with small diagenetic foci (see Figure 2).

London (Aiello & Molleson 1993). People were buried in this crypt between AD 1729 and 1859, and coffin plates linked with parish registers have allowed the sex and age of 389 individuals to be established independently (Molleson & Cox 1993). The accumulation of osteons and fragments in the cortical bone of the femur midshaft showed a relationship with known age that was at least as good as the more commonly used age estimation methods based upon morphological changes to the pubic symphysis joint. Like most others, this study used thin sections and light microscopy, which show the osteons clearly in well preserved material, but there is considerable potential for using the simpler sectioning technique and wider applicability of SEM-BSE in a wider range of archaeological material.

Back scattered electron imaging shows very clearly the effects of diagenetic change (Bell 1990; Bell et al. 1991; Bell & Jones 1991; Bell et al. 1996; Hagelberg et al. 1991; Hillson & Baond 1996). The most frequent change is cracking. Cracks generally run around the osteons, between the osteocytes lacunae and Haversian

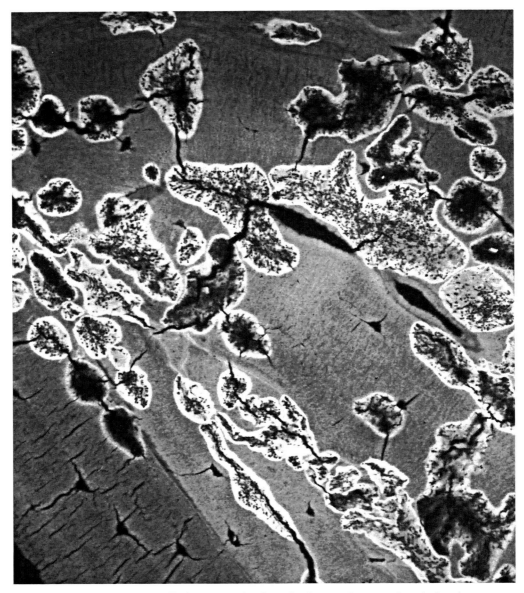

Fig. 2: Bone from a sheep mandibular ramus, also from the Overton Down earthwork. Specimen preparation and microscopy as in Figure 1. Field width 2.25 mm. Prominent diagenetic foci mark much of the surface. Each focus is 10-20 μm across its narrowest dimension. Most have a bright and therefore heavily mineralised margin (see text). Within this margin, most show very variable mineralization, including many small voids 1 μm or so across which may represent the positions of micro organisms. Other foci, have areas of much lower density inside. Several are linked by cracks and, in some cases, the foci seem to have exploited pre-existing cracks.

canals – presumably they follow along lines of weakness (Figures 1 & 2). They are seen as dark lines in back scattered electron images, because they create voids which are filled with the embedding resin. Some at least may be artefacts of the preparation techniques used for microscopy, but others are clearly a part of the diagenetic change (below). Another common change is so-called *diagenetic foci* (Figures 2 & 3). These are irregular areas of altered mineralisation, often 100 μm or so across, although they may be divided up into multiple lobes. Sometimes they follow along the structure of the original bone, but in other specimens they cut across it. Often they follow cracking (Figure 2) and this shows that some of the cracks were produced early on in the process of diagenetic change. The foci vary in structure, but most are surrounded by a heavily mineralised margin. Inside this, they may show generally lower mineralisation than the surrounding bone, so they make an irregular dark patch in the image. Others show patches of heavy mineralisation inside, with voids which might represent the positions taken by micro-organisms during decomposition of the bone. It is assumed that the diagenetic foci are the result of the activities of micro-organisms, but this is poorly understood. What is clear, however, is that most diagenetic foci represent a re-organisation of the mineral content of the bone, rather than removal of it, because they usually show both hyper- and hypo-mineralised areas. In addition, elements of the original structure may still be visible, passing through the foci. The SEM-BSE imaging is particularly suitable for showing diagenetic foci and, during the past 20 years, has been the main means of investigating them. Bones from both anatomy or zoology museum collections and archaeological sites may show diagenetic foci, and they seem to be produced in a wide variety of burial conditions. Diagenetic changes are particularly problematic for the light microscopy of bone thin sections, because they scatter light and make it very difficult to see any elements of structure. SEM-BSE images are much less affected, but some specimens are so altered that virtually no original structure remains. Not surprisingly, the microstructural intactness of a specimen is a good indicator of the survival of organic material (Hedges & Millard 1995; Hedges & Wallace 1978) within bone, or dentine. Enamel, on the other hand, seems unaffected by such changes and nearly always shows clear details of its original structure. In fact, the features of enamel may be more clearly visible in some archaeological specimens than they are in the fresh tissue. This is particularly noticeable for example in the remains from Spitalfields (Figure 4) and it is often true of fossilised teeth. The factors involved in this enhancement of structure are unknown.

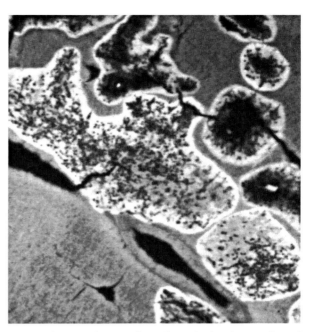

Fig. 3: Close up of an area from Figure 2, showing a variety of diagenetic foci.

Enamel development

Enamel develops as a series of layered structures that are best shown by light microscopy of ground and polished thin sections. In fact, these sections are anything but thin, because it is rarely possible to polish them thinner than 60 μm. In practice, most tooth sections are around 100 μm, because these show the required structures most clearly. The usual plane of section is through the highest point of the cusp of a tooth crown. Freshly extracted teeth can usually be sectioned directly, but archaeological teeth often need the support provided by embedding with resin (e.g. PMMA). The initial cuts are made using a sectioning machine with a rotating circular blade which is loaded with industrial diamond. The specimens are usually mounted for cutting using wax, and two parallel cuts about 250 μm apart are made. The slice of tooth is detached and then polished down to a thickness of 100 μm. The most important thing in polishing seems to be to use a good quality, heavy polishing jig. It is important to get section preparation right, because this has a fundamental effect on the ability of the microscopist to see the necessary structures. Experience has shown that it is difficult to make really successful sections of teeth, and a great deal of practice and acquired skill are needed.

The developmental structures (Osborn 1981; Osborn & Ten Cate 1983; Ten Cate 1985) seen in enamel are the result of light scattering effects along planes of increased porosity, or the effects of boundaries between areas of different refractive index. These effects are cumulative through the thickness of the section. As a result, they are usually seen best in relatively thick sections of 100 μm, and under low magnification. Higher magnification objective lenses have a shallower depth of focus which makes the effects harder to see. On the other hand, because the lines that are seen under the microscope are the cumulative effect of light passing through several layers of developmental structures, great care needs to be taken to make sure they are not the artefact of such factors as plane of section. One way to do this is to check key areas with confocal light microscopy (Table 2), where the image plane is a very thin region within the section – typically 1 μm or so. Confocal microscopes are not a practical way to count developmental structures through the full thickness of enamel, but they can be used as a valuable check on the reality of the smaller structures which are seen in more conventional light microscopy.

Under the microscope, the main units of enamel structure (Boyde 1989; Hillson 1996; Osborn & Ten Cate 1983) are bundles of crystals known as *prisms* (Figure 4). These are about 4 μm in diameter and radiate out towards the surface of the tooth crown from the enamel dentine junction. Crossing these prisms is a regular series of dark smudges, also at a spacing of about 4 μm. These are known as the *prism cross striations*, and for many years they have been thought to be circadian – that is they represent a roughly 24 hourly rhythm. The evidence for this was circumstantial, but it has now been firmly established using archaeological material (see

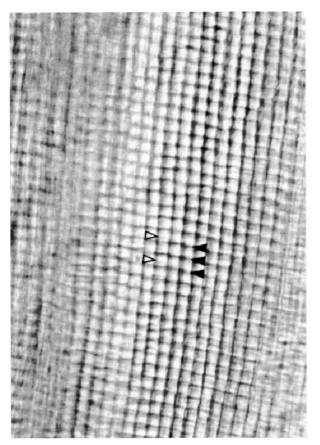

Fig. 4: Thin section (around 100 µm thick) of enamel from a tooth excavated from the crypt at Christ Church, Spitalfields in London (see text), examined in transmitted light microscope. Field width 88 µm. The almost vertical lines represent the prisms of the enamel. The dark shadows that appear to divide them into segments are the prism cross striations – three are marked out with black arrows. The open arrows mark the phenomenon called "doubling" which is seen in many sections (marks between the main striations). The shadowy lines running diagonally across the prisms are part of the system of increments that includes the brown striae of Retzius.

below). At low magnifications, it is possible to see a series of additional lines cutting diagonally across the prisms (Figure 5). These are dark, fuzzy bands, which are not sharply defined through much of the enamel thickness. Where they angle up towards the surface, however, they become sharper and are clearly regularly spaced at 30-40 µm or so apart. In this better defined region, it is possible to count prism cross striations between them. The count is constant in all the teeth of one individual, but it varies between individuals from 7 to 10 cross striations (Fitzgerald & Rose 2000). The average count of all studies so far is 9. These coarser lines are known as the *brown striae of Retzius*, after the famous Swedish anatomist Anders Retzius who published the first good description of them. They are fuzzy and ill defined because they are caused by the light scattering effect of thin porous planes running through the enamel. These planes mark out successive layers of enamel matrix formation during the development of the tooth. By following them, it is possible to see that the very first layers are formed underneath the point at which each tooth cusp will develop, and are dome-shaped. Domes of gradually increasing size are laid down on top of one another until the full height of the cusp is formed (Figure 6). After this, the layers become sleeve-like and overlap one another down the side of the tooth crown. It is useful to distinguish between the dome-like layers of the *cuspal enamel* and the sleeve-like layers of the *lateral enamel*. Where each brown stria of Retzius angles up to meet the surface in the lateral enamel, there is a microscopic groove in the surface (Figures 5 & 7). Even with modest magnification, it is possible to see these grooves, running across the sides of the crown with a spacing that varies from 100 µm or so near the cusp tip, to 30 µm near the base of the crown (Figures 7 & 9). They are most clearly seen, however, with a scanning electron microscope operated in the secondary electron (or Everhart-Thornley detector) mode. The specimen needs to be oriented carefully relative to the electron beam and detector, because the grooves are very shallow indeed. They are also affected by abrasion of the crown surface, particularly the toothbrush abrasion seen in modern teeth. These patterns of surface grooves have been given a variety of names, but are usually known as *perikymata*.

Both the brown striae of Retzius and the perikymata vary in prominence through any one tooth (Figures 5, 7 & 9). Sequences of particularly prominent striae and perikymata can be matched between different teeth from one individual, showing that these variations are caused by some systemic physiological fluctuation, although the precise nature of this is still unclear. The different teeth in the permanent dentition form in a sequence, starting with the first molar, followed by the incisors, canine, premolars, second molar and third molar (Figure 8). The periods of crown formation for different teeth overlap so that, for example, a set of prominent striae seen in the cuspal enamel of a canine might be visible in the lateral enamel of the first molar. In this way, it is possible to build up a series of matches, linking all the teeth in the permanent dentition, up to the completion of the second molar. The third molar crown usually starts to form after this point and so cannot be matched in. The starting point of the sequence is the initiation of enamel formation in the first molar. In many babies, this starts a little before they are born (although there is some variability relating to the length of gestation and different parts of the tooth start to form earlier than others). The point of birth is marked in some individuals by a marked brown stria of Retzius – known as a *neonatal line* – amongst the first formed dome-shaped layers of cuspal enamel. Not everyone shows these, presumably because of timing differences in the initiation of tooth formation, or some variation in the

Ancient bones and teeth on the microstructural level 149

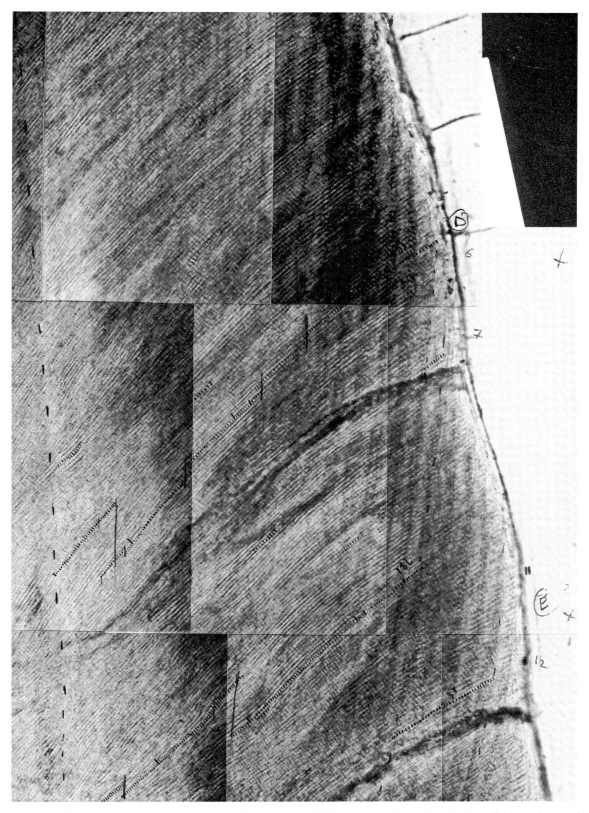

Fig. 5: Thin section of enamel in a lower first incisor from Individual 2431, buried at Christ Church, Spitalfields, again examined by transmitted light microscopy. Field width about 770 µm. The picture is a photomontage, made up from several fields of view and the marks and notes show the way in which counts of prism cross striations were made. The surface of the crown is to the left and the brown striae of Retzius are running almost vertically up to meet it. In this part of the crown, they are regularly spaced and quite sharply defined, but the deeper part of the enamel shows them less clearly. The prisms cross the picture diagonally, towards the surface and this particular specimen shows many very clear cross striations which were used to arrive at a counted sequence for this individual. The two defects described in the text can be seen as depressions in the surface, numbered 6/7 and 11/12.

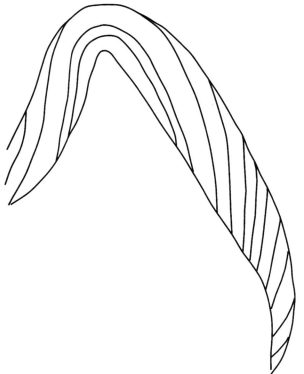

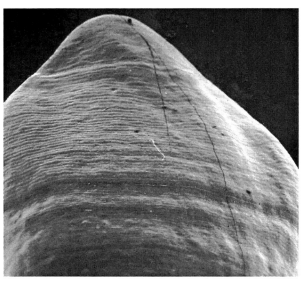

Fig. 7: Epoxy resin replica of the surface of a post-Medieval tooth from London, examined in a Hitachi S570 scanning electron microscope, using the Everhart-Thornley detector. Field width 6 mm. The tooth is practically unworn and the perikymata are very sharply defined, crossing the crown. It is also possible to see very clearly marked enamel hypoplasia defects of the furrow type.

Fig. 6: Tracing of a canine section, showing the trend of brown striae of Retzius in the enamel. The first increments of enamel matrix are dome-shaped and the later increments are like overlapping sleeves, down the crown side.

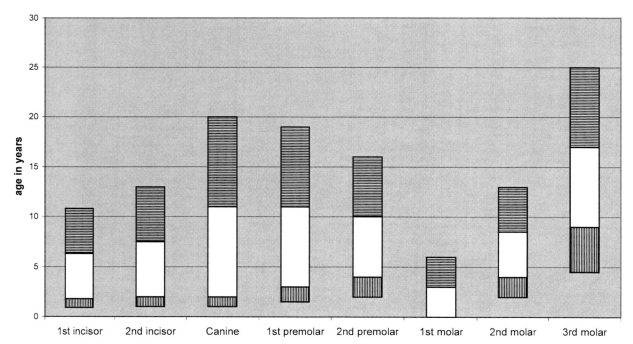

Fig. 8: Formation times for the crowns of lower permanent teeth. The shaded parts of the bars denote the variation in ages at which the crowns are initiated and completed. Figures come from (Schour & Massler 1940; Schour & Massler 1941) and are only approximate.

physiological effect of birth. Neonatal lines are usually clearly shown in deciduous dentitions (milk teeth), all of which start to form well before birth, but both brown striae and cross striations are more difficult to see in deciduous enamel and it has not so far proved possible to match sequences between dentitions. For permanent teeth, however, once a sequence of matching prominent brown striae has been established, it is possible to calibrate it by making counts of cross striations between prominent striae.

This has been checked in the dentitions of the Spitalfields children (above) (Antoine et al. 1999). From within the sample of named known-age individuals, 8 children were selected whose tooth crowns were still forming at the point of death (younger than 6 years of age for most individuals). The developing teeth were dissected from the jaws and microscope sections were prepared for all the different types of upper and lower permanent teeth. Matched sequences of cross striations and brown striae were built up, and the total counts of cross striations between birth and death were compared with the known age at death from the coffin plates. A good example of this is the little boy labelled Individual 2431, who died on the 14th December 1822 aged two years, six months and three weeks (or 2.56 years, which equals 1216 days). Examples of both permanent upper and lower incisors, upper and lower canine, upper and lower first premolar, upper and lower first molar were sectioned (Figure 5). A neonatal line was recognised under the first formed cusp of the first molar and cross striation counts from this to the last formed the increments of the enamel totalled 1190. This is only 26 cross striations short of the expected total 1216 which would match the age of the child. The teeth of this child have a constant count of 9 cross striations between each pair of brown striae of Retzius, so 26 cross striations represents only just under 3 brown striae. The last formed enamel matrix would not, in any case, be fully mineralised so it is expected that some of this delicate material would be lost in archaeological specimens. Bearing this in mind, the cross striation count matches the known age very closely. If cross striations departed significantly from a circadian rhythm, a much greater difference would be expected.

The Spitalfields specimens illustrate something else as well. Development of teeth may be punctuated by a number of interruptions that leave traces as defects in the structure of the enamel. These are commonly known as Developmental Defects of Enamel (DDE), or *enamel hypoplasia* (Hillson & Bond 1997). There is clinical and laboratory evidence that fever and nutritional deficiencies can cause such defects. They take several forms, but the most common is a simple groove or furrow that runs around the circumference of the crown, following the line of the perikymata. Several furrows of this type are

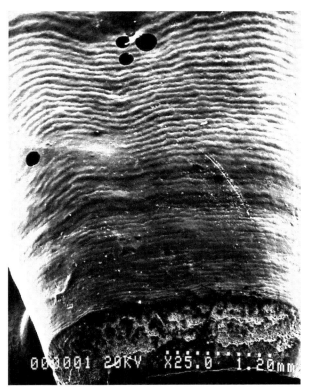

Fig. 9: Epoxy resin replica of the surface of the lower first incisor from Individual 2431, microscopy as for Figure 7. Field width 2.85 mm. The tooth had never erupted into the mouth, but was dissected from the jaw, so the perikymata are very clearly marked with no sign of wear (except for a post-mortem scratch). At the base of the crown, the irregular surface of the last-formed enamel matrix can be seen – it is likely that a certain amount of this less well mineralised tissue has been eroded away. The perikymata show a gradual decrease in spacing down the crown. This is a usual part of the geometry of crown growth. In the middle of the picture, the two defects described in the text can be seen (see also Figure 5).

shown in Individual 2431 from Spitalfields (Figure 9). One particularly characteristic group of furrow defects consists of two steps, each including two perikymata, with three perikymata in between (Hillson et al. 1999). Matching defects with these particular characteristics can be seen in the incisors, canines and first molars in the parts of the crown that would have been forming at the same age. There is no other combination of defects exactly like this anywhere else on the tooth crowns of Individual 2431. The spacing between the grooves of the perikymata can be measured along the edge of the surface, seen in the sections. Figure 10 shows the spacing measurements for both the lower first incisor and lower first molar in the 22 perikymata that include the pair of defects. They range in both teeth from 70 to 80 μm in the perikymata higher up the crown, down to 60 μm or so lower down the crown. This is expected, as the way in which teeth develop brings about a progressive decrease in perikymata spacing. The spacing between the pairs of perikymata in the two steps of the defect increase to over

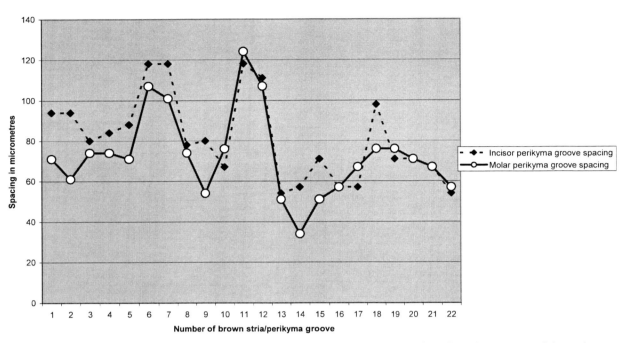

Fig. 10: Spacings between the grooves of the perikymata (equivalent to the brown striae of Retzius), measured down the section of the lower first incisor from Individual 2431. The perikyma grooves are numbered from just above the defects. The two grooves which make the "higher step" of the defect are numbered 6 and 7, and the grooves of the lower step are numbered 11 and 12. The spacings 5/6 and 6/7 are longer than the intervening spacings for both the first incisor and first molar, as the the spacings 11/12 and 12/13.

100 μm and, in between the two steps, the perikymata spacing returns to normal. At the same time, in the section underneath (Figure 5), the counts of cross striations continue evenly, at 9 between each pair of brown striae, and the spacing between brown striae remains even. The only change seen within the section is a slight increase in prominence of the brown striae associated with the perikymata of the steps in the defect. From the cross striation counts, it is possible to say that the higher step of the defect represents a disturbance to growth starting at one year seven months of age, during December 1821, just one year before the death of the child. This disturbance affected two perikymata/brown striae, 9 days apart and then the disturbance seems to have stopped for 27 days after that. The second step represents a similar disturbance 5 weeks after the first, in January or February 1822, again disrupting two perikymata/brown striae, 9 days apart. This sequence does not yield any information about the nature of the event which caused these growth disruptions, although it is likely from their timing that they represent minor episodes of childhood fever, rather than a longer term illness or nutritional deficiency.

Dental enamel defects of this kind are well known for humans, but what is less well-known is that they can be seen in a whole range of other mammals. Both very character around both perikymata and defect are particularly prominent in pig teeth and, if the covering of cement is stripped away, they can be seen clearly in horse (Figure 11), cattle and elephant teeth.

Layering in dental cement

One of the most common applications of histology in the archaeological and museum collections is the examination of layering in cement which is thought to represent an annual growth rhythm. In animals such as humans and carnivores, the cement generally coats only the roots and, in smaller creatures may be a millimetre or less in greatest thickness. It acts as an anchoring tissue for the attachment of the ligamentous fibres that bind the tooth into its bony socket. In the teeth of the largest animals of this group, such as bears, the cement forms a much more substantial coating. The thickest cement coatings, however, are found in those animals where it covers the crown as well as the root. The tall crowns of cattle, horse and some toothed whales emerge gradually from the jaws so that, in the earlier parts of the sequence, the sides of the crown need to attach to the binding ligament that holds the tooth, and the root only becomes involved later on. Annual bands were first recognised, in fact, in the dentine of the large canines of fur seals and elephant seals. Here, the annual rhythm is very strong, as the males do not feed at all during the

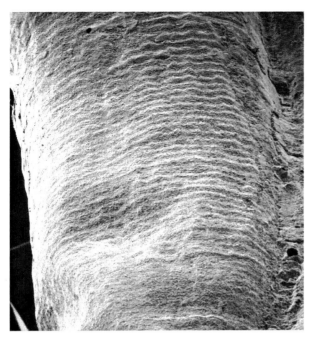

Fig. 11: Epoxy resin replica of the surface of a horse molar, from which the cement has been removed mechanically. Microscopy as for Figure 7. Field width 6.63 mm. Clear perikymata can be seen running across the enamel surface.

mating season when they haul out onto beaches and guard a harem of females. The tall canines of the males grow persistently over 20 years or so, and it is possible to see really prominent bands of poorly mineralised dentine marking the periods of mating (Scheffer 1950). These can be made out with the naked eye on simple cut surfaces of the teeth. Similarly prominent layerings can be seen in the large, persistently growing teeth of sperm whales, which follow an annual round of migration and again go through a period during which they feed very little. A whole variety of smaller toothed whales, and tusked marine mammals such as walrus and dugong similarly show dentine layering and it is the basis of age estimation methods employed by such organisations as the International Whaling Commission (Perrin & Myrick 1980). There is also layering visible in sections of the cement in all these animals but, where the dentine can be used, it often gives more consistent results. Cement layers are used more in the smaller seals (Mansfield & Fisher 1960), where it is more difficult to use dentine layers in the same way, but there is some discussion about what particular part of the cement layering sequence represents the breeding/moult interval.

Cement layering has also been of particular interest in land mammals, where dentine much more rarely shows evidence of gross annual or seasonal layering (Klevezal 1996). The main focus has been hibernating carnivores, such as the bears, and ungulates such as deer. These layers cannot be seen with the naked eye, and the normal procedure is to demineralise the tooth and then section the organic component that remains using a microtome. This results in a very thin section (ideally <5 μm thick) which can be stained in a way that shows a clear alternation of lightly and darkly stained layering in the cement. Counts of these layers are routinely used for age estimation in wildlife management. They may well yield good age estimates, but there are still significant difficulties. One is that there is no consensus as to what is causing the variation in the stain reaction – is it variation in the density of the cells of cement (cementocytes), in the density or organisation of the collagen fibres which play a prominent part in the architecture of cement, or in the organisation of the non-collagen organic component. Other potential factors are added when thin sections of intact, undemineralised tissue are used to show the layers, where they may be due to any of the above factors and also the pattern of mineralisation of the tissue. Thin sections, however, typically contain rather a thick slice of tissue (above). Where the layers are thin, and on the sides of the roots of small teeth they may be only a few micrometres apart, it is unlikely there would be adequate resolution in this type of section to see them properly. In addition to all this, relatively few studies have been done on known age animals where the method could be tested and, where they have been carried out, there has been variable success (Hillson 1990). It is understandable that it would be difficult to find groups of wild animals whose age at death is precisely and independently known.

It may be that human teeth may help to answer these questions. Their sequence of development and histology are better known than those of any other mammal. The long human life span makes it possible that many cement layers could accumulate, so the test would be a better one than provided by the short lives of many wild animals. Also, it is easy to accumulate large collections of known-age teeth from dental surgeries at which patients can be asked to donate their extracted teeth. The first question, however, is whether or not humans have a pattern of layering in any way similar to that of large ungulates and carnivores. Surprising though it may seem, this does seem to be the case and the largest study (Charles et al. 1986; Charles et al. 1989; Condon et al. 1986) has suggested that there is at least a moderate relationship between the number of layers and the age at extraction, even if there was not a direct one-to-one relationship between layers and years in all specimens. This study used demineralised, microtome sectioned and stained preparations of recently extracted teeth. For archaeological and museum specimens, however, this type of preparation often does not work well because of the poor survival of the organic component and the diagenetic changes (above) which may damage large areas of the tissue. A more widely applicable alternative is SEM-BSE imaging of cut, polished surfaces as described for bone above. This has the added advantage

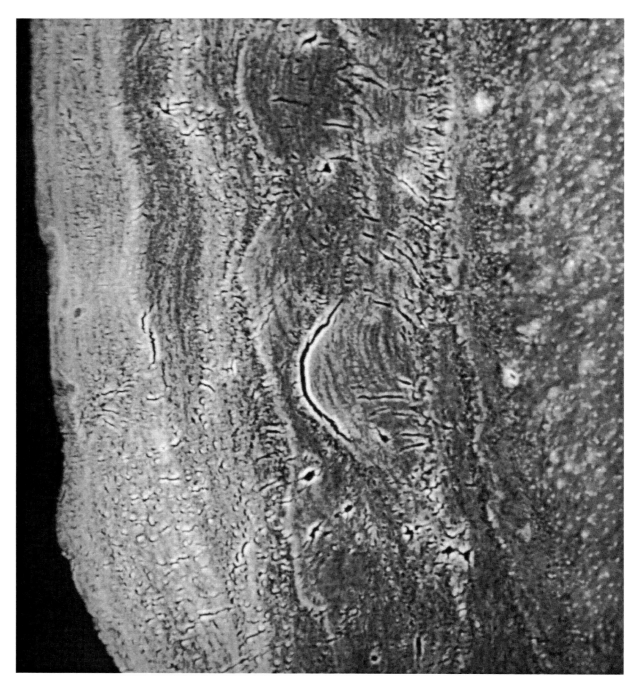

Fig. 12: Section of an extracted lower 1st premolar from a 48 year old individual. Specimen preparation and microscopy as in Figure 1. Field width 0.5 mm. On the right of the picture is the dentine, marked by a pattern of bright dots which represent highly mineralised material filling in the tubules which pass through the tissue. The irregular bright line about one quarter the way across the right had side is the cement-dentine junction and then, to the left of this, is a succession of dark and bright lines which mark out the layering in the cement. The small dark void are the lacunae of the cementocytes and a pattern of cracks runs through the cement. In most cases, these follow the orientation of the larger, so-called extrinsic, collagen fibres in the cement. A consistent count of 36 well defined layers could be made for this tooth, although irregularities can clearly be seen in the picture. The roots of 1st premolars are completed between 12 and 13 years of age so, if the tooth was extracted at 48 years, it would be expected that there would be 35-36 years worth of cement formation. This fits well with the layer count in this specimen.

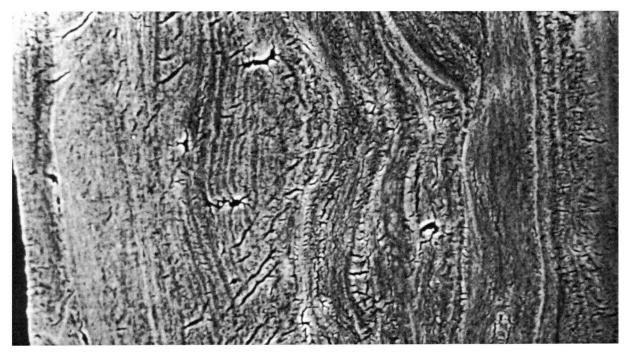

Fig. 13: Higher magnification image of a section from the upper third molar in the same individual as for Figure 12. Field width 300 μm. The cement-dentine junction is at the right of the picture, and the root surface at the left. 38 layers were counted. 3rd molar roots are completed at some time between 18 and 25 years of age (although these teeth are very variable), so with an extraction at 48 years it would be expected that 23-30 years worth of cement formation would be present on the root. This is a little bit under the count of 38 layers.

that it potentially reduces the number of factors involved in forming the image. SEM-BSE images, where the specimen and microscope are set up to optimise compositional contrasts, mostly show the mineralisation of the tissue. It needs to be borne in mind that the orientation of crystals might affect the rate of polishing in ways that could create surface irregularities which are also shown, but the image is still largely concerned with organisation of the mineral component and it should be clearer what is causing the layering seen. As part of a small pilot project, five teeth extracted from one individual aged 48 (Figures 12 and 13) were sectioned and examined in SEM-BSE mode. Layering showed as alternating bright and dark lines, but in many places these were irregular. Strongly developed layers were counted at various points on the sections (there were additional more weakly developing layers in many cases) and the mean count recorded. Roots take some years to form, and it is not clear at which point it should be assumed that the cement layering starts. The conventional starting point for counts is the completion of the roots and, if this is taken from a range of developmental studies, the counts of layers for different teeth were not far away from what might be expected for the time over which cement would be deposited on the root surface (see caption of Figure 12).

Conclusions / Schlussfolgerung

The microstructure of bone and dental tissues provides a great deal of detailed evidence about the state of preservation of archaeological specimens, and their developmental history. Methods and microscopy and preparation techniques need to be carefully chosen, so that they are appropriate for archaeological material.

Die mikrostrukturelle Organisation von Knochen- und Zahngewebe birgt zahlreiche Hinweise sowohl bezüglich des Erhaltungsgrades archäologischer Funde, als auch derer Entwicklungsgeschichte. Die jeweilige methodische Vorgehensweise hinsichtlich der Mikroskopier- und Präparationstechniken muss jedoch mit Bedacht so gewählt werden, dass sie dem archäologischen Fundgut angemessen ist.

References

Aiello L.C. & Molleson T., 1993.
Are microscopic ageing techniques more accurate than macroscopic ageing techniques. *Journal of Archaeological Science* **20**: 689-704.

Antoine D.M., Dean M.C. & Hillson S.W., 1999.
The Periodicity of Incremental Structures in Dental Enamel Based on the Developing Dentition of Post-Medieval Known-age Children. In: Mayhall J. T. & Heikinnen T. (eds.). *Dental Morphology '98*: 48-55. Oulu: Oulu University Press.

Bell L.S., 1990.
Palaeopathology and diagenesis: an SEM evaluation of structural changes using backscattered electron imaging. *Journal of Archaeological Science* **17**: 85-102.

Bell L.S., Boyde A. & Jones S.J., 1991.
Diagenetic alteration to teeth in situ illustrated by backscattered electron imaging. *Scanning* **13**: 173-183.

Bell L.S. & Jones S.J., 1991.
Macroscopic and microscopic evaluation of archaeological pathological bone: backscattered electron imaging of putative Pagetic bone. *International Journal of Osteoarchaeology* **1**: 179-184.

Bell L.S., Skinner M.F. & Jones S.J., 1996.
The speed of post mortem change to the human skeleton and its taphonomic significance. *Forensic Science International* **82**: 129-140.

Boyde A., 1972.
Scanning electron microscope studies of bone. In: Bourne G. H. (ed.). *Biochemistry and physiology of bone*: 259-310. London: Academic Press.

Boyde A., 1984.
Methodology of calcified tissue specimen preparation for scanning electron microscopy. In: Dickson G.R. (ed.) *Methods of calcified tissue preparation*: 251-307. Amsterdam: Elsevier.

Boyde A., 1989.
Enamel. In: Berkovitz B.K.B., Boyde A., Frank R.M., Höhling H.J., Moxham B.J., Nalbandian J. & Tonge C.H. (eds.). *Teeth*: 309-473. New York, Berlin & Heidelberg: Springer Verlag.

Boyde A., Hendel P., Hendel R., Maconnachie E. & Jones S.J., 1990.
Human cranial bone structure and the healing of cranial bone grafts: a study using backscattered electron imaging and confocal microscopy. *Anatomy & Embryology* **181**: 235-251.

Boyde A. & Jones S.J., 1984.
Back-scattered electron imaging of skeletal tissues. *Metabolic Bone Disease & Related Research* **5**: 145-150.

Boyde A., Maconnachie E., Reid S.A., Delling G. & Mundy G.R., 1986.
Scanning electron microscopy in bone pathology: review of methods, potential and applications. *Scanning Microscopy*, **1986/IV**: 1537-1554.

Charles D.K., Condon K., Cheverud J.M. & Buikstra J.E., 1986.
Cementum annulation and age determination in Homo sapiens. I. Tooth variability and observer error. *American Journal of Physical Anthropology* **71**: 311-320.

Charles D.K., Condon K., Cheverud J.M. & Buikstra J.E., 1989.
Estimating age at death from growth layer groups in cementum. In: Iscan M.Y. (ed.). *Age markers in the human skeleton*: 277-316. Springfield: Charles C. Thomas.

Condon K., Charles D.K., Cheverud J.M. & Buikstra J.E., 1986.
Cementum annulation and age determination in Homo sapiens. II. Estimates and accuracy. *American Journal of Physical Anthropology* **71**: 321-330.

Fitzgerald C. & Rose J., 2000.
Reading between the lines: dental development and subadult age assessment using the microstructureal growth markers of teeth. In: Katzenberg M.A. & Saunders S.R. (ed.). *Biological anthropology of the human skeleton*:163-186. New York: John Wiley & Sons, Inc.

Hagelberg E., Bell L.S., Allen T., Boyde A., Jones S.J. & Clegg J.B., 1991.
Analysis of ancient bone DNA: techniques and applications. In: Eglinton G. & Curry G.B. (ed.). *Molecules through time: fossil molecules and biochemical systematics*: 399-408. London: The Royal Society.

Hedges R.E.M. & Millard A.R., 1995.
Bones and groundwater: towards the modelling of diagenetic processes. *Journal of Archaeological Science* **22**: 155-164.

Hedges R.E.M. & Wallace C.J.A., 1978.
The survival of biochemical information from archaeological bone. *Journal of Archaeological Science* **5**: 377-386.

Hillson S.W., 1990.
Teeth. 1st paperback Ed. Cambridge: Cambridge University Press.

Hillson S.W., 1996.
Dental anthropology. Cambridge: Cambridge University Press.

Hillson S.W., Antoine D.M. & Dean M.C., 1999.
A Detailed Developmental Study of the Defects of Dental Enamel in a Group of Post-Medieval Children from London. In: Mayhall J.T. & Heikinnen T. (ed.). *Dental Morphology '98*: 102-111. Oulu: Oulu University Press.

Hillson S.W. & Bond S., 1996.
A scanning electron microscope study of bone, cement, dentine and enamel. In: Bell M., Fowler P. & Hillson S.W. (ed.) *The experimental earthwork project, 1960-1992*, York: 185-194. Council for British Archaeology.

Hillson S.W. & Bond S., 1997.
Relationship of enamel hypoplasia to the pattern of tooth crown growth: a discussion. *American Journal of Physical Anthropology* **104**: 89-104.

Klevezal G.A., 1996.
Recording structures of mammals : determination of age and reconstruction of life history. Rotterdam: A.A. Balkema.

Mansfield A.W. & Fisher H.D., 1960.
Age determination in the harbour seal Phoca vitulina L. *Nature* **186**: 92-93.

Molleson T. & Cox M., 1993.
The people of Spitalfields: the Middling Sort. York: Council for British Archaeology.

Osborn J.W., 1981.
Dental anatomy and embryology. In: Rowe A.H.R. & Johns R.B.(ed.). Oxford : Blackwell Scientific Publications.

Osborn J.W. & Ten Cate A.R., 1983.
Advanced Dental Histology. 4th Ed. Bristol: John Wright.

Perrin W.F. & Myrick A.C., 1980.
Growth of odontocetes and sirenians: problems in age determination. Proceedings of the International Conference on determining age of odontocete Ceteans (and Sirenians), La Jolla, California, September 5-19, 1978. Cambridge : International Whaling Commission.

Reid S.A. & Boyde A., 1987.
Changes in the mineral density distribution in human bone with age: image analysis using backscattered electrons in the SEM. *Journal of Bone and Mineral Research* 2: 13-22.

Robling A.G. & Stout S.D., 2000.
Histomorphometry of human cortical bone: applications to age estimation. In: Katzenberg M.A. & Saunders S.R. (ed.). *Biological anthropology of the human skeleton*: 187-213. New York: John Wiley & Sons, Inc.

Scheffer V.B., 1950.
Growth layers on the teeth of Pinnipedia as an indication of age. *Science* 112: 309-311.

Schour I. & Massler M., 1940.
Studies in tooth development: the growth pattern of human teeth. *Journal of the American Dental Association* 27: 1778-1931.

Schour I. & Massler M., 1941.
The development of the human dentition. *Journal of the American Dental Association* 28: 1153-1160.

Ten Cate A.R., 1985.
Oral histology: development, structure and function. St Louis: C V Mosby.

Ubelaker D.H., 1986.
Estimation of age at death from histology of human bone. In: Zimmerman M.R. & Angel J.L. (ed.). *Dating and the age determination of biological materials*: 240-247. London: Croom Helm.

Williams P.L. & Bannister L.H., 1995.
Gray's anatomy: the anatomical basis of Medicine and Surgery. 38th, London: Churchill Livingstone.

Interpreting the trace-element components of bone – a current perspective from the Laboratory for Archaeological Chemistry

James H. Burton, T. Douglas Price, Department of Anthropology,
University of Wisconsin-Madison

Abstract / Zusammenfassung

Although strontium and barium enter bone in proportion to their dietary levels, the conventional interpretation that less strontium and barium imply more meat in the diet is unwarranted. Ba/Ca and Sr/Ca tend to reflect the principal calcium source in the diet rather than varying in proportion to variations in the diet.

Although Ba/Ca and Sr/Ca don't vary greatly with changes in diet, they do have significant geographic variation. This lack of local dietary variation together with the geographic differences implies that Ba/Ca and Sr/Ca may have potential in provenance studies of humans through the analysis of dental enamel, which retains the elemental composition obtained during one's early childhood years.

Obgleich Strontium und Barium in einer bestimmten Relation zu deren Konzentrationen in der Nahrung in den Knochen eingelagert werden, ist die konventionelle Interpretation nicht gerechtfertigt, weniger Strontium und Barium sprächen für einen höheren Fleischanteil in der Nahrung. Die Ba/Ca- und Sr/Ca- Verhältnisse des Knochens spiegeln eher die hauptsächliche Ca-Quelle in der Nahrung wider, als dass sie in Relation zu den entsprechenden Verhältnissen in den Nahrungsmitteln variieren würden.

Obgleich die Ba/Ca- und Sr/Ca-Verhältnisse auch bei verschiedenen Ernährungsmustern im Skelett nicht sonderlich stark schwanken, ist dagegen deren geographische Variabilität erheblich. Das Fehlen nahrungsabhängiger, gemeinsam mit signifikanter geographischer Variabilität impliziert, dass die Ba/Ca- und Sr/Ca-Verhältnisse im Skelett potentiell für Herkunftsuntersuchungen geeignet sind, insbesondere durch Analyse des Zahnschmelzes, welcher die Elementzusammensetzung während der frühen Kindheit archiviert.

Keywords: Strontium, Bone chemistry, Paleodiet, Barium
Strontium, chemische Zusammensetzung des Knochens, Paläoernährung, Barium

Introduction

Archaeologists began to take a significant interest in the trace-elemental composition of bone, and specifically in strontium (Sr), during the 1970's when researchers began to recognize that carnivores have less Sr in their bones than herbivores (e.g., Brown 1973; Gilbert 1975; Kavanaugh 1979; Schoeninger 1979; Parker & Toots 1980; Katzenberg 1984). If one could use this principle to assess the relative amount of meat versus vegetation consumed by prehistoric people, then one could by extrapolation examine substantial anthropological issues such as the relative amount of hunting versus gathering as well as changes in subsistence strategies from hunting and gathering to agriculture. One of the primary activities of the Laboratory for Archaeological Chemistry over the last 15 years has been to investigate the sources of variation in the trace-element composition of human remains and to develop anthropological applications of this information. These investigations include examination of strontium as an indicator of plants versus meat and other possible dietary inferences such as seafood consumption, the use of barium for drawing such inferences, and the study of non-dietary sources of variation in these elements such as geographic variation and post-mortem (diagenetic) alteration. We are currently examining the potential of elemental abundances to provide locational data, analogous with and in addition to the use of $^{87}Sr/^{86}Sr$ ratios for provenience information.

Strontium and the plant/meat ratio of the diet

Nuclear testing during the Cold War led to the recognition that radioactive ^{90}Sr, produced by such testing, was entering the environment, with specific concern that it would accumulate in calcium-rich foods such as milk and subsequently in the human skeleton where it initiates lethal disease processes. This concern led to extensive investigations of strontium distribution in the environment (e.g., Hodges et al. 1950; Comar et al. 1952; Comar et al. 1955; Alexander et al. 1956; Comar et al. 1956;Hartsook et al. 1956; Turekian & Kulp 1956; Comar et al. 1957; Thurber et al. 1958; Burton & Mercer 1962), which found that strontium does indeed move through the food-chain and accumulate in bone, also that biological processes discriminate against strontium absorption in preference for calcium. Thus the ratio of strontium to calcium in an organism is only about one fifth the Sr/Ca of the diet. Herbivores have Sr/Ca equal to 20% of the Sr/Ca of the plants they consume, and carnivores feeding on these herbivores likewise have a similar reduction to 20% of the Sr/Ca of the herbivores, i.e., to about 4% that of the herbivore diet (Elias et al. 1982). This trophic-level reduction of the amount of strontium relative to calcium is commonly called "biopurification". Because intermediate diets would consequently have intermediate Sr/Ca, which would be tracked by the bone, analysis of bone Sr/Ca should reflect the relative amount of meat versus plants. Our studies, consisting of both feeding experiments (Price et al. 1986) and examination of modern food webs (Price et al. 1985; Burton et al. 1999), demonstrated that bone does faithfully track dietary strontium levels, i.e., bone Sr/Ca is consistently one-fifth that of the diet.

Nonetheless, application of this principle to determine the percentage of meat in the diet proved problematic. Three significant issues became apparent: 1> insensitivity to the dietary plant/meat ratio for intermediate diets (Burton & Wright 1995), 2> regional variation in strontium levels, precluding direct comparison between populations in different areas without adjusting for intrinsic geographic variability (Burton et al. 1999), 3> post-mortem changes in the strontium levels in bone through equilibration with the depositional matrix (Price et al. 1992).

Insensitivity to the plant/meat ratio

Probably the most confounding aspect of strontium dietary studies is that even though bone strontium levels demonstrably and consistently reflect dietary strontium levels, and bones of carnivores have consistently less strontium than bones of herbivores, intermediate diets do not yield proportionally intermediate strontium levels in bone (Burton & Wright 1995). That is to say, a diet of half meat and half vegetation will not produce bone strontium levels halfway between that of a pure carnivore and a pure herbivore. The resolution of this paradox is revealed by actually calculating the composition of such a diet, the Sr/Ca ratio of which will be almost identical to the Sr/Ca of the purely vegetarian diet. Bones tracking such a diet will faithfully reflect the diet Sr/Ca, but will nonetheless resemble herbivores. In animals almost all of the calcium and strontium is not in meat but resides in bone, which is not normally a large part of human diets. Whatever dietary component has the highest calcium generally determines the dietary Sr/Ca, with little influence from low calcium foods. Because meat has very little calcium compared to most vegetation, the low Sr/Ca of meat has little influence on Sr/Ca of the whole diet unless it comprises by far the majority of the diet (>80%, in which case it becomes a significant calcium source). Thus carnivores have low Sr/Ca but omnivores generally have high Sr/Ca regardless of the amount of meat they consume. We wish to emphasize that this is not due to a failure of bone to track dietary Sr/Ca, rather because it does track the diet.

Geographic variation

Another problem, albeit the most amenable to a practical solution, is the intrinsic variation in strontium levels arising from different strontium levels in different regions irrespective of diet. For example carnivores of the Yosemite, California food-chain study of Elias et al. (1982) appropriately have 1/5 the strontium of Yosemite herbivores, but the Yosemite carnivores have as much strontium as we found in pure herbivores of northern Wisconsin, i.e. five-times the strontium of Wisconsin carnivores (Burton et al. 1999). Direct comparison of Sr/Ca between the two locations cannot yield inferences about the degree of biopurification of dietary calcium without first adjusting for this large geographic difference. Fortunately such an adjustment for geographic variation can be made by analyzing, where possible, the same species, or at least species from the same trophic-level, from each region and comparing the study samples to these rather than making a direct comparison between regions.

Diagenesis

While strontium may not be useful as a quantitative measure of the amount of meat in the diet, the extensive feeding experiments and studies of modern food webs unequivocally demonstrate that bone does quantitatively reflect the dietary Sr/Ca and does reflect the principal dietary source of calcium. Nonetheless, in studies of fossil bone there remains another, more intractable problem: post-depositional, or 'diagenetic', alteration of the biological level of strontium. Numerous studies, including our own (e.g., Lambert et al. 1985; Sillen 1986; Price et al.1992; Ezzo 1992), have resulted in many ways of trying to assess and ameliorate post-depositional changes in bone. With few exceptions, proposed measures of diagenetic alteration suffer from the problem that they are proxies rather than direct measures of alteration of strontium levels. Bone may be altered in its cellular structure as indicated microscopically, or in its crystallinity, as revealed by X-ray diffraction and infrared spectroscopy, without necessarily being badly contaminated by diagenetic strontium. Conversely, bone may exhibit unaltered, biological values for elements such as calcium (Ca) and phosphorus (P) and the absence of non-biological elements such as zirconium (Zr), yet have accumulated Sr from the soil.

Researchers have explored several procedures, mainly various methods of washing by acids, to reduce diagenetic contamination. By monitoring $^{87}Sr/^{86}Sr$, as mentioned above, or various proxies such as the Ca/P ratio, we can assess the effectiveness of these washing methods for reducing post-depositional contaminants. The results of these experiments provide no consensus, however, on optimal washing methods. Some propose using relatively strong acid to aggressively remove soluble contaminants (Price et al. 1992); others advocate washing with weakly acid buffers (Sillen 1986, 1989) to avoid further recrystallization, which could fix contaminants within the residual bone material. Most use the bone that remains after aggressive washing, but some use the last washes and avoid residual bone that might contain insoluble diagenetic fluorides (Sillen 1986, 1989). We suspect that the actual optimal method depends upon the specific depositional contexts, the age of the bones, and the abundance or scarcity of initial material and thus varies on a case-by-case basis. The optimal treatment of many grams of bone from modern fauna will not likely be the same as that required by a milligram of early hominid material. We currently use multiple washes by pure water and 5% acetic acid (considered to be a relatively 'aggressive' method) and analyze the remaining bone material rather than the washes. Even so, diagenesis remains a significant problem and often these methods may fail to significantly remove contamination.

Strontium and seafood

Because seawater is relatively high in mineral salts, including strontium, some suspected that seafood would thus elevate bone strontium (Schoeninger & Peebles 1981). We examined animals from modern marine food webs as well as human remains from sites where seafood was evidently a major component of diet (e.g., coastal Peru and the Aleutian Islands). We found, ironically, that people consuming significant seafood had slightly less, not more, strontium in their bones than had those who lived inland, without significant access to seafood. This paradox was resolved by noting that modern food webs in the marine environment exhibit biopurification just as do terrestrial food webs. Those consuming 'seafood', generally implying large amounts of fish (Peru) or flesh of marine mammals (Aleutians) were consuming food from the highest trophic levels and hence quite 'biopurified'. Thus their Sr/Ca was similar to bones of people with terrestrial meat-rich diets and hence were slightly lower than bone Sr/Ca produced by a typical terrestrial diet with at least moderate amounts of vegetation (Burton & Price 1999).

Barium

Because barium (Ba) is an alkaline-earth element like strontium, we added Ba to our analytical protocols and found that it did, indeed, behave like strontium. This similarity had been noted by Elias et al. (1982) in their examination of a food chain in California, where they found that, presumably because the ionic-radius of barium is larger than that of strontium, biological systems more strongly discriminate against barium. We found in our own food chain studies a strong correlation between strontium and barium (Burton et al. 1999), with barium dropping by a factor of ten (bone Ba/Ca = x10% diet Ba/Ca) with each trophic level in contrast to a factor of five (x20%) for strontium. Thus we initially proposed that barium would be more efficient that strontium as a trophic indicator. While barium clearly exhibits a greater degree of biopurification, it, too, is subject to the constraints discussed above for strontium. It does reveal the extent of biopurification of dietary calcium, but mainly reflects the major calcium source rather than a quantitative measure of plants versus meat and is also affected by geographic variation and diagenetic contamination

We discovered that barium in one significant aspect differs from strontium, i.e. regarding marine resources. While seafood produces bone Sr/Ca comparable to that of terrestrial carnivores, it lowers bone Ba/Ca by more than an order of magnitude (Burton & Price 1990, 1991). We interpret this as a result of the high sulfate concentrations of the oceans coupled with the extremely low solubility of barite (barium sulfate). Barium is essentially insoluble in seawater and is thus unavailable in the marine food web. Marine Ba/Ca is several orders of magnitude lower than typical terrestrial levels. We initially thought this could be a sensitive monitor of marine resources, but as is the case with Sr/Ca, it, rather than being a quantitative marine/terrestrial measure, mainly reveals, when such is the case, that the ocean is the major calcium source.

Ba and Sr as geographic variables in teeth

The facts that bone Sr/Ca and Ba/Ca can be relatively unresponsive to significant changes in diet, typically varying locally by less than a factor of four, and that differences between some locations can exceed several orders of magnitude can be restated more simply as "variation among different regions might exceed variation within the regions", i.e. the 'provenience postulate'. In other words, bones from different locations could chemically differ enough such that the composition could be used as a determinant of the location from which the bone comes. Because bone remodels and continues to incorporate Sr and Ba, the Sr/Ca and Ba/Ca ratios reflect the chemistry of the local, adult diet (at least in prehistoric cases). Archaeologists normally know the burial context of bones, which parsimoniously can be assumed to be close to the place of residence and hence the 'origin' of the bone. Dental enamel of non-deciduous teeth is, in contrast to bone, deposited during early childhood and does not remodel or later incorporate barium or strontium. Thus, the Ba/Ca and Sr/Ca ratios of dental enamel reflect the place of residence during early childhood rather than adulthood and thus have utility in the study of human mobility much as we currently exploit the isotopic ratio of $^{87}Sr/^{86}Sr$, which is independent of biological processes but varies with geological context.

Our initial studies of deer populations showed significant Ba/Ca and Sr/Ca differences among deer from various locations although they are completely herbivorous (Burton et al. 2002). Knipper (2002) recently examined sheep bones and teeth from various European locations and similarly found geographic variation exceeded the intrinsic local variation. We are currently evaluating Ba/Ca and Sr/Ca in dental enamel for their potential in providing data on human provenience. We recently examined dental enamel from Tarascan elites (Burton et al. 2003, Cahue 2001) to determine whether they matched either those from the Tarascan administrative center of Tzintzantzun or local non-elites in the western Tarascan area. Although Sr/Ca did not differ between the regions and thus was not useful for such discrimination, the Ba/Ca results from the elites matched the local non-elites and significantly differed from those from the administrative center, consistent with local origin of the elites. Although Ba/Ca and Sr/Ca are already providing useful results, we suggest that ideally they be used as a supplement to, rather than a substitute for, $^{87}Sr/^{86}Sr$ measurements.

Summary / Zusammenfassung

Strontium and barium enter bone in proportion to dietary Sr/Ca and Ba/Ca, respectively. Although meat has lower strontium and barium than plants, meat also has low calcium and thus does not affect dietary Sr/Ca and Ba/Ca or, consequently, bone Sr/Ca or Ba/Ca. Bones tend to have Sr/Ca and Ba/Ca characteristic of the principal calcium source rather than respond sensitively to variations in diet. On the other hand, there can be substantial geographic variation in environmental Sr/Ca and Ba/Ca, which are hence reflected in dietary and bone Sr/Ca and Ba/Ca. Because dental enamel develops during the first years of life and does not chemically remodel, it retains the calcium and associated strontium and barium deposited during childhood. Thus these elements have potential for studying human mobility, especially when used as an adjunct to $^{87}Sr/^{86}Sr$ data.

Strontium und Barium werden in den Knochen in Relation zu den Sr/Ca- und Ba/Ca-Verhältnissen in der Nahrung eingelagert. Obgleich Fleisch weniger Strontium und Barium enthält als pflanzliche Nahrung, enthält es doch gleichzeitig wenig Calcium, so dass dieser Nahrungsbestandteil keinen Einfluß auf die Sr/Ca- und Ba/Ca-Verhältnisse in der Nahrung hat, und konsequenterweise auch nicht auf jene des Knochens. Die Sr/Ca- und Ba/Ca-Verhältnisse im Knochen sind eher auf die hauptsächliche Calcium-Quelle in der Nahrung bezogen, als dass sie empfindlich gegenüber unterschiedlichen Ernährungsmustern wären. Andererseits kommen substantielle geographische Variationen in bezug auf habitattypische Sr/Ca- und Ba/Ca-Verhältnisse vor, welche dann ihrerseits zu entsprechenden Variationen dieser Werte in der Nahrung und im Knochen führen. Da der Zahnschmelz während der frühen Kindheit präzipitiert und keinem weiteren Umbau unterliegt, archiviert dieses Gewebe das Calcium sowie assoziiertes Strontium und Barium, welches in diesem Entwicklungszeitraum inkorporiert wurde. Sr- und Ba-Analysen haben daher ein Forschungspotential in bezug auf menschliche Mobilität, insbesondere ergänzend zu $^{87}Sr/^{86}Sr$-Daten.

References

Alexander G.V., Nusbaum R.E. & MacDonald N.S., 1956.
The relative retention of strontium and calcium in bone tissue. *Journal of Biological Chemistry* **218**: 911-919.

Brown A.B., 1973.
Bone strontium content as a dietary indicator in human skeletal populations. Ph.D. dissertation. Department of Anthropology, University of Michigan, Ann Arbor.

Burton J.D. & Mercer E.R., 1962.
Discrimination between Strontium and Calcium in their Passage from Diet to the Bone of Adult Man. *Nature* **193**: 846-847.

Burton J.H. & Price T.D., 1990.
The ratio of barium to strontium as a paleodietary indicator of consumption of marine resources. *Journal of Archaeological Science* **17**: 547-557.

Burton J.H. & Price T.D. 1991.
Paleodietary applications of barium values in bone. In: Pernicka E. & Wagner G.A. (eds). *Proceedings of the 27th International Symposium on Archaeometry, Heidelberg, 1990*: 787-795. Berkhauser Verlag, Basel.

Burton J.H. & Price T.D., 1999.
Evaluation of bone strontium as a measure of seafood consumption. *International Journal of Osteoarchaeology* **9**: 233-236.

Burton J.H., Price T.D., Cahue L. & Wright L.E., 2003.
The use of barium and strontium abundances in human skeletal tissues to determine their geographic origins. *International Journal of Osteoarchaeology* **13**: 88-95.

Burton J.H., Price T.D. & Middleton W.D., 1999.
Correlation of bone Ba/Ca and Sr/Ca due to biological purification of calcium. *Journal of Archaeological Science* **26**: 609-616.

Burton J.H. & Wright L.E., 1995.
Nonlinearity in the relationship between bone Sr/Ca and diet: Paleodietary Implications. *American Journal of Physical Anthropology* **96**: 273-282.

Cahue L., 2001.
The effect of environmental change and economic power on the diet of Tarascan elites. Ph.D. dissertation. Department of Anthropology, University of Michigan, Ann Arbor.

Comar C., Lotz W. E. & Boyd G. A., 1952.
Autoradiographic studies of calcium, phosphorous, and strontium distribution in the bones of the growing pig. *American Journal of Anatomy* **90**: 113-125.

Comar C., Whitney I. B. & Lengemann F.W., 1955.
Comparative utilization of dietary Sr90 and calcium by the developing rat fetus and growing rat. *Proceedings of the Society for Experimental Biology* **88**: 232-236.

Comar C., Wasserman R. H. & Nold M.N., 1956.
Strontium-calcium discrimination factors in the rat. *Proceedings of the Society for Experimental Biology* **92**: 859-863.

Comar C., Russell R.S. & Wasserman R.H., 1957.
Strontium-calcium movement from soil to man. *Science* **126**: 485-496.

Elias R.W., Hirao Y. & Patterson C.C., 1982.
The circumvention of the natural biopurification of calcium along nutrient pathways by atmospheric inputs of industrial lead. *Geochimica et Cosmochimica Acta* **46**: 2561-2580.

Ezzo J.A., 1992.
A test of diet versus diagenesis at Ventana Cave, Arizona. *Journal of Archaeological Science* **19**: 23-37.

Gilbert R.I., 1975.
Trace element analyses of three skeletal Amerindian populations at Dickson Mounds. Ph.D. dissertation, University of Massachusetts, Amherst.

Hartsook E., Cowan R. L., Chandler P. T. & Whelan J. B., 1956.
The relative retention of strontium and calcium in bone tissue. *Journal of Biological Chemistry* **218**: 911-919.

Hodges R. M., MacDonald N. S., Nusbaum R., Stearns S., Ezmirlian F., Spain P. & McArthur C. 1950.
The strontium content of human bone. *Journal of Biological Chemistry* **185**: 519-524.

Katzenberg M.A., 1984.
Chemical Analysis of Prehistoric Human Bone from Five Temporally Distinct Populations in Southern Ontario. Toronto: Archaeological Survey of Canada / National Museum of Man.

Kavanaugh M., 1979.
Strontium in bone as a dietary indicator. M.A. thesis. University of Wisconsin, Madison.

Knipper C., 2002.
Mobility, diet and diagenesis – Trace elemental analyses of faunal remains from southern Germany. In: 33rd International Symposium on Archaeometry. 22-26 April 2002, Amsterdam. Program and Abstracts 160.

Lambert J.B., Simpson S.V., Szpunar C.B. & Buikstra J.E., 1985.
Bone diagenesis and dietary analysis. *Journal of Human Evolution* **14**: 477-482.

Parker R.B. & Toots H., 1980.
Trace elements in bones as paleobiological indicators. In: Behrensmeyer A.K. & Hill A.P. (eds). *Fossils in the Making*: 197-207. Chicago: University of Chicago Press.

Price T.D., Connor M & Parsen J.D., 1985.
Bone strontium analysis and the reconstruction of diet: strontium discrimination in white-tailed deer. *Journal of Archaeological Science* **12**: 419-442.

Price T.D., Swick R.W. & Chase E. P., 1986.
Bone chemistry and prehistoric diet: strontium studies of laboratory rats. *American Journal of Physical Anthropology* **70**: 365-375.

Price T.D., Blitz J., Burton J.H. & Ezzo J., 1992.
Diagenesis in Prehistoric Bone: Problems and Solutions. *Journal of Archaeological Science* **19**: 513-529.

Schoeninger M.J., 1979.
Diet and status at Chalcatzingo: some empirical and technical aspects of strontium analysis. *American Journal of Physical Anthropology* **51**: 295-310.

Schoeninger M.J. & Peebles C.S., 1981.
Effect of mollusc eating on human bone strontium levels. *Journal of Archaeological Science* **8**:391-397.

Sillen A., 1986.
Biogenic and diagenetic Sr/Ca in Plio-pleistocene fossils of the Omo Shungura Formation. *Paleobiology* **12**: 311-323.

Sillen A., 1989.
Diagenesis of the inorganic phase of cortical bone. In: Price T.D. (ed). *The Chemistry of Prehistoric Bone*: 211-229. Cambridge: Cambridge University Press.

Thurber D.L., Kulp J.L., Hodges E., Gast P.W. & Wampler J.M., 1958.
Common strontium content of the human skeleton. *Science* **128**: 256-257.

Turekian K.K. & Kulp J.L., 1956.
Strontium content of human bones. *Science* **124**: 405-407.

Bone Collections are DNA data banks

Carles Lalueza-Fox, Secció Antropologia, Dept. Biologia Animal,
Facultat de Biologia, Universitat de Barcelona

Abstract / Zusammenfassung

Information about some of the most relevant ancient DNA studies is presented, quoting those that fulfil the main authentication criteria of independent replication, cloning of PCR products, phylogenetic sense, amino acid analysis and quantitative PCR that have been proposed by some researchers. Also, data about cloning of PCR products have been compiled revealing that there is no direct relationship between age and nucleotide misincorporation rate (with maximum error values of up to 6 nucleotides per 1000 bp). Conclusions that emerge from this study are that there have been more efforts made to prove the authenticity of nonhuman ancient DNA results than there have been to prove human results. In addition, it was determined that some of the more extraordinary claims, like the retrieval of DNA from very old material (e.g., insects in amber or Miocene magnolia leaves), nuclear genes (e.g., Y-chromosome markers or β-globin genes) or very old human DNA (e.g., from the 60,000 year-old Lake Mungo remains) are not supported by any authentication criteria.

Informationen aus einigen besonders relevanten Untersuchungen über konservierte DNA in archäologischen Skelettfunden wurden zusammengetragen, wobei insbesondere solche Berücksichtigung fanden, welche die wesentlichen Authentifizierungskriterien erfüllen, wie sie von einer Reihe von Wissenschaftlern vorgeschlagen worden sind: unabhängige Replikation, Klonierung der PCR-Produkte, phylogenetische Aussagekraft, Aminosäureanalyse und quantitative PCR. Anhand einer Kompilation von Daten zur Klonierung von PCR Produkten konnte gezeigt werden, dass kein direkter Zusammenhang zwischen dem Alter der Probe und der Rate falsch inkorporierter Nukleotide besteht (mit maximalen Fehlern von bis zu sechs Nukleotiden pro tausend bp). Es stellt sich heraus, dass bislang mehr Anstrengungen unternommen wurden, um die Authentizität von nicht humaner konservierter DNA zu belegen, als im Falle von humaner DNA. Darüberhinaus wurde festgestellt, dass im Falle einiger als besonders sensationell bezeichneter Ergebnisse, wie der Gewinnung von DNA aus sehr altem Material (z. B. in Bernstein eingeschlossenen Insekten oder miozänen Magnolienblättern), von nukleären Genen (z. B. Y-chromosomaler Marker oder β-Globingenen), oder auch sehr alter humaner DNA (z. B. aus den 60 000 Jahre alten Funden vom Lake Mungo) von keinerlei Authentifizierungskriterien unterstützt werden.

Keywords: ancient DNA, authentication criteria, cloning, PCR
alte DNA, Authentifizierungskriterien, Klonierung, PCR

Introduction

Ancient DNA studies are defined as the retrieval of genetic material from past remains. They are related to different scientific fields, including palaeontology, human population genetics, molecular phylogeny and forensics. The emergence of this new field is associated with the technical improvements in biomolecular tools, especially the Polymerase Chain Reaction (PCR), in the last two decades. Its main achievements are probably the recovery of DNA from an extinct human species, the Neandertals (Krings et al. 1997, Ovchinnikov et al. 2000, Krings et al. 2000, Schmitz et al. 2002), and the recovery of the first functional extinct genome, the mitochondrial genome, from the giant ratite birds of New Zealand, the moas (Cooper et al. 2001). These studies and many others from the field have been conducted on bone, teeth or dried tissue held in various zoological, anthropological and natural history museums. Ancient DNA has thus provided a new use for many museum specimens.

However, the field is plagued by numerous technical problems, to the point that some of the most extraordinary claims, like the retrieval of DNA from a dinosaur or from amber-entombed insects, have been proven false or have been seriously criticised. The proposal of some authentication criteria (Cooper & Poinar 2000) addresses these problems. If all or most of these criteria were routinely followed, the ancient DNA field would be soundly established. Among the different criteria recommended, I have selected the most stringent, and compiled most of the ancient DNA studies that have been published with a view to testing which are the most solid and which criteria are the most widely followed.

Authentication criteria

The main five authentication criteria proposed are:
1- Replication of results in an independent laboratory.
2- Cloning of PCR products.
3- Phylogenetic sense.
4- Amino acid analysis.
5- Quantitative PCR.

Other and additional criteria are not considered in this review, but are assumed in all ancient DNA studies. Among these are: undertaking multiple extractions, inclusion of PCR and extraction blanks in molecular reactions, working in proper conditions (with dedicated ancient DNA laboratories that are physically isolated from the main laboratories, with sterile, positive air conditions and UV lights), external cleaning of samples and noncontamination of samples during the study.

The potential of the PCR implies a high risk of contamination, particularly from anterior PCRs. A standard amplification can create billions of DNA copies in aerosols called amplicons, which are perfectly suitable as template DNAs for posterior reactions. The proposal of methodological standards is not ludicrous. Failure to adhere usually means that it will be impossible to discount intra-laboratory contamination as a DNA source. Even worse, despite claims for scientific rigour, some journals continue to publish papers that simply do not fulfil any of these suggestions, thus exposing the entire field to discredit.

Independent replication. In my opinion, the replication of sequences in a second laboratory is the most stringent criterion since it eliminates the possibility of intra-laboratory contamination which is by far the most important source of contamination. The problem lies in the fact that it needs the collaboration of two different laboratories in a difficult field.

Cloning. Cloning PCR products into bacteria is a strategy designed to obtain consensus sequences from multiple amplifications and to overcome the problems posed by template degradation. Damage in the original DNA implies not only fragmentation but also random oxidation of bases because in a damaged nucleotide, the Taq DNA polymerase can insert a wrong nucleotide while copying the template. If this occurs in the very first cycles of the PCR, then the majority of the final products can have an incorrect base. The cloning of different PCR products not only helps to differentiate between original nucleotides and wrong bases, but also to unravel possible mixing of sequences (see Table 1). Of course, the frequency of misincorporation is correlated to the initial number of DNA templates. In well-preserved samples and in particular cases (Fig. 1), cloning will be relatively noninformative.

Phylogenetic sense. If a sequence from an extinct species is similar, but not identical to sequences from extant, related species, then the results are said to have phylogenetic sense. For instance, Greenwood et al. (1999) found an insertion in the 18S rDNA nuclear gene from a mammoth that was not described in other mammals. Further sequencing of the Asian and African elephants showed a similar insertion. In contrast, studies of insects in amber revealed that the sequences retrieved, presumably from ten or hundred million-year-old specimens, showed almost no difference between them and that of modern insects. Thus, those results had no phylogenetic sense and are false.

The application of this criterion to human sequences is less helpful. Assuming we are studying samples from a different continent, it may be expected that sequences retrieved would cluster among mtDNA lineages from that area and not from the researcher's homeland. But, phylogenetic criteria cannot be applied, for instance, to prove European sequences obtained by European researchers. Therefore, it can be considered a weak authentication criterion for work on human samples.

Amino acid analysis. The same taphonomic factors that destroy DNA also affect amino acids from bones. For one, racemisation, or the change from the amino acid L-form, constituent of all living animals, to its symmetrical D-form, is also fuelled by the same factors that degrade DNA, such as pH, temperature or chemical enzymatic reactions.

A second point to note is that different amino acids change at different rates. Aspartic acid, for instance, is an amino acid that changes very fast, making it ideal for tracking recent taphonomic processes. Thus, samples with low L-aspartic content will be suggestive of bad DNA preservation (it has been suggested that samples with D/L asp values higher than 0.10 are not expected to contain endogenous DNA).

Quantification. This is a system that estimates the number of initial DNA molecules in a sample by using a quantitative PCR. It has been suggested that a very low number of DNA template copies will result in inconsistent results, even with the cloning of PCR products.

Material and Methods

The main ancient DNA papers available have been compiled, with the aim not to point out possible flawed papers, but rather to discover tendencies in the authentication criteria over the years. Only papers with published sequences have been reviewed. Forensic studies, like the identification of the Romanov family, have not been included. For nonhuman species, five criteria have

	16,223	16,278	16,298	16,303
Direct sequencing (Barcelona laboratory)	T	-	C	-
Independent replication (Oxford laboratory) Cloning results				
Clone1	T	-	-	-
Clone2	T	-	-	A
Clone3	T	-	-	-
Clone4	T	-	-	-
Clone5	T	-	-	-
Clone6	T	-	-	-
Clone7	T	-	-	-
Clone8	-	G	-	-
Clone9	-	G	-	-
Clone10	-	G	-	-
Clone11	-	G	-	-
Clone12	-	-	C	-
Clone13	-	-	C	-
Clone14	-	-	C	-
Clone15	-	-	C	-

Tab. 1: Results of PCR cloning in an ancient Central Asian sample. The direct sequence of this sample for the 16,209-16,356 fragment in Barcelona was (16,223 [T] and 16,298 [C]); therefore, the sample was tentatively attributed to a C haplogroup (Asian). However, the cloning of the PCR product in the Ancient Biomolecules Centre in Oxford (undertaken by M.T.P. Gilbert, directed by A. Cooper) showed that the sample was, in fact, a mixing of three different sequences: sequence a) 16,223 [T]; sequence b) 16,278 [G] and sequence c) 16,298 [C]. All these sequences are European or West Eurasian. Posterior amplification and sequencing of the +13,262 AluI site confirmed that the sample did not have any C haplogroup sequence.

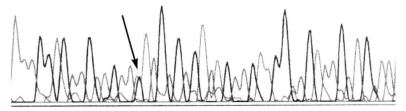

Fig. 1: Segment of a direct sequence from Central Asian ancient teeth (fragment L16,209-H16,401, reverse strand). The arrow marks a heteroplasmy at the 16,239 position. In this case, the cloning of the PCR product would not solve the uncertainty in this position, since it is expected that approximately half of the clones would show a T and the other half a C in this position (there is also a T in np16,223, with no heteroplasmy). The typing of this position was posteriorly clarified with an additional amplification and sequencing of the 16,209-H16,356 fragment, that showed again the T in np 16,223, but no change in 16,239 with respect to the reference sequence.

been considered: independent replication, cloning, phylogenetic sense, amino acid analysis and quantitative PCR. For human species, phylogenetic sense, as explained above, has not been considered. These criteria have been considered positively even when only partially followed (e.g., cloning of some products or replication of partial sequences).

Data from ancient DNA studies that have cloned PCR products have been analysed statistically to determine if there is a relationship between cloning results, error rates and other variables, such as the antiquity of the sample. To clarify this issue, an error rate has been estimated, counting the number of nucleotide substitutions per 1000 bp. Multiple substitutions (shared by more than one clone), are counted only once since they are considered as a single mutational event in the PCR process.

Results

The results for nonhuman studies are shown in Table 2 and the results for human studies in Table 3. They are categorised according to publication dates, starting with the earliest.

Different patterns can be observed. In nonhuman species, the phylogenetic sense is a criterion that strongly supports the authenticity of most of the results obtained. The only exceptions are the magnolia leaf and amber insects, which had sequences that contradicted the supposed ages of the DNA. Through the years, a consistent trend towards increasing the number of authentication criteria may be observed. Between 1984 and 1994, only 18.2% of the total possible authentication cells have been fulfilled, while in the following eight years (1994-2002), the percentage increased to 60%. Of all the cri-

Specimen	Retrieved DNA	Dating	Independent replication	Cloning PCR products	Phylogenetic sense	Amino acid analysis	Quantitative PCR	Reference
Quagga	cyt c mtDNA	140 BP			X			Higuchi et al. 1984
Marsupial wolf	cytb/12S mtDNA	>100 BP			X			Thomas et al. 1989
Magnolia leaf	rbcL chloroplastDNA	17 MYR						Golenberg et al. 1990
Saber Tooth Cat	12S mtDNA/FLA 1	14000 BP		X	X			Janczewski et al. 1992
Termite in amber	18S/16S mtDNA	25-30 MYR						DeSalle et al. 1992
Equus hemionus	16S mtDNA	25000 BP			X			Höss and Pääbo 1993
Weevil in amber	18S/ITS	120-135 MYR						Cano et al. 1993
Mammoth	16S mtDNA	9700->50000 BP			X			Höss et al. 1994
Mammoth	cytb mtDNA	>47000 BP			X			Hagelberg et al. 1994
Cave Bear	HVR1 mtDNA	40000 BP		X	X			Hänni et al. 1994
Blue antelope	cyt b mtDNA	200 BP			X			Robinson et al. 1996
Ground Sloth	12S/16S mtDNA	>10000 BP	X		X		X	Höss et al. 1996
Steller´s Sea cow	cytb mtDNA	?			X			Ozawa et al. 1997
Feldhofer 1	HVR1/2 mtDNA	30000-100000 BP	X	X	X	X	X	Krings et al. 1997/1999
Mammoth/Sloth/Cave Bear	Nuclear genes	13000-33000 BP		X	X			Greenwood et al. 1999
Mezmaiskaya	HVR1 mtDNA	29000 BP	X	X	X			Ovchinnikov et al. 2000
Myotragus balearicus	cytb mtDNA	7500 BP	X	X	X	X		Lalueza-Fox et al. 2000
Vindija	HVR1/2 mtDNA	42000 BP		X	X	X		Krings et al. 2000
Megaladapis edwardsi	cytb mtDNA	Pleist./Holocene			X			Montagnon et al. 2001
Moa (Dinornis/Emeus)	mtDNA genome	1200-600 BP	X	X	X	X	X	Cooper et al. 2001
Dodo	12S cytb mtDNA	>400 BP	X	X	X			Saphiro et al. 2002
Myotragus balearicus	cyt b mtDNA	10000 BP	X	X	X			Lalueza-Fox et al. 2002b

Tab. 2: Authentication criteria used in some of the most relevant papers on nonhuman ancient DNA in recent years.

Specimen	Retrieved DNA	Dating	Independent Replication	Cloning PCR products	Amino acid analysis	Quantitative PCR	Reference
Europeans	Y-chromosome	200-300 BP					Hummel and Herrmann 1991
Pacific Islanders	HVR1 mtDNA	2,700-1,600 BP					Hagelberg and Clegg 1993
Tyrolean Ice Man	HVR1 mtDNA	5,200 BP	X	X		X	Handt et al. 1994
Europeans	Beta-globin gene	12,000 BP					Béraud-Colomb et al. 1995
Europeans	X-Y-chromosome	200-8,000 BP					Faerman et al. 1995
Amerindians	HVR1 mtDNA	>100 BP				X	Lalueza-Fox 1996
Amerindians	HVR1 mtDNA	600-1,000 BP		X		X	Handt et al. 1996
Amerindians	HVR1 mtDNA	1,530-150 BP					Monsalve et al. 1996
Amerindians	HVR1 mtDNA	1,000-650 BP					Parr et al. 1996
Europeans	X-Y-chromosome	2,000 BP					Faerman et al. 1997
Amerindians	HVR1 mtDNA	700 BP					Stone and Stoneking 1998
Xinese	HVR1 mtDNA	2,000 BP					Oota et al. 1999
Italian Meso/Neolithic	HVR1 mtDNA	6,000-14,000 BP	X	X	X	X	Di Benedetto et al. 2000
Xinese	HVR1 mtDNA	2,500 BP					Wang et al. 2000
Asian Paleo/Neolithic	HVR1 mtDNA	6,500-25,000 BP					Oota et al. 2001
Mungo Lake	HVR1 mtDNA	60,000 BP		X			Adcock et al. 2001
Amerindians	HVR1 mtDNA	1,400-300 BP					Lalueza-Fox et al. 2001
Upper Paleolithic	HVR1 mtDNA	25,000 BP	X	X	X	X	Caramelli et al. 2003
Amerindians	HVR1 mtDNA	1000 BP	X	X			Lalueza-Fox et al. 2002a

Tab. 3: Authentication criteria used in some of the most relevant papers on human ancient DNA in recent years.

teria, quantitative PCR is the least used. There is also a surprising lack of criteria in the most sensational of the claims (the magnolia leaf and the amber insects), despite having been published in scientific journals with the highest profile in the field.

In the human species, the number of criteria used has been minimal. Many studies do not follow any criteria, and only three studies in more than ten years fulfil three or more criteria. The most extraordinary claims, like the retrieval of nuclear DNA, including sex chromosomes and ß-globin genes, the analysis of very old specimens, including the 60,000-year-old Lake Mungo remains, do not follow any of the four proposed authentication criteria. In these cases, there is no clear effort towards improving the reliability of these studies.

It was also discovered that independent replication and amino acid analysis are the less frequently used criteria.

The results of cloning are shown in Table 4. It can be seen here that a general relationship between age and DNA damage does not exist. Modern samples show error rates of around 2 and 3.5 per 1000bp, but since these substitutions are mostly singletons (i.e., not shared by more than one clone) they are therefore mainly attributable to cloning artefacts. Some ancient samples show values similar to these figures, indicating good DNA preservation, even after thousands of years. Multiple substitutions seem to be correlated with high error values and therefore, to high degradation levels. One interesting observation is that there is no relation between age and error rate or age and excess of transitions, contrary to some suggestions (Figures 2a and 2b). From the cloning results, it can be concluded that age is not the main factor in DNA damage, but that local taphonomic proccesses, like temperature and pH, are probably more important in the degradation or preservation of genetic material.

Discussion

Ancient DNA is a reliable technique if properly used. However, it must be kept in mind that it is a destructive approach, and the chances of success can certainly be small. For instance, there are four Neandertal sequences already published, with information from up to 30 specimens worked on by several laboratories. This represents around a ten percent success rate in samples that are around 30,000 to 40,000 years old. There are around 300 Neandertal specimens described, meaning that in the best of scenarios, we can reasonably expect to have around 30 Neandertal sequences after some years (and with some luck). We can also expect that more recent samples, like the Upper Paleolithic Cromagnons, could have a significantly higher success rate, but how much higher remains a question. In any case, a museum must be aware of these limitations and assume responsibility for the specimens' destruction. This adds to the fact that previous amino acid analysis also involves the destruction of some samples.

To complicate the "destructive" technology required by ancient DNA studies, teeth, which are the best DNA receptacles since the hard enamel matrix protects the pulp cavity against external destructive agents, are, at the same time providers of interesting anthropological information, including evidence related to diet, pathology, development, sexual dimorphism and phylogeny. Their destruction during DNA retrieval therefore often poses a difficult research choice, even when permission is obtained. To partially overcome these problems, some institutions have a policy of obtaining casts of the specimens to be subjected to ancient DNA research. In a project undertaken at *deCODE* Genetics (Reykjavik, Iceland) on the ancient remains of the first Icelandic settlers, the National Museum of Iceland agreed to collaborate on the project only after casts of all teeth to be analysed were made.

Another important issue involves physical conditions. Sampling for ancient DNA analysis needs the active collaboration of the institutions where the materials are preserved. In an ideal scenario, the curators would send the samples to the laboratories where the work is to be carried out, thus preventing anyone working in the molecular laboratory from coming into contact with the remains, thereby eliminating a potential source of contamination. Of course, this means that the curators should have the ability to take samples safely and in turn, having equipment like drilling machines, sterile gear, lab coats, face masks and gloves for handling the specimens (materials which are widely accessible and relatively cheap).

On the other hand, institutions willing to collaborate with ancient DNA researchers should take into consideration not only the problems involved, but also the authentication criteria proposed. One way would be that they provide samples only upon provision of a short list of the relevant papers from the requesting researchers and then make the decision as to whether to provide samples based on a written proposal. It must be emphasised that ancient DNA studies need not only collaboration between museums and laboratories, but also among molecular centres.

Despite all its technical problems, ancient DNA, if done properly, can provide new, unexpected and eventually crucial data for resolving some of the hottest anthropological and evolutionary controversies. At the same time, it offers museums the opportunity to be, not dead places where materials are held, but dynamic centres of new research.

Sample	Date (approx)	Region (mtDNA)	N (clones)	Total (bp)	TS	TV	single	Multiple	Total	Error (/1000bp)
Tataupa	Modern	protein genes	24	7308	12	3	15	1	15	2,05
Kiwi	Modern	protein genes	11	3672	12	0	11	1	12	3,27
Cassowary	Modern	protein genes	16	4513	10	3	13	0	13	2,88
Taino	-500	CRI	11	1947	3	2	5	0	5	2,57
Amerindian (vc15a)	-600	CRI	34	4005	5	5	10	0	10	2,50
Dinornis	-600	protein genes	70	33620	44	10	48	6	54	1,61
Emeus	-1200	protein genes	67	26929	52	5	56	1	57	2,12
Mullerornis	-3350	protein genes	9	1467	5	1	6	0	6	4,09
Ice Man	-5000	CRI	54	13847	22	10	27	5	32	2,31
Myotragus (1)	-6000	CYTB	6	330	1	1	1	1	2	6,06
Borgo Novo	-6050	CRI	127	10973	33	4	27	9	36	3,28
Mezzocorona	-6400	CRI	111	9568	44	4	39	9	48	5,02
Myotragus (2)	-10000	CYTB	24	3262	14	5	15	4	19	5,82
Villabruna	-14000	CRI	132	11313	33	13	30	16	46	4,07
Mezmaiskaya	-29000	CRI	7	1720	5	0	5	0	5	2,91
Feldhofer	-30000	CRI	219	15401	53	7	40	20	60	3,90
Feldhofer	-30000	CRII	92	10209	24	5	27	2	29	2,84
Paglicci	-25000	CRI	19	3923	5	1	5	1	6	1,53

Tab. 4: Results of PCR cloning in some of the most relevant ancient DNA studies, quoting the number of clones (N [clones]), total nucleotides sequenced (Total [bp]), transitions observed (TS), transversions (TV), single substitutions observed in the clones (single), multiple substitutions observed in more than one clone (multiple), total substitutions observed in the clones (Total), and error misincorporation rate per 1000 nucleotides (Error [/1000 bp]).

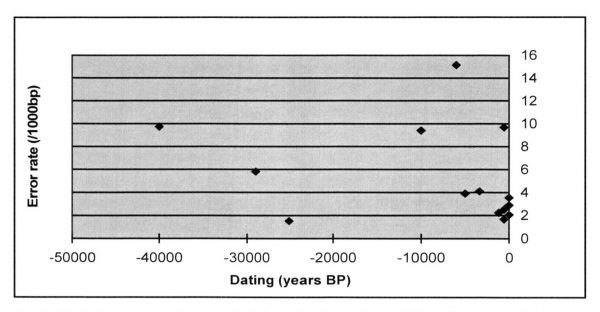

Fig. 2a: Relation between age and error rate in cloning results of several ancient DNA studies (number of misincorporations per 1000 bp). The error rates of the oldest samples are within the range found in some modern DNA samples cloned.

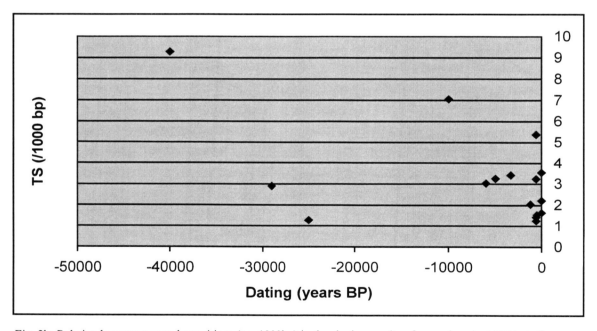

Fig. 2b: Relation between age and transitions (per 1000bp) in the cloning results of several ancient DNA studies.

Future research on ancient DNA is oriented towards the nuclear genome. With the knowledge generated by the new genomic era, more and more information about the genes involved in fundamental aspects such as growth, brain development or punctual phenotypic traits will be available. However, with the present technical approaches, only a few, extremely well-preserved samples will be amenable to ancient nuclear DNA retrieval. For instance, up to now, all attempts to obtain a single copy nuclear gene from the *Myotragus balearicus* samples that yielded the mitochondrial DNA (Lalueza-Fox et al. 2002a) have failed, but access to the nucleus will open a new era in aDNA. It will become possible, for example, to generate transgenic animals by inserting an ancient gene into an extant, related species, to see the effect on the phenotype of the ancient gene. The retrieval of the MC1R gene (melanocortin receptor 1, involved in mammal pigmentation) from the extinct *Myotragus*, for instance, could provide some clue to the pigmentation of this bovid through a transgenic ancient-extant species experiment. Such experiments, however, would raise new ethical issues and should be socially discussed.

Bibliography

Adcock G.J., Dennis E.S., Easteal S., Huttley G.A., Jermiin L.S., Peacock W.J. & Thorne A., 2001.
Mitochondria DNA sequences in ancient Australians: implications for modern human origins. *Proceedings of the National Academy of Sciences USA* **98** (2): 537-542.

Béraud-Colomb E., Roubin R., Martin J., Maroc N., Gardeisen A., Trabuchet G., Goossens M., 1995.
Human ß-globin gene polymorphisms characterized in DNA extracted from ancient bones 12,000 years old. *American Journal of Human Genetics* **57**: 1267-1274.

Cano R.J., Poinar H.N., Pieniazek N.J., Acra A. & Poinar G.O., 1993.
Amplification and sequencing of DNA from a 120-135-million-year-old weevil. *Nature* **363**: 536-538.

Caramelli D., Lalueza-Fox C., Vernesi C., Lari M., Casoli A., Mallegni F., Chiarelli B., Dupanloup I., Bertranpetit J., Barbujani G. & Bertorelle G., 2003.
Evidence for a genetic discontinuity between Neandertals and Cromagnons. *Proceedings of the National Academy of Sciences USA* (in press).

Cooper A. & Poinar H.N., 2000.
Ancient DNA: Do it right or not at all. *Science* **289**: 1139.

Cooper A., Lalueza-Fox C., Anderson S., Rambaut A., Austin J. & Ward R., 2001.
Complete mitochondrial genome sequences of two extinct moas clarify ratite evolution. *Nature* **409**: 704-707.

DeSalle R., Gatesy J., Wheeler W. & Grimaldi D., 1992.
DNA Sequences from a Fossil Termite in Oligo-Miocene Amber and Their Phylogenetic Implications. *Science* **257**: 1933-1936.

Di Benedetto G., Nasidze I.S., Stenico M., Nigro L., Krings M., Lanzinger M., Vigilant L., Stoneking M., Pääbo S. & Barbujani G., 2000.
Mitochondrial DNA sequences in prehistoric human remains from the Alps. *European Journal of Human Genetics* **8**: 669-677.

Faerman M., Filon D., Kahila G., Greenblatt C.L., Smith P., Oppenheim A., 1995
Sex identification of archaeological human remains based on amplification of the X and Y amelogenin alleles. *Gene* **167**(1-2):327-332.

Faerman M., Kahila G., Smith P., Greenblatt C., Stager L., Filon D., Oppenheim A., 1997.
DNA analysis reveals the sex of infanticide victims. *Nature* **385**(6613): 212-213.

Golenberg E.M., Giannasi D.E., Clegg M.T., Smiley C.J., Durbin M., Henderson D. & Zurawski G., 1990.
Chloroplast DNA sequence from a Miocene Magnolia species. *Nature* **344**: 656-658.

Greenwood A., Capelli C., Possnert G. & Pääbo S., 1999.
Nuclear DNA sequences from Late Pleistocene megafauna. *Molecular Biology and Evolution* **16**(11): 1466-1473.

Hagelberg E. & Clegg J.B., 1993.
Genetic polymorphisms in prehistoric Pacific islanders determined by analysis of ancient bone DNA. *Proceedings of the Royal Society of London* **252**: 163-170.

Hagelberg E., Thomas M.G., Cook C.E., Sher A.V., Baryshnikov G.F. & Lister A.M., 1994.
DNA from mammoth bones. *Nature* **370**: 333-334.

Handt O., Richards M., Trommsdorff M., Kilger C., Simanainen J., Georgiev O., Bauer K., Stone A., Hedges R., Schaffner W., Utermann G., Sykes B. & Pääbo S., 1994.
Molecular Genetic Analysis of the Tyrolean Ice Man. *Science* **264**: 1775-1778.

Handt O., Krings M., Ward R.H. & Pääbo S., 1996.
The retrieval of ancient human DNA sequences. *American Journal of Human Genetics* **59**: 368-376.

Hänni C., Laudet V., Stehelin D. & Taberlet P., 1994.
Tracking the origins of the cave bear (Ursus spelaeus) by mitochondrial DNA sequencing. *Proceedings of the National Academy of Sciences USA* **91**: 12336-12340.

Higuchi R.G., Bowman B., Freiberger M., Ryder O.A. & Wilson A.C., 1984.
DNA sequence from the quagga, an extinct member of the horse family. *Nature* **312**: 282-284.

Höss M. & Pääbo S., 1993.
DNA extraction from Pleistocene bones and analysis by a silica-based purification method. *Nucleic Acid Research* **21**: 3913-3914.

Höss M., Pääbo S. & Vereshchagin N.K., 1994.
Reproducible mammoth DNA sequences. *Nature* **370**: 333.

Höss M., Dilling A., Currant A. & Pääbo S., 1996.
Molecular phylogeny of the extinct ground sloth Mylodon darwinii. *Proceedings of the National Academy of Sciences USA* **93**: 181-185.

Hummel S. & Herrmann B., 1991.
Y-Chromosome-Specific DNA amplified in ancient human bone. *Naturwissenschaften* **78**: 266-267.

Janczewski D.N., Yuhki N., Gilbert D.A., Jefferson G.T. & O'Brien S.J., 1992.
Molecular phylogenetic inference from saber-toothed cat fossils of Rancho La Brea. *Proceeding National Academy of Sciences USA.* **89**: 9769-9773.

Krings M., Stone A., Schmitz R.W., Krainitzki H., Stoneking M. & Pääbo S., 1997.
Neandertal DNA sequences and the origin of modern humans. *Cell* **90**: 19-30.

Krings M., Geisert H., Schmitz R.W., Krainitzki H. & Pääbo S., 1999.
DNA sequence of the mitochondrial hypervariable region II from the neanderthal type specimen. *Proceedings of the National Academy of Sciences USA* **96**:5581-5585.

Krings M., Capelli C., Tschentscher F., Geisert H., Meyer S., von Haeseler A., Grossschmidt K., Possnert G., Paunovic M. & Pääbo S. 2000.
A view of Neandertal genetic diversity. *Nature Genetics* **26**(2): 144-6

Lalueza-Fox C., 1996.
Ancient mtDNA analysis in extinct aborigines from Tierra del Fuego/Patagonia. *Ancient Biomolecules* **1**: 43-54.

Lalueza-Fox C., Bertranpetit J., Alcover J.A., Shailer N. & Hagelberg E., 2000.
Mitochondrial DNA from *Myotragus balearicus*, an extinct Bovid from the Balearic Islands. *Journal of Experimental Zoology (Molecular and Developmental Evolution)* **288**: 56-62.

Lalueza-Fox C., Luna-Calderón F., Calafell F., Morera B. & Bertranpetit J., 2001.
MtDNA from extinct Tainos and the peopling of the Caribbean. *Annals of Human Genetics* **65**: 137-151.

Lalueza-Fox C., Gilbert M.T.P., Martinez-Fuentes A.J., Calafell F. & Bertranpetit J., 2002a.
MtDNA from pre-Columbian Ciboneys from Cuba and the colonization of the Caribbean. (submitted).

Lalueza-Fox C., Saphiro B., Bover P., Alcover J.A. & Bertranpetit, J., 2002b.
Molecular phylogeny and evolution of the extinct bovid Myotragus balearicus. *Molecular Phylogenetics and Evolution* (in press).

Monsalve V., Cardenas F., Guhl F., Delaney A.D. & Devine D.V., 1996.
Phylogenetic analysis of mtDNA lineages in South American mummies. *Annals of Human Genetics* **60** (4): 293-303.

Montagnon D., Ravaoarimanana B., Rakotosamimanana B. & Rumpler Y., 2001.
Ancient DNA from Megaladapis edwardsi (Malagasy subfossil): preliminary results using partial Cytochrome b sequence. *Folia Primatologica* **72**: 30-32.

Oota H., Saitou N., Matsushita T. & Ueda S., 1999.
Molecular genetic analysis of remains of a 2,000-year-old human population in China –and its relevance for the origin of the modern Japanese population. *American Journal of Human Genetics* **64**: 250-258.

Oota H., Kurosaki K., Pookajorn S., Ishida T. & Ueda S., 2001.
Genetic Study of the Paleolithic and Neolithic Southeast Asians. *Human Biology* **73** (2): 225-231.

Ovchinnikov I.V., Götherstrom A., Romanova G.P., Kharitonov V.M., Lidén K., & Goodwin W., 2000.
Molecular analysis of Neanderthal DNA from the Northern Caucasus. *Nature* **404**: 490-493.

Ozawa T., Hayashi S. & Mikhelson V. M., 1997.
Phylogenetic position of Mammoth and Steller's sea cow within Tethytheria demonstrated by mitochondrial DNA sequences. *Journal of Molecular Evolution* **44**:406-413.

Parr R.L., Carlyle S.W. & O'rourke D.H., 1996.
Ancient DNA analysis of Fremont Amerindians of the Great Salt Lake Wetlands. *American Journal of Physical Anthropology* **99**: 507-518.

Robinson T.J. & Bastos A.D., 1996.
Mitochondrial DNA sequence relationships of the extinct Blue Antelope *Hippotragus leucophaeus*. *Naturwissenschaften* **83**: 178-182.

Saphiro B., Sibthorpe D., Rambaut A., Austin J., Wragg G.M., Bininda-Emonds O.R.P., Lee P.L.M. & Cooper A., 2002.
Flight of the Dodo. *Science* **295**: 1683.

Schmitz R. W., Serre D., Bonani G., Feine S., Hillgruber F., Krainitzki H., Pääbo S. & Smith F. H., 2002.
The Neandertal type site revisited: Interdisciplinary investigations of skeletal remains from the Neander Valley, Germany. *Proceedings of the National Academy of Sciences USA* **99**: 13342-13347.

Stone A.C. & Stoneking M., 1998.
MtDNA analysis of a prehistoric Oneota population: implications for the peopling of the New World. *American Journal of Human Genetics* **62**: 1153-1170.

Thomas R.H., Schaffner W., Wilson A.C. & Pääbo S., 1989.
DNA phylogeny of the extinct marsupial wolf. *Nature* **340**: 465-467.

Wang L., Oota H., Saitou N., Jin F., Matsushita T. & Ueda S., 2000.
Genetic structure of a 2500-year-old human population in China and its spatiotemporal changes. *Molecular Biology and Evolution* **17** (9): 1396-1400.

Part V:

The Socio-Cultural Aspect and Modern Implications

Bioarchaeological Collections and the cultural heritage

Helmut Bender, Archäologie der Römischen Provinzen, Universität Passau

Abstract / Zusammenfassung

Explanations about the increase in size of cattle during the Roman period in the northwestern provinces of the Imperium Romanum as well as the origin of the Bajuvarii have been and still remain controversial in archaeological science. The hope of my archaeological guild is that with the aid of modern natural-scientific methods these questions will be resolved in the near future.

Erklärungen zur Grössenzunahme von Rindern während der römischen Zeit in den nordwestlichen Provinzen des Römischen Reiches und Untersuchungen zur Herkunft der Bajuwaren werden in der archäologischen Wissenschaft sehr kontrovers diskutiert. Es ist die Hoffnung der Fachwissenschaft, daß mit Hilfe der modernen Naturwissenschaften diese Fragen in naher Zukunft gelöst werden können.

Keywords: Roman Empire, Early Middle Ages, cattle size, Bajuvarii, significance of bioarchaeological collections
Römisches Reich, Frühmittelalter, Größe von Hausrindern, Bajuwaren, Bedeutung bioarchäologischer Sammlungen

If, in the following attempt to clarify certain problems, I say something from an archaeological perspective about the descent of the Bajuvarii, ancestors of contemporary Bavarians, and I contribute observations about the size of Roman cattle, this juxtaposition of topics is purely accidental, in no way deliberate, and results only from an interest in the history of prior research.

I would like to restrict my observations to both these themes and to remain within a chronological period that is familiar to me, namely, the Roman Empire and the Early Middle Ages (Czysz et al. 1995), from the first to the seventh century A.D. I shall begin with animals and in the second part discuss the Bajuvarii in order to conclude with two notices: first, calling attention to the great significance of the natural-science collections of the State of Bavaria, and second, suggesting future tasks and potentials for research.

The phenomenon of an increase in size of domestic animals – in particular cattle, horses and domestic fowl – during the Roman period, in the northwestern provinces from Britain to Hungary, has become widely known among specialists and need not be discussed again today (Fig. 1). In his fine book "Römische Tierhaltung und Tierzucht" (Peters 1998), Joris Peters has thoroughly discussed this subject and attempted to provide explanations for the phenomenon (Breuer et al. 1999, 2001). Among unitary or multiple factors offered to explain an increase in cattle-size is the view that cattle were imported deliberately to improve the stock through cross-breeding. Another view points to improvements in animal-husbandry, as well as supple-

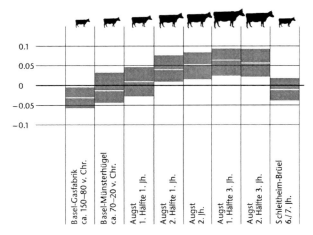

Fig.1: Increase in size of cattle-bones from excavations in northern Switzerland; starting in Pre-Roman times: Basel-Gasfabrik 150 – 80 BC to Early Middles Ages: Schleitheim-Brüel 6th/7th century AD (L. Zürcher in: G. Breuer, Augusta Raurica 2002/1, p. 2).

ments of more energy-rich nutrition. Another explanation – more favorable climatic conditions – appears in many, somewhat more popular articles. But how did things really work? Which factors were really decisive? Was it one or several of the ones already mentioned? The whole question is even more fundamentally complex. One finds in J. Peters's book, for example, that alongside the larger Roman cattle, smaller-sized, presumably native breeds still co-existed, perhaps because they were superior milk-producers (Peters 1998, 278; Pucher & Schmitzberger 2001). These questions are of interest not only to the natural-scientist but also to the archaeologist and the historian. Why is it that after the end of the Roman period, roughly the fifth century onward, the sizes of cattle, pigs, and sheep seem to decrease, but not that of horses (Breuer et al. 2001)? Until just a few years ago, one was satisfied simply to record these phenomena and to suggest explanations that were partly subjective, partly factually-based. Excavations provided the raw material. As you are well aware, during investigations of Roman settlements such material accumulates in large quantities (Becker & Benecke 2001). After one cleans the finds, then identifies the animal bones, establishes their sizes, makes statistical counts, generates projections as to the total animal population, one can then perhaps reconstruct the fauna during a particular time-period. Frequently, after these procedures – I have personally experienced this – one must then re-bury or even discard the evaluated animal bones because the storage facilities of museums and institutes seem to be already filled (Boessneck 1990, 138). Rarely are animal-artifacts considered to be worthy of conservation.

Lack of storage-space has been the reason offered to justify the disposal of animal bones under professional supervision. But is that a valid justification? One has always proceeded otherwise with the usual archaeological artifacts. Nobody would come up with the idea that excavated evidence--artifacts like ceramics, glass, metal, and stone-- after being catalogued, identified, drawn, and so on, should then be selectively thrown away! Animal bones, too, are irreplaceable documents (and the same, of course, is true for the archaeobotanical samples: Kreuz 2002, 477); they should be treated like any other artifact, whether written or material (for example Ginella & Heigold-Stadelmann 1999). How gladly would we have certain find-complexes with animal bones available so that, with the help of completely new natural-scientific methods, we could address the questions raised earlier! These methods have already been mentioned during the course of this colloquium: Ancient DNA-analysis (Pusch et al. 2001) or, as it was recently described in a paper on human bones, strontium isotope ratios (Teschler-Nicola & Prohaska et al. 2001; Schweissing 2002).

Now a few thoughts on the Bajuvarii, which lead us to anthropology, a field in which I admit to feeling rather less competent. For at least four hundred years, roughly since the Renaissance, scholarship has wrestled with this problem (Fischer 1988, 1995; Geisler 1998). It is a problem of great significance because within it seems to lie the origin of self-esteem and ethnic pride among contemporary Bavarians. Three main theories have been advanced concerning the Bajuvarii: that they were autochthonous; or, that they were a mixture of various ethnic groups; or, that they were immigrants from some other place of origin. Precisely on the basis of archaeological finds, a consensus has been reached in the past decade; namely, that the Bajuvarii arose on Bavarian soil only in the late fifth century, the result of an intermingling of very different ethnic groups: Romanized, autochthonous natives mixed with Alamanni, Thuringians, Ostrogoths, and Lombards. In the book "Römer und Bajuwaren an der Donau" [Romans and Bajuvarii on the river Danube] Thomas Fischer (Fischer 1988, 47-48) has presented the following conclusions (Translation of the German Text): "The presence of Germanic people from Bohemia is shown in Friedenhain-Přešťóvice, in the Late-Roman fort cemeteries, and in the grave-row cemetery [Reihengräberfeld] of Straubing-Bajuwarenstraße. This group thus constitutes the link, missing until now, between Late Antiquity and the Early Middle Ages, located precisely in a region considered to be the heartland of the Bavarian stem-duchy in the area of its capital Regensburg. So it would not be mistaken to see, in the group Friedenhain-Přešťóvice, the "Baiovarii," or "The Men from Bohemia." One should not, however, consider them original Bavarians as such, but rather one of many groups out of which the Bajuvarii people was formed."

Here in Germany, anthropology seems to play a less helpful role in the postwar period following the unfortunate years of the Third Reich: to differentiate, for example, a "Roman" skeleton from an already "Romanized" Alamannic or Lombard skeleton can be, in my opinion, a very subjective process. To my knowledge, blind-testing criteria, without archaeological-historical preconceptions, were not established beforehand. As a result, the classical case of "three settlement groups" in the cemetery of Neuburg on the Danube has since become obsolete (Ziegelmayer 1979; Schröter 1990). The theory became untenable because both the preceding archaeological investigations of the find materials, and their arrangement into three horizontally-graphed settlement groups were guided by the archaeologist's preconception. The immigration of a tribal element from Bohemia to the Danube region – a group we designate from its eponymous find-sites as the "Friedenhain-Přešťóvice Group" (Fig. 2) – appears to be the most-clearly recognizable through its archaeological remains, for its ceramics, discovered in the two important Late-Antique cemeteries, Azlburg I and II, near Straubing (Fig. 3), have close parallels in southern Bohemia. Important (ceramic) forms came finally from the cemetery of Straubing-Bajuwarenstraße (Geisler 1998). This area, occupied for one or two generations at the end of the fifth century, seemed to record an ethnogenesis involving Romanized indigenous folk, Alamanni, Thuringians, Ostrogoths, Lombards, and Friedenhain-Přešťóvice people. In the cemeteries Azlburg I and II (Prammer 1989, 90-98), according to the now widely accepted descent-theory of the Bajuvarii, could be found the tribal element from Bohemia that contributed to the ethnogenesis of the Bajuvarii.

A beautiful theory, until anthropology renewed its investigation of the skeletons and began an analysis of

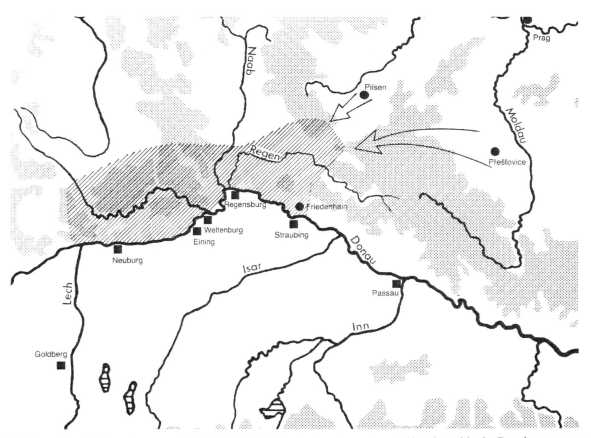

Fig.2: Sites with of pottery from so called Friedenhain-Přeštovice-type in southern Bohemia and in the Danube area around Ratisbon and Passau (Fischer 1988, p. 34 Fig. 28).

strontium isotope ratios! I thank Mike Schweissing (Schweissing 2002), Gisela Grupe and Günther Moosbauer (pers. communication), archaeological specialist of the materials, for making the first results accessible to me. According to the analysis, the Azlburg-people had not immigrated at all; they must be considered largely as autochthonous at a great percentage. On the other hand, the group living in Friedenhain on the northern side of the Danube (Fig. 3) should be recognized as exogenous. The people living in Straubing on the southern side, in their external appearance and in their material goods partially fit the new trend: the Romans looked like Germanic people! Fortunately, the bones of these human beings were preserved, as was formerly the usual practice, so that now, through application of newly available natural-scientific methods, older questions can be brought closer to a more objective solution.

Through my choice of examples, namely Roman cattle and Bajuvarii, I wished to make several points. One is that the human and animal bones preserved in the Staatssammlung für Anthropologie (Glowatzki 1977; Ziegelmayer 2000) und Palaeoanatomie (Boessneck 1990) represent a bioarchaeological potential whose value should not be underestimated. Through newer natural-scientific research methods, it will be possible

Fig. 3: Area around Straubing. 1 Friedenhain, 2 St. Peter, late Roman fortlet, 4 Late Roman cemetery Azlburg I, 5 Late Roman Cemetery Azlburg II, 6 Cemetery of the Early Middle Ages Straubing-Bajuwarenstrasse (after H. Geisler catalogue "Die Bajuwaren. Ausstellung Mattsee/Rosenheim 1988", p. 63 Fig. 30 [Fischer & Geisler 1988]).

in the near future to answer questions about the descent of the Bajuvarii and the increasing size of domesticated animals during the Roman Empire. Since, in this context, the same methods can be applied to the investigation of human and animal bones, both ancillary disciplines, anthropology and palaeoanatomy, can work purposefully together on research projects. No one can presently imagine what new techniques and new analytical methods will arise in the future to stimulate even broader questions and their answers. It would be greatly advantageous if both research disciplines, linked together in the second millennium, could share dwindling financial resources for innovative research projects and for joint project applications.

As an archaeologist I find this prospect enormously exciting and glance a bit enviously at these new starting points in natural-scientific archaeology. Of course, I mean that in a wholly positive sense and I do think that even traditional archaeological and historical science will become involved. But herein lies another problem. Although archaeologists could until now take it for granted that the human and animal bones from their excavations would be properly treated and, as is usual among natural scientists, that the results would appear promptly in catalogue- and analytical manuscripts, these conditions are no longer possible. There is insufficient personnel to carry out these services to the usual extent. Further, suitable room is unavailable to display and to process larger find-complexes. Will bones, once again, have to be thrown away because storage space is unavailable? In my presentation, I hope to have shown that to discard bones is unacceptable and that every bone-find should be treated like any other archaeological or historical source because none of us can know what new questions will be posed in the future to the bioarchaeological archive (to mention only two examples: Jacomet & Schibler 2001; Bender 1997). The legislator is obligated and requested to be concerned hereafter about suitable conservation and to make possible innovative research. "Do it right or not at all" (Cooper & Poinar 2000) is the motto of my contribution to the questions that have engaged us throughout this workshop.

Acknowledgements

This lecture, given at the symposium of the 100th anniversary of the anthropological collection, is delivered for publication unchanged; only the references and the bibliography are supplemented. For help and discussion I would like to thank especially G. Grupe, G. Moosbauer and J. Peters; translation of my German text by L. Okamura, Columbia (Missouri).

References

Becker C. & Benecke N., 2001.
Archaeozoology in Germany. Its Course of Development. *Archaeofauna* **10**: 163-182.

Bender H., 1997.
Agrargeschichte Deutschlands in der römischen Kaiserzeit innerhalb der Grenzen des Imperium Romanum. In: Lüning J. et al. (eds.). *Deutsche Agrargeschichte. Vor- und Frühgeschichte*: 263-374. Stuttgart: Ulmer.

Boessneck J., 1990.
Institut für Palaeoanatomie, Domestikationsforschung und Geschichte der Tiermedizin. In: von den Driesch A. (ed.). *200 Jahre tierärztliche Lehre und Forschung in München*: 131-139. Stuttgart / New York: Schattauer.

Breuer G. et al., 1999.
Grössenveränderungen des Hausrindes. Osteometrische Untersuchungen grosser Fundserien aus der Nordschweiz von der Spätlatènezeit bis ins Frühmittelalter am Beispiel von Basel, Augst (Augusta Raurica) und Schleitheim-Brüel. *Jahresberichte Augst und Kaiseraugst* **20**: 207-228.

Breuer G. et al., 2001.
Veränderung der Körpergrösse von Haustieren aus Fundstellen der Nordschweiz von der Spätlatènezeit bis ins Frühmittelalter. *Jahresberichte aus Augst und Kaiseraugst* **22**: 161-178.

Breuer G., 2002.
Grosse Haustiere dank den Römern. *Augusta Raurica* 2002/1: 2-4.

Cooper A. & Poinar H.N., 2000.
Ancient DNA: Do It Right or Not at All. *Science* **289**: 1139.

Czysz W. et al., 1995.
Die Römer in Bayern. Stuttgart: Theiss.

Fischer Th., 1988.
Römer und Bajuwaren an der Donau. Bilder zur Frühgeschichte Ostbayerns. Regensburg: Pustet.

Fischer Th., 1995.
Von den Römern zu den Bajuwaren. In: Czysz W. et al. (eds.). *Die Römer in Bayern*: 405-411. Stuttgart: Theiss.

Fischer Th. & Geisler H., 1988.
Herkunft und Stammesbildung der Baiern aus archäologischer Sicht. In: Dannheimer H. & Dopsch H. (eds.). *Die Bajuwaren. Von Severin bis Tassilo 488-788*: 61-68. München, Salzburg.

Geisler H., 1998.
Das frühbairische Gräberfeld Straubing-Bajuwarenstraße I. Katalog der archäologischen Befunde und Funde. Internationale Archäologie 30. Rahden/Westf.: Leidorf.

Ginella F. & Heigold-Stadelmann A., 1999.
Ein Beitrag zur Nahrungswirtschaft und zur Verpflegung römischer Truppen im Legionslager Vindonissa/Windisch (CH). Archäozoologische Auswertung der Tierknochen aus der Grabung Vindonissa-Feuerwehrmagazin 1976. *Gesellschaft Pro Vindonissa. Jahresbericht* 1999: 3-28.

Glowatzki G., 1977.
Zur Geschichte der Anthropologischen Staatssammlung München. In: Schröter P. (ed.). *75 Jahre Anthropologische Staatssammlung München 1902-1977*: 15-18. München.

Jacomet St. & Schibler J., 2001.
Umwelt und Ernährung. In: Furger A. et al. (eds.): *Die Schweiz zur Zeit der Römer. Multikulturelles Kräftespiel vom 1. bis 5. Jahrhundert*. Archäologie und Kulturgeschichte der Schweiz 3: 107-130. Zürich: Neue Zürcher Zeitung.

Kreuz A., 2002.
Zufall, Müll und Mißgeschicke. Zu Herkunft und archäobotanischer Auswertung vor- und frühgeschichtlicher Pflanzenfunde. *Antike Welt* **33**: 475-479.

Moosbauer G., in preparation.
Straubing in der Spätantike. Romanen, Germanen an der Schwelle zum Frühen Mittelalter.

Peters J., 1998.
Römische Tierhaltung und Tierzucht. Eine Synthese aus archäozoologischer Untersuchung und schriftlich-bildlicher Überlieferung. Passauer Universitätsschriften zur Archäologie 5. Rahden/Westf.: Leidorf.

Prammer J., 1989.
Das römische Straubing. Ausgrabungen-Schatzfund – Gäubodenmuseum. Bayerische Museen 11. München / Zürich: Schnell & Steiner.

Pucher E. & Schmitzberger M., 2001.
Belege zum Fortbestand keltisch-norischer Rinder während der mittleren bis späten römischen Kaiserzeit vom Michlhallberg im Salzkammergut. In: Grabherr G. *Michlhallberg. Die Ausgrabungen in der römischen Siedlung 1997-1999 und die Untersuchungen an der zugehörigen Straßentrasse.* Schriftenreihe des Kammerhofmuseums Bad Aussee 22: 241-269. Bad Aussee.

Pusch C.M. et al., 2001.
Molekulare Paläobiologie. Ancient DNA und Authenzität. *Germania* **79**: 121-141.

Schröter P., 1990.
Exkurs: Zum spätrömischen Gräberfeld von Neuburg a.d. Donau. In: Fischer Th. Das Umland des römischen Regensburg. Münchner Beiträge zur Vor- und Frühgeschichte 42: 388-391. München: Beck.

Schweissing M., 2002.
Strontiunisotopenanalyse (87Sr/86Sr). Eine archäometrische Applikation zur Klärung anthropologischer und archäologischer Fragestellungen in Bezug auf Migration und Handel (unpubl. Ph. D. Fakultät für Biologie, Ludwig Maximilians Universität München).

Teschler-Nicola M., Prohaska Th. et al, 2001.
$^{87}Sr/^{86}Sr$-Isotopenverhältnis in (prä)historischen menschlichen Skelettresten. Indikator für individuellen Ortswechsel. *Archäologie Österreichs* **12/1-2**: 70-74.

Ziegelmayer G., 1979.
Die anthropologischen Befunde. In: Keller E. Das spätrömische Gräberfeld von Neuburg an der Donau. Materialhefte zur Bayerischen Vorgeschichte. Reihe A 40: 71-113. Kallmünz/Opf: Laßleben.

Ziegelmayer G., 2000.
Zur Geschichte des Instituts für Anthropologie und Humangenetik. Institutskolloquium am 21. November 2000. München.

Diversity conservation: Rare domestic farm animal breeds

Hans Hinrich Sambraus, Lehrgebiet für Tierhaltung und Verhaltenskunde,
Technische Universität München

Abstract / Zusammenfassung

Farm animal breeds, as we know them in Europe today, have existed since the middle of the 18th century. With the exception of the horse, individuals belonging to the same breed are rather similar to each other, also in respect to their performance. However, breeds with higher performance, such as in production of milk, meat, eggs, wool etc., were very quickly preferred over those with lower performance, with the result that the older local breeds have become extinct or almost extinct. Several older local breeds, however, definitely had their advantages: they were robust, modest, of high production quality and good use in terms of landscape protection. The worth of breeds, however, should not be judged solely on production value. They have been created by man and are therefore, part of the human cultural heritage, and consequently, should not be destroyed, just as one would not destroy an old painting or building. Local breeds are frequently similar to free-range types morphologically. However, all breeds change their appearance with time, and with the exception of their colour, the animals may become taller or smaller, stouter or more slender. It will become harder for future scientists to identify a specific breed from archaeological finds.

Haustierrassen im heutigen Sinne gibt es seit Mitte des 18. Jahrhunderts. Die Individuen einer Rasse sind, abgesehen vom Pferd, im Aussehen recht einheitlich; sie haben auch ungefähr dieselbe Leistung. Rassen mit höherer Leistung (Milch, Fleisch, Eier, Wolle usw.) wurden bald denen mit quantitativ geringerer Leistung vorgezogen. Dabei konnte es aber sein, dass diese Landrassen sich durch Qualitätsmerkmale auszeichneten: Sie sind robust und anspruchslos, haben eine gute Produktqualität und lassen sich z. B. gut im Landschaftsschutz einsetzen. Rassen sollten aber nicht nur nach ihrer Leistung beurteilt werden. Sie wurden vom Menschen geschaffen und sind damit ein Kulturgut. Man sollte sie ebensowenig zerstören wie ein altes Bild oder Gebäude. Landrassen ähneln morphologisch häufig der Wildform. Allerdings ändern alle Rassen im Verlaufe der Zeit, abgesehen von der Färbung, ihr Erscheinungsbild. Die Tiere können größer oder kleiner, stämmiger oder schlanker werden. Die Wissenschaftler kommender Zeiten werden es oft schwer haben, Ausgrabungsfunde einer bestimmten Rasse zuzuordnen.

Key Words: Domestication, Farm Animals, Rare Breeds, Breeding Aim, Yield, Cultural Heritage, Landscape Protection
Domestikation, Nutztiere, Gefährdete Rassen, Zuchtziel, Leistung, Kulturgut, Landschaftsschutz

Introduction

There were no farm animal breeds, as we know them in Europe today until the middle of the 18th century. Animals belonging to one species differed from region to region, and one or the other larger island or mountain valley system was populated by animals with particular characteristics. One determinant was food – where the basic fodder supply was adequate, one could "afford" larger animals; thus, in the areas with poorer soils or of a higher location, the animals were more slightly built. It is conceivable that in certain areas, animals belonging to one species tended to have a particular colour, due either to the personal preferences of the owner, or to the authorities' ideals, which the owners tried to put into practice.

Only one species of animals has been bred for many centuries, and that is the horse. Simultaneously, only a particular type of horse was named a "breed," namely, the thoroughbred animals of the nobility. Working horses, such as those used in farming, were not seen as a breed at this time.

The term "breed"

The term "breed" is applied to a group of domestic animals, which as a result of their mutual breeding history, appearance, certain physiological and ethological features and performance, are very similar to each other. On closer examination, this general formulation is only valid for some features and then, only to a certain extent. The heaviest animals most certainly possessed common origins. The colouring of an animal can indeed be helpful in its classification, but does occasionally lead to errors.

In spite of the desired unity and the breeding aims as formulated by the breeding societies, significant differences have emerged within each breed in the past. This becomes apparent in expressions such as "type" and "line". Leading breeding authorities or influential breeders often sought their own types for their area. Some experts can recognize such animals from their appearance alone.

The word "breed"

The expression "race" has been used in animal breeding circles for several centuries. Its origins lie in the Arabic "Ras", meaning here, amongst other things, mountain range, that is, a section of the whole mountain range. For domestic animals, the term was first used in France (as "race"), and in the beginning, only for horses. Not until the 18th century did the term "race" gain acceptance amongst other species, and only in the 19th century did the word "Rasse" first come in common use in the German-speaking areas.

The beginning of the formation of breeds in England

Until the middle of the 18th century, breeding was, to a great extent, empirically done. The development of systematic influence began with Robert Bakewell (1725-1795) and his cattle breeding. This development was triggered off by the increasing demand for animal products in the towns as well as the introduction of forage farming. The animals could now be better fed and the owner achieved a better price for the product.

Bakewell's breeding procedure was based on three principles.

a) There must be a clear idea about the breeding aim.

b) Only such individuals, which best fulfil the requirements shall be used for reproduction and mated with each other.

c) For further breeding, only the offspring that best meets the breeding aims shall be used.

In the past, the owner of the animal proceeded in a completely different manner. A farmer was proud to have a cow in his barn that was strong enough to work hard. One of his other cows had a high milk yield, and yet another delivered fleshy calves. However, as they all belonged to the same reproduction group, the special characteristics of the individual animals disappeared in the next generation. With Bakewell, a period of building up of breeds in domestic animal breeding began. He left no written notes behind, but did however, have a number of intelligent scholars, the best known of whom were the brothers Charles and Robert Colling. They produced, within a relatively short space of time, the cattle breed "Shorthorn" (Fig. 1). This breed was soon superior in yield to all other cattle breeds. This is obvious as the Shorthorn was crossed into almost all other breeds on the European continent, if it had not already been taken over as a pure breed.

In the same pattern, further breeds of cattle appeared, also in the 18th century, in Great Britain. Like the Shorthorn, they too, received worldwide recognition: Hereford, Aberdeen-Angus, Ayrshire, Jersey and Guernsey.

The formation of breeds amongst individual animal species

The European continent took on the British breeding methods rather slowly. From 1800 onwards, time and effort was dedicated to cattle breeding, and in the first half of the 19th century, intensive breeding of pigs was carried out. This occurred through crossing them with Chinese masked pigs. With pigs, it is occasionally claimed that the belt (e.g., Angeln Saddleback pig and the Schwäbisch-Hall pig) was brought to central Europe by the Chinese pigs. This is very doubtful. The belt appears on all domestic animals and is, in general, a sign of domestication that appears as a mutation repeatedly, on pigs as well as in other species, like cattle, goats and rabbits. In contrast to this, the saddled profile of the head (Fig. 2) was neither known nor desired on European pigs until 1800. The reason for this is that up to the end of the 19th century, pigs were kept by herdsmen. They had to, to a certain extent, root their food out of the ground and to do this, needed long, pointed snouts. Such a snout is not necessary for pigs that are fed purely from a trough.

Finally, in 1890, breeding goats with the aim of increasing their yield began. The "German Improved" goats, as they were then known, had three characteristics to fulfil:
1. High milk yield
2. Short hair
3. Hornlessness

Short hair was demanded for hygiene, and hornless animals because the risk of an accident occurring to the people working with them was reduced.

Some breeding rules apply only to cattle, pigs, sheep and goats, and not, however, to horses. For horses, two of these rules are
– breeding for unity of colour
– being free of cross-breeding

For most horse breeds, all basic colours are allowed, namely, black, white, chestnut and brown. White mark-

Diversity conservation

Fig. 1: Shorthorn cattle have existed since the 18th century (Photo: Sambraus)

Fig. 2: It is obvious that the pig breed "Middle White" originates from east Asian animals (Photo: Sambraus)

ings are also accepted, but dappling excludes them. Crossbreeding is allowed, as it is believed that horses without improved blood become ungainly. As a result of this, Thoroughbreds were crossed in repeatedly (English or Arabian Thoroughbreds). These breeds are considered to be improved. There are only a few breeding societies of particular horse breeds which allow only one colour: Lipizzaners are, apart from very rare exceptions, white while Friesians are always black. For some other horse breeds there is a dominating colour, but this, however, is dictated by the fashion. It is often difficult with individual horses to name the breed. There are, however, certain types:
- cold blood
- heavy blood horse
- blood horse
- Thoroughbred
- special breeds

Fig. 3: Achal-Tekke are extremely slender and have long legs. Nevertheless, they do not count as Thoroughbreds (Photo: Sambraus)

This classification is, at least, valid for large horses with a height of at least 148 cm at withers. All horses with a height of less than 147.3 cm at withers are termed ponies (14 hands and 2 inches = 147.3 cm). The type depends on its use. For heavy haulage work, strong, well-muscled draft horses were and are required, while for farm work and coaches, heavy blood horses were used. Riding horses were supposed to be elegant, and here, light horses, known in the past as half-breed horses, were suitable.

The term "Thoroughbred" was only used for the English and Arabian Thoroughbreds. These are light, long-limbed, sensitive and intelligent animals. Other breeds of this type, such as the Akhal-Tekke (Fig. 3), fall under the classification of "special breeds," like the Trotter.

Breeding procedure

In the past, one male was kept for a few females. Even if this male used as/put to stud in the surrounding farms, the number of his offspring was still not great. The situation changed c.60 years ago with the introduction of artificial insemination. The sperm could be diluted and ejaculates split up, so that for example, for cattle, a single ejaculate would be enough to fertilize 100 cows. Later, deep-freezing the samples allowed them to be accessible on a worldwide scale.

Nowadays, there is more required than the parent animals' high yield in order to use them for breeding. A bull's genetic make-up, for example, is first tested by examining hundreds of his female offsprings before he is taken on for breeding on a larger scale. This way, the selection of the few required male animals is incredibly large and one can reckon with rapid progress in breeding.

Progress in breeding

Progress in breeding usually means changes in the appearance of a breed. This is not true for colour, but for size, weight and the proportion of the different parts of the body to each other. A higher milk yield can be achieved by changing the genetics or the feeding. More milk can also be expected from larger cows, which explains why the Holstein-Friesian is so much higher and heavier compared to the Old Black Pied. This affects many things, including even the way in which the cow barn is built.

Breeds are in no way static things. The breeding aim can alter, and with it, the appearance of the breed. For cattle, the height at withers has been measured by the German Agriculture Society (DLG) since the end of the 19th century. It has neither remained constant, nor progressed in one particular direction, but has gone up and down, and often for economic reasons. To mention one example: after World War Two, many families owned a single cow, just to be self-sufficient. For one family, a smaller animal was adequate, and this so-called "economic type," which was significantly lighter than animals of the same breed before and after this time, was encouraged during this period.

The advances in breeding

Performance testing led not only to a high rate of competition within a breed but also meant that the individual breeds could now be better compared with each other. Many things can be verified by size and number, such as the amount of milk, layer of fat, weight gain and so on. In this way, some breeds proved themselves as being better than others. According to the saying, "The better one is the good one's enemy," it led to many breeds being suppressed, with the result that in the 20th century, only a few animals of some specific breeds were left (Tab. 1).

In pigs, a large difference in the breeds can be detected. Worldwide, there are only two breeding aims for pigs: a large percentage of valuable parts (ham, shoulder, chops) and not too much fat. Sheep are still kept in a relatively natural state, even when climatic conditions vary considerably. As a result, while there are only six pig breeds of economic significance worldwide, there are dozens of different breeds of sheep.

The decline in the number of breeds occurred mainly between 1950 and 1980. A significant breakthrough occurred in the 1960s, which was a decade of unbelievable change in the world of agriculture.

In 1951, the four most common cattle breeds in Germany made up 86.7% of the cattle raised, but by 1997 it had climbed to 96.6 %. For pigs, there are similar figures to hand: in 1951, the four top breeds accounted for 96.0 % of the animals raised, and as early as 1968, it had reached 99.7 %. For sheep, however, the percentage of the four most common breeds has decreased, from 89.1% in 1955 to 69.9% in 1994. During that time, several other breeds had come from abroad to central Europe.

Advantages of endangered breeds

At present, as already mentioned, breeds whose production can easily be measured in quantity are preferred. There are, however, characteristics, which are more difficult to define. For the endangered breeds, they can be summed up as being "robust and undemanding". At a closer look, this can be defined, amongst other things, as:
– good constitution
– longevity
– high fertility
– few problems at birth
– good mothering qualities
– resistance to illness
– high quality of products
– suitability to local conditions
– being satisfied with lower quality fodder
– suitability for raising in Third World countries.

Heath sheep, an example of such a breed, are satisfied with heather; something that few other breeds could survive on. Studies have shown that native breeds kept under the same conditions suffer significantly less from endoparasites than high yielding breeds. To mention another example: the White Mountain Sheep does not have particularly high quality wool, but rather, smooth wool. There have been attempts to improve the quality of the wool by crossbreeding these sheep with Merinos in Austria. As expected, they were successful, but the animals became prone to illness. This was due to the fact that rain ran off the smooth coat of the original type like rain runs off the plumage of a hen. Merino wool, on the other hand, soaks rain up like blotting paper, causing the animals to suffer often from hypothermia and to become more prone to illnesses.

Unpopular breeding aims

For every species of animals and every breeding trend, there are particular breeding aims at the forefront. There is nothing final, however, about these aims. They can change in the course of time. Here are three examples of this:

a) The Finnish native sheep's existence was threatened during the 1960s. Its wool was only of average quality, well outdone by the Merino sheep, and it was therefore replaced to a large extent, by other breeds. However, quite suddenly, cotton and synthetic fibres in turn, replaced sheep's wool and sheep were now being kept almost purely for their meat. The Finnish native sheep are extremely fertile and the ewes produce three or sometimes even four lambs in one litter. Due to their high fertility, they were crossbred into many other breeds and the demand for this breed was increased significantly this way.

Breed	1951	2000
German Landrace	66.4	64.5
Improved	6.5	12.8
Belgian Landrace	-	0.1
Piétrain	-	18.9
Angeln Saddleback	13.3	0.0
Schwäbisch-Hall	9.8	0.3
German Weide Pig	1.7 ⎫ 27.2	- ⎫ 0.3
Cornwall	1.4	-
Berkshire	1.0	-
Large White	-	0.9
Duroc	-	0.3
Leicoma	-	1.7
Hampshire	-	0.2
Bentheim Black and White	-	0.1

Tab. 1: Percentage of herd-book pigs in Germany 1951 and 2000

b) Since the First World War, the Piétrain pig (Fig. 4) has possessed, proportionately, a lot of meat and only a little fat. Immediately following the Second World War, fat was in demand and the Piétrain almost died out, so much so that in Belgium, the apparently single significant breeder left found himself being laughed at for being behind the times. With the rapid economic upturn, however, low-fat meat was in demand again, and the Piétrain breed experienced an amazing boom.

c) In the USA, Texas Longhorn cattle were a common breed in the first half of the 19th century. Slim and long-legged, they could easily be driven thousands of miles from the Southwest to the slaughterhouses in the Northeast. With the introduction of the railroad, the compact short-legged British breeds were preferred. The Texas Longhorn nearly died out, until it was discovered that their dry meat was very suitable in hamburger production. The Texas Longhorn's existence is now secure.

Fig. 4: The Piétrain pig has a high muscle content and little-fat. Unfortunately PSE-meat is quite common (Photo: Sambraus)

Cultural assets

Breeds cannot be measured purely on the basis of economic criteria. Cultural assets are exhibited for their cultural, not economic value. Likewise, mankind has created the domestic animal breeds and they are often identified with particular areas. The Lüneburg Heath is unimaginable without the Grey Horned Heath, and Eringer cattle, used for cow fights, exists nowhere else except in Wallis, Switzerland. The Hungarian Puszta would not be complete without Hungarian Steppe cattle (Fig. 5) and Racka sheep. The loss of a domestic animal breed *is* simultaneously the loss of a part of a unique farming culture.

The appearance of a breed

Most breeds are marked by a particularly striking optical appearance. Just as Pinzgau cattle have their charac-

Fig. 5: The grey Steppecattle with their extremely long horns are typical of the Hungarian Puszta (Photo: Sambraus)

teristic pigmented side plates, the Holstein-Friesian are characteristically chequered black and white and Simmenthal cows always have their characteristic white heads. That is how it is now, and that is how it was in the first half of the 19th century. Farmers are conservative by nature and tend to hang onto familiar ideas. Thanks to this mentality (and in opposition to official breeding strategy), certain breeds still exist. Against the orders of the authorities, the farmers resisted plans to "purify" various breeds with the argument that: "My grandfather owned this breed; my father owned them; I own them, and my son shall own them too". It was due to this attitude that we still have breeds like the "Carinthian Spectacled" today.

There are only a few examples of breeds that have completely changed their appearance and then took on another name. The "German improved land race pig," a larder of the old sort, was bred to create a new "modern meat type pig" but continued to be called "German land race". Where horses are concerned, the warm-blooded types of the various breeding areas (Holstein, Hannoveran, Westphalian etc.) have become very similar due to the exchange of breeding animals. Apart from this, a different type of horse has been evolving for the past few decades. The farm working horse has become a light, rideable, saddle horse. While each breeding society continues to have its own brand, the breed, however, is called "German saddle horse."

Reasons for using a different breeding method

As far as appearance is concerned, individual animals of one breed are reasonably similar. Their appearance represents a particular performance. If an animal does not compare with the visual picture of a known breed, one must assume that it is a crossbreed. Its particular performance can hardly be estimated.

Crossbreeding can easily be achieved where the areas of distribution of the various breeds clash. For instance, in Upper Bavaria, the Fleckvieh (= Simmenthal) is kept, in Allgäu, the Braunvieh. The small area from where the Murnau-Werdenfelser cattle originate lies between these two areas, and south of it is the area of the Tirolian Grauvieh. Finally, a little further to the north are the Franconians.

The situation is quite different in Iceland. Only one breed of horse, cattle, sheep and goat has been kept in this country for over a thousand years. Only through imports can crossbreeding take place. Apart from the fact that imports are illegal, word would soon get round the whole country with warnings of crossbreeds with poorer performance. That is why the Icelanders can, as it were, afford to accept all colourings (Fig. 6). They have all the colourings of the various cattle breeds known on the European continent and several more. At present, however, there is a change taking place amongst the Icelandic cattle. Until a few decades ago, although genetically hornless animals did occur, they were the exception. For some time now, in connection with the change to loose housing, animals are being selected more and more for their hornlessness.

Conclusion

Endangered breeds should be maintained for genetic as well as cultural reasons. At the same time, an endangered breed is not something static. Size, weight and the

type can change and so can the number of breeds. While there were 30 breeds of cattle in Bavaria in the 19th century, only four were left after the Second World War. Many breeds have been imported from other countries in the past two decades and there are now more than 20 different breeds of cattle again. Not even the presence or lack of horns leads to identification because through the destruction of the hornbuds, horn growth can be stopped. Palaeontologists of a later period will have difficult job in trying to identify the breed to which an animal belonged.

Fig. 6: In contrast to most European breeds of cattle, Icelandic cattle have many different colours. They have also been bred to lack horns (Photo: Sambraus)

Summary / Zusammenfassung

Breeds of livestock, as we now know them, have existed since the middle of the 18th century. Except in horses, the term "race" means not only do the animals look like one another, but also that they show similar agricultural production values. For example, some breeds of cattle are characterized by high milk yield, others by high daily weight gain, and still others by their high capacity to be used as work animals. The differentiation of livestock breeds on the European continent took place during the 19th century, at the beginning for cattle, in the first half for pigs, in the second half for sheep, and at the end for goats.

The broad similarity of individuals within each breed allowed the different breeds to be easily distinguished. This allowed some breeds to be further strengthened at the expense of others. In so doing, quantitative traits were used almost exclusively: high milk yield, high daily weight gain, high egg production. Although many older local breeds did not offer these traits, they did have other, less observable advantages. Local breeds are generally robust, require low maintenance and show specialized product qualities, for example, a unique flavour of meat, or they are good mothers. The worth of domesticated animals should not be judged solely by their production values. Through their creation by man, they have become part of our history in culture. In the same way that one does not destroy old paintings, chop down old trees, or demolish old buildings, one should not let the old breeds go extinct. Many are symbolic of particular regions or are necessary to maintain local microhabitats.

Rassen im heutigen Sinn gibt es bei Haustieren erst seit Mitte des 18. Jahrhunderts. Bei den landwirtschaftlichen Nutztieren, nicht bei den Pferden, bedeutet der Begriff "Rasse" im allgemeinen nicht nur, dass die Tiere einander im Aussehen weitgehend gleichen; sie sind sich auch in der Leistung ähnlich. Bei den Rindern bedeutet das zum Beispiel, dass einzelne Rassen eine hohe Milchleistung haben, bei anderen die täglichen Zunahmen der Masttiere sehr hoch sind, oder dass sie kräftige Arbeitstiere abgeben. Die Rassebildung fand auf dem europäischen Kontinent bei Rindern Anfang des 19. Jahrhunderts statt, bei Schweinen in der ersten Hälfte des 19. Jahrhunderts, bei Schafen in der zweiten Hälfte des 19. Jahrhunderts und bei der Ziege Ende des 19. Jahrhunderts.

Durch die weitgehende Einheitlichkeit der Individuen einer Rasse war ein guter Vergleich der Rassen untereinander möglich. Dadurch wurden einzelne Rassen immer mehr bevorzugt, die Bestände anderer Rassen gingen dagegen zurück. Allerdings achtet man bei der Wahl fast ausschließlich auf quantitative Merkmale: Viel Milch, hohe tägliche Zunahmen, viele Eier. Viele alte Landrassen waren in dieser Hinsicht nicht so gut, sie hatten allerdings andere Vorteile, die man kaum beachtete. Landrassen sind im allgemeinen robust und anspruchslos; sie haben eine besondere Produktqualität oder z. B. gute Muttereigenschaften. Haustierrassen sollten aber nicht nur nach ihrer Leistung beurteilt werden. Sie wurden vom Menschen geschaffen und stellen ein Kulturgut dar. Ebensowenig wie man ein Bild zerstört, einen alten Baum fällt oder ein altes Gebäude einreißt, sollte man die alten Rassen zu Grunde gehen lassen. Viele von ihnen sind kennzeichnend für bestimmte Regionen oder erhalten durch ihr Nahrungsspektrum bestimmte Landschaften.

Part VI:

Current Projects of the Collection of Anthropology and Palaeoanatomy

Vertebrate food webs and subsistence strategies of Meso- and Neolithic populations of central Europe

Gisela Grupe[1], Živko Mikić[2], Joris Peters[1], Henriette Manhart[1]

[1]Staatssammlung für Anthropologie und Paläoanatomie, München
[2]Philosophische Fakultät, Anthropologische Sammlung, Universität Belgrad

Abstract / Zusammenfassung

Vertebrate trophic webs at the Mesolithic site at Vlasac in the Iron Gates region and at a Neolithic site in southern Bavaria have been reconstructed by stable isotope analysis of human and animal bone collagen. The analyses reveal that trophic relationships were maintained among nonhuman vertebrates in both terrestrial and aquatic food chains and that sympatric Mesolithic and Neolithic humans were located at the top of the food chains isotopically, most probably also due to the consumption of freshwater fish. For the Iron Gates region, this interpretation is strongly supported by archaeological evidence and could thus serve as a model for other European inland sites. Our findings suggest that the consumption of freshwater fish and other aquatic vertebrates by humans may have been frequently underestimated because the preservation and recovery of such bone remains is usually difficult or impossible for taphonomic reasons. Despite the maintenance of trophic relationships at the sites, stable nitrogen isotopic ratios in vertebrates from the Iron Gates region show a consistent and marked shift in terms of more positive values in comparison to the Bavarian site. As a consequence, any valid interpretation of stable isotope ratios in terms of palaeodiet or palaeoecosystem analysis must be based on the comparative analysis of animal bone finds from the same historical period, and at least, from the same restricted geographical and ecologically defined area.

Für den mesolithischen Fundort Vlasac am Eisernen Tor und für einen neolithischen Fundort in Südbayern wurden die Nahrungsnetze zwischen verschiedenen Wirbeltierspezies anhand stabiler Kohlenstoff- und Stickstoffisotope aus dem Kollagen tierischer und menschlicher Skelettfunde rekonstruiert. Innerhalb der nichtmenschlichen Wirbeltiere waren die Trophiestufen sowohl für das terrestrische als auch das aquatische Nahrungsnetz gewahrt. Zeitgleiche mesolithische und neolithische menschliche Skelettfunde befanden sich gemäß ihrer Isotopie jeweils an der Spitze der Nahrungskette, höchst wahrscheinlich auch aufgrund des regelmäßigen Konsums von Flussfischen. Für die Region des Eisernen Tores wird diese Interpretation nachhaltig durch die archäologischen Befunde unterstützt, so dass diese Fundstelle als Modellsituation für andere Fundplätze im europäischen Binnenland herangezogen werden kann. Unsere Ergebnisse legen nahe, dass der Konsum von Süsswasserfischen oder anderen aquatischen Wirbeltieren durch den Menschen häufig unterschätzt worden ist, da taphonomische Prozesse die Überlieferung und auch die Aufdeckung von Knochen dieser Tiere behindern. Obgleich die unterschiedlichen Trophiestufen für beide untersuchten Regionen klar herausgearbeitet werden konnten, waren die stabilen Stickstoffisotopien der Wirbeltiere vom Eisernen Tor im Vergleich zu den südbayerischen Funden deutlich und in sich konsistent in den Bereich positiverer Werte verschoben. Daraus folgt, dass jede valide Interpretation von Isotopendaten im Sinne einer Rekonstruktion früher Ernährungsgewohnheiten oder früher Ökosysteme auf der vergleichenden Analyse von zeitgleichen Tierfunden basieren muß, welche darüber hinaus zumindest aus der selben begrenzten geographischen, ökologisch definierten Region stammen müssen.

Keywords: food webs, Mesolithic, Neolithic, isotopes, carbon, nitrogen
Nahrungsnetze, Mesolithikum, Neolithikum, Isotope, Kohlenstoff, Stickstoff

Introduction

By the end of the Mesolithic, southeastern Europe was inhabited by hunter-gatherer populations that already preferred select animal and plant species for consumption and the manufacture of artefacts. According to De Laet (1994), such a subsistence strategy might be viewed as a sort of "pre-Neolithic" lifestyle. By the middle of the 9th millennium calBC, the pre-pottery Neolithic B had evolved from the pre-pottery Neolithic A (ca 9500–8500 calBC) in southeastern Anatolia and northern Syria, from where it spread via Asia minor and the Levante to reach the southeastern region of the European continent as a fully developed Neolithic culture in the 7th millennium BC. The cultural change spread further along the River Danube via the Balkans to reach southern and central

Germany at the turn of the 6th and the 5th millennium calBC. In a joint project between the State Collection of Anthropology and Palaeoanatomy, the Anthropological Institute at the University of Belgrade and the Geobio-Center of the University of Munich, a study into the effects of the respective subsistence strategies on the local palaeoecosystem was undertaken, the first results of which are presented in this paper. The still-open question as to whether the spread of the Neolithic culture was the result of several successive waves of emigrating farmers or of adoption by neighbouring hunter-gatherer populations might probably be attempted at by first understanding the various local ecological parameters that would have had a marked impact on this cultural change. Such palaeoecosystem analyses not only focus on human dietary behaviour, but also specifically on the local vertebrate trophic webs: the introduction of domestic livestock into a new region will have an impact on the local game populations as soon as the species have to compete over naturally occurring food resources. One might also recall an early competition between man and his domestic animals over food resources, a phenomenon well-known in later periods in history.

Another aspect frequently neglected – most often simply due to the paucity of relevant archaeological finds – is the exploitation of freshwater food resources by humans and their omnivorous domestic animals. Humans, much more than many other higher vertebrates, are biologically dependent on constant access to drinking water and consequently, even temporary human occupation sites are always located in the immediate vicinity of springs, lakes or rivers. There are hardly any logical reasons why humans should give up collecting high quality and easily accessible food from such aquatic systems, even after the introduction of a self-sustaining farming lifestyle.

This ongoing project aims at the reconstruction of vertebrate food webs by stable isotope analyses of bone collagen and bone structural carbonate of human and animal bone finds from three areas relevant to the Neolithic transition in Europe. The first region is Anatolia, place of origin of the new lifestyle, represented by the sites of Göbekli Tepe (1st half of the 9th millennium calBC), Nevalı Çori (8600–7500 calBC), and Gürcütepe (7500–7000 calBC). Bone analyses on finds from these sites are in progress and have yet to be completed. Next, finds from the Iron Gates region on the Balkan, a very important transition zone and also centre of early animal husbandry, together with some Mesolithic and Neolithic finds from southern Germany (cf. materials section) have already been completely investigated in terms of stable isotopes from bone collagen.

Stable isotopes measured in bone collagen are ^{13}C and ^{12}C, and ^{15}N and ^{14}N, respectively, and they have served for palaeodietary analyses in numerous cases since the carbon and nitrogen isotope ratios in collagen reflect the protein part of the daily diet of the individual consumer (Ambrose & Norr 1993). Although stable isotopes of carbon and nitrogen differ by one mass unit only, the difference is still considerable in view of the generally low atomic weight. Molecules containing the respective light isotope are more volatile, and kinetic isotope effects occurring in the course of the transport of carbon or nitrogen through the geo-, bio- and hydrosphere add to the establishment of reservoirs within the various compartments of an ecosystem, which are defined by restricted ranges in isotopic ratios. These ratios are expressed in the δ-notation as δ^{13}C and δ^{15}N respectively, related to PDB (δ^{13}C) or AIR (δ^{15}N) standards.

Both the Iron Gates area and southern Germany are C_3-plant-dominated regions of the temperate climates with no access to the open sea, and therefore, the collagen of terrestrial mammals will exhibit δ^{13}C-values of around −21‰. A certain variability is brought about by gross and microclimatic specifities and plant cover, e.g., the well-known canopy effect, which causes considerably lowered δ^{13}C-values (Ambrose 1993). Freshwater fish from the archaeological record have by far less frequently been analysed (e.g., Dufour et al. 1999, Katzenberg & Weber 1999), but the majority of publications reveal a considerable variability with generally lower carbon isotopic ratios (Schulting & Richards 2002). More important for the reconstruction of food webs in temperate climates are δ^{15}N-ratios which are excellent indicators of the trophic position of an individual. The well-known trophic level effect averages +4‰, but may vary roughly between 2.5 and 6‰, depending on local parameters (Ambrose 1993). δ^{15}N in bone collagen therefore permits the differentiation between herbivores, omnivores, primary and secondary carnivores (Schwarcz & Schoeninger 1992, Ambrose 1993). For mammalian sucklings, mother's milk is a pure, animal-derived nutritive source; hence weaning practices can be reconstructed by the age-dependent variability of δ^{15}N (Bocherens et al. 1997, Balasse et al. 1999, Dittmann & Grupe 2000). Again, freshwater organisms exhibit high variability in terms of collagen δ^{15}N, the values of which mainly tend to exceed terrestrial ones (Bonsall et al. 1997, Katzenberg & Weber 1999).

Excellent palaeoecosystem analyses, which require the investigation of both human and animal skeletal remains from the same site and time, have been published by Weber et al. (2002) on Lake Baikal and by Ambrose (1986) on Africa. Pleistocene environments have been frequently addressed by Bocherens and his co-workers (e.g., Bocherens 1997, Bocherens et al. 1999). Where the Neolithic transition in Europe is concerned, there are only a few studies available, and among them the pioneering work on Denmark by

Tauber (Tauber 1981), who was able to show a change from a marine to a terrestrial subsistence at the turn of the Neolithic. A similar result was achieved by Lubell et al. (1994) for Portugal. British sites have been investigated by, for instance, Richards and Hedges (1999a, b). The Mesolithic human finds from the Ofnet Cave in Bavaria, which are also subject of this paper, have already been analysed by Bocherens et al. (1997); however, no appropriate animal bone material for comparison was available at that time. As will be shown in this paper, the interpretation of the Ofnet people as having been predominantly successful hunters will have to be questioned. Extensive analyses of human bone finds from the Iron Gates have been performed by Bonsall and co-workers (Bonsall et al. 1997, 2000); however, again without adequate consideration of animal bones. For the Lepenski Vir site (see below), a single otter find as well as three bovid and three fish bones were analysed, without the species level being determined. Also, the bovids were "thought to be from domestic cattle" only (Bonsall et al. 1997). We hope to be able to show through our data that palaeodietary analyses for humans, but more, palaeoecosystem reconstructions will not be possible and cannot result in robust interpretations as long as the isotopic characteristics of the contemporary and sympatric vertebrate food webs are insufficiently known.

Material

The data set that is yet available consists of 140 human and 82 animal bone samples, *in toto* 222 individuals. The majority of the finds came from the Iron Gates region in Yugoslavia, namely, the Mesolithic site at Vlasac and the proto- and early Neolithic strata of Lepenski Vir. Both sites are located only about three kilometres apart in the same valley on the banks of the Danube. A well-protected area that is surrounded by high mountains and endowed with special microclimatic features (Srejovic 1975, Nemeskeri 1978), Lepenski Vir's abundance of wild boar and red deer remains indicates that the habitat was wet, well-watered and densely forested (Bökönyi 1978). The Mesolithic burial site at Vlasac is horizontally stratified into an early (9000–7600 calBC) and a later (7600–6500 calBC) phase. Altogether, 117 human individuals forming a considerable variety of inhumation types were excavated from 84 graves (Nemeskeri 1978). Animal bone remains largely represented kitchen refuse. Fish constitute the overwhelming majority of the animal remains; among them catfish that weighed up to over 100 kilograms, carps that weighed up to 20 kilograms apiece, pike and sturgeon. Although Mesolithic people at Vlasac appear to have been specialized catfish fishermen, fish from the Danube ranked only second as a source of meat according to Bökönyi (1978). The only domestic animal at Vlasac was the dog which was represented by a surprisingly high number of 160 individuals. Other vertebrate species identified were the aurochs, chamois, red deer, roe deer, wild cat, pine marten, badger, lynx, brown bear, red fox, beaver, squirrel, hedgehog, several bird species and European pond turtle. The majority of species was probably hunted for both their meat and pelts (Bökönyi 1978). The animal bone specimens analysed in our study have not been published to date since they have only recently been found, mingled with bone artefacts from the site. Unfortunately, our samples did not yield bones from herbivores.

The site of Lepenski Vir is stratified into proto-LV (about 7000 calBC), early Neolithic LV I (6300–6000 calBC) and late Neolithic LV I (6000–5600 calBC). Later Neolithic phases of this site are not addressed in this paper. The faunal spectrum of early Neolithic Lepenski Vir is different from the later Neolithic periods in that – just like Vlasac – the only domestic animal is the dog, and fish remains (57% of the total assemblage) outnumber game, which in turn is dominated by red deer (36% of the assemblage). While game remains are abundant in the faunal record, domestic animals like cattle, pig, sheep and goat do not show up before the Neolithic phase of LV III (Bökönyi 1970). A total of 170 human individuals have been excavated from all phases of Lepenski Vir (Mikic 1992). Burial types at Lepenski Vir are even more diverse than in Vlasac.

Since both sites are located in very close proximity of each other, cultural exchange between the two populations is plausible. Interestingly, Vlasac was a permanently settled area in the early occupation phase but turned into a seasonal settlement in the later phase. The opposite holds true for Lepenski Vir: according to archaeological records (Nemeskeri 1978), the seasonal camp in the course of proto-LV turned into a permanent one during LV I.

No such horizontally stratified burial sites exist for the Neolithic transition in southern Germany where the River Danube would have facilitated early contacts between the populations of the Balkan and present-day Bavaria. Stable isotope data on 25 burials from the Mesolithic Ofnet Cave (Schmidt 1913; 5500 calBC), which had been established previously (Bocherens et al. 1997), were included into this study. In addition, collagen $\delta^{13}C$ and $\delta^{15}N$ values were measured for the single epipalaeolithic find from Neuessing (Gieseler 1977; 16 200 calBC), and for another five late Neolithic Bell Beaker individuals from the Weichering site (Weinig 1992; 2500–2000 BC). The analyses of another ca. 75 human Neolithic skeletons, covering all periods from the early to late Neolithic in southern Bavaria are in progress. Animal bone samples were available for the nearby Neolithic site of Pestenacker (ca. 3800–3400 calBC, Schönfeld 1991) which revealed a wide variety of both wild and domestic animals including many freshwater vertebrates, thanks to extensive sifting in the course of the excavation. A few additional animal bone finds considered here came from the nearby Neolithic site at Unfriedshausen (Vagedes 1998). Altogether, the Neolithic animal bone series is made up of sheep, cattle, goat, pig, dog, wild horses, red deer, roe deer, wild boar, red fox, brown bear, otter, beaver, mallard, crane, European pond turtle, pike and some cyprinids. Since despite a common feeding behaviour of a species (e.g., herbivores or carnivores, grazers or browsers) individual dietary preferences do exist, and therefore, at least three to five different individuals per species were sampled whenever available. Tables 1 – 5 summarize all specimens analysed.

Methods

Collagen was retrieved from the bones by gelatinisation: after cleaning under running tap water and air drying, about 250 mg of whole bone was homogenized and the powder demineralised for 20 minutes in 10 mL 1M HCl. The acid was then removed by centrifuging and the pellet was washed with distilled water until pH equilibrium was achieved. Humic substances were removed by incubating the demineralised sample for 20 hours in 10 mL 0.125M NaOH, which was then removed again by centrifuging and washing in distilled water. For gelatinisation, the sample was kept for 10 to 15 hours in 10 mL hot (90° C) 0.001M HCl (pH 3) and finally passed through a glass filter (pore size 5 µm). The gelatine was lyophilised until further processing.

Aliquots of the gelatine extracts were forwarded to amino acid analysis to check for their preservation. About 1 mg of gelatine was hydrolysed in 1 mL 6N HCl for 24 hours. After evaporation of the acid, the sample was diluted in 1 mL distilled water plus 1 mL Li-citrate buffer (pH 2.2). 20 µL of this stock solution was analysed by an amino acid analyser (Pharmacia Alpha plus, Li-system) for the amino acid profile. Beyond proteinogenetic amino acids, a selection of their breakdown products and also of microbial amino acids was included into the spectrum.

$\delta^{15}N$ and $\delta^{13}C$ were measured online by mass spectrometry (Thermo Finnigan Delta plus), coupled with a CHN analyser (Thermo Finnigan NA2500). About 0.3 mg of lyophilised gelatine was forwarded into Sn-capsules and burnt under oxidizing conditions at 1500 °C in the CHN analyser. The resulting gases were forwarded in the form of CO_2 and N_2, respectively, by use of He (4.6) as carrier gas to the mass spectrometer. Isotopic ratios were measured against a laboratory standard that had been calibrated against IAEA standards NBS 19 and NBS 20 for CO_2, and against N1 and N2 for N_2 respectively. Isotopic ratios were expressed in the conventional δ-notation against PDB ($\delta^{13}C$) and AIR ($\delta^{15}N$) standards. Measurement error never exceeded 0.15‰. Molar C/N ratios served as an additional check for the integrity of the samples.

Prior to these molecular analyses, thin sections of a selection of human and animal bone samples were prepared and inspected under transmitted and polarized light (Herrmann et al. 1990) to get an impression of the overall structural integrity of the finds. Since isotopic analysis is an invasive method, and the finds are of considerable scientific value, sampling for histology was dependent on the availability of suitable specimens. In addition, since a selection of Neolithic herbivores had already been sectioned in the course of another project (cf. contribution by Karola Dittmann, this volume), the images were made available to this project to minimize destruction of the finds.

Results

All isotopic data, the gelatine yields in percent of specimen weight, percentage of carbon and nitrogen in the gelatine, and the respective molar C/N-ratios are listed in Tables 1 – 5. Histological preservation of the finds varied from excellent to fairly bad, and as expected, gelatine yields were quite variable. Amino acid chromatograms of the specimens were typical for collagen type I, with no or only minor presence of microbial amino acids. Also, molar C/N ratios, with very few exceptions only, varied within the physiological range of between 2.9 and 3.6 (Ambrose 1993), and despite a considerable variability of carbon and nitrogen percentages (Fig. 1). The measurement data presented in this paper are therefore considered biological rather than diagenetic signals and are suitable for interpretation of palaeodiets.

Discussion

The data set from the Iron Gates (Fig. 2) serves as a very good model for trophic relationships, and in particular for the human exploitation of freshwater food resources, which was clearly evident from the archaeological finds of fish bones and fishing devices (Srejovic 1975). All carnivores cluster around $\delta^{15}N$-values of about 10 ‰, but do differ slightly in $\delta^{13}C$ values. The most carnivorous species analysed is the lynx, which at the same time, exhibits the lowest $\delta^{13}C$-value of all the terrestrial vertebrates in the data set, presumably due to its preference for more densely wooded areas. It is noteworthy that the domestic dogs at Vlasac are hardly, if ever, distinguishable from pine martens and wolves. In terms of diet, the dogs may have benefited from the

Vertebrate food webs and subsistence strategies

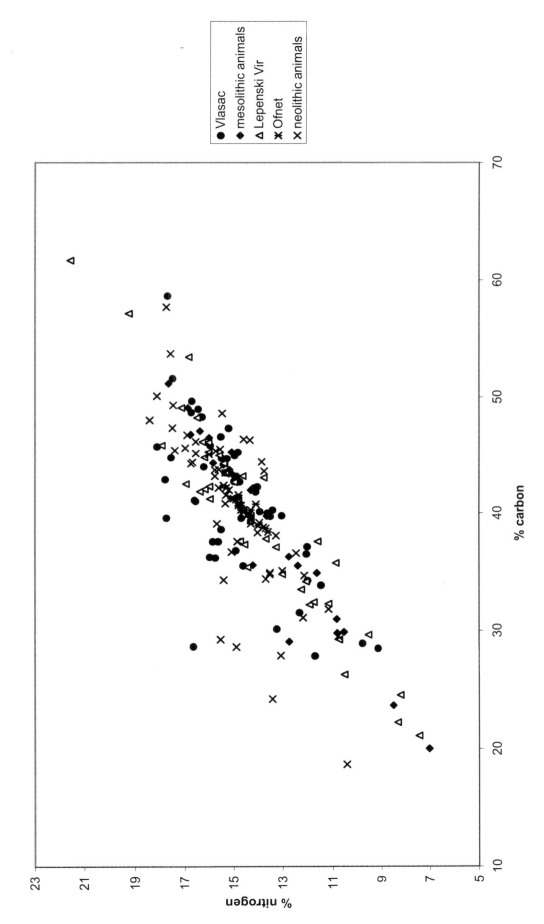

Fig. 1: Carbon and nitrogen yields (in %) in gelatine recovered from Mesolithic and Neolithic bone finds. Specimens with low percentages nevertheless had molar C/N ratios in the accepted physiological range.

Vlasac									
no	age	sex	cal BC	% gelatine	%C	%N	molar C/N	$\delta^{13}C$	$\delta^{15}N$
4a	young adult	male	7600-6500	3.52	46.63	15.51	3.51	-19.19	15.07
4b	old adult	male	partial	3.99	48.33	16.33	3.45	-19.13	15.05
5	13-16 y	nd	7600-6500	3.43	47.36	15.2	3.63	-20.26	14
6	old adult	male	9000-7600	3.14	41.07	16.54	2.9	-19.71	15.15
6a	neonate	nd	9000-7600	2.67	34.26	12	2.66	-19.62	16.82
7	15-16 y	nd	9000-7600	2.72	40.16	13.94	3.36	-20.34	12.19
10	neonate	nd	9000-7600	3.68	43.49	15.34	3.33	-20.34	14.91
11b	young adult	female	7600-6500	3.54	48.32	16.27	3.46	-20.01	14.51
12	neonate	nd	?	4.47	44.77	15.48	3.37	-19.48	18.31
12a	neonate	nd	7600-6500	3.19	39.6	14.69	3.14	-20.04	13.4
12b	neonate	nd	7600-6500	3.35	43.18	14.89	3.38	-19.66	17.3
16	old adult	male	partial	3.46	42.19	14.17	3.47	-20.04	12.26
17	young adult	male	7600-6500	3.94	39.62	17.75	2.6	-20.31	13.79
18c	9-10 y	nd	9000-7600	1.91	33.87	11.47	3.44	-21.03	12.86
21	neonate	nd	9000-7600	1.95	28.93	9.8	3.44	-20.28	16.84
23	old adult	male	7600-6500	1.49	48.71	16.72	3.4	-19.31	14.23
24	young adult	female	6647-6625	3.17	42.91	17.8	2.8	-18.73	14.39
27	old adult	female	7600-6500	2.27	51.61	17.5	3.44	-19.4	14.32
32	old adult	female	9000-7600	1.32	45.89	15.94	3.36	-20.22	13.22
35a	neonate	nd	9000-7600	2.76	41.87	14.11	3.46	-20.09	15.03
36	young adult	female	partial	2.5	39.75	13.51	3.43	-19.09	14.97
36 (I)	neonate	nd	partial	1.45	40.03	13.62	3.43	-19.35	16.45
38	old adult	female	9000-7600	1.56	36.83	14.91	2.88	-18.81	15.31
42	neonate	nd	7600-6500	5	36.19	15.73	2.68	-20.06	14.05
44	old adult	male	7600-6500	2.84	48.99	16.43	3.48	-19.66	13.33
47	young adult	female	9000-7600	2.05	45	14.96	3.51	-19.26	15.03
48	old adult	female	9000-7600	1.7	37.58	15.61	2.81	-18.39	14.72
50A (I)	neonate	nd	9000-7600	2.59	42.72	14.76	3.38	-20.8	15.21
51	10-12 y	nd	9000-7600	0.56	28.51	9.15	3.63	-21.93	9.87
53	8-9 y	nd	9000-7600	1.24	44.73	15.27	3.42	-19.43	15.02
54	old adult	male	7032-6401	0.37	31.53	12.32	2.99	-20.04	14.22
55	young adult	female	9000-7600	3.15	58.66	17.71	3.87	-19.64	15.19
58b	neonate	nd	7600-6500	1.87	39.8	13.04	3.56	-20.47	13.67
59	neonate	nd	9000-7600	1.53	42.75	14.98	3.33	-20.64	14
61	neonate	nd	9000-7600	3.33	42.29	14.04	3.51	-20.15	14.84
62	neonate	nd	9000-7600	2.82	40.29	13.41	3.5	-20.1	15
63	old adult	male	9000-7600	1.39	44.84	17.56	2.98	-19.26	14.26

Tab. 1a: Datings, biological and analytical data for the human skeletal finds from Vlasac.

Vlasac									
65a	young adult	female	partial	2.43	41.16	16.59	2.9	-19.69	14.06
66a	neonate	nd	partial	0.41	27.88	11.7	2.78	-20.73	10.12
67	young adult	female	9000-7600	1.64	45.78	18.13	2.95	-18.69	13.83
69	old adult	male	partial	1.58	37.59	15.83	2.77	-19.42	14.41
71	old adult	female	9000-7600	1.38	37.15	12.01	3.59	-19.6	14.25
72	old adult	female	9949-8843	2.55	49.68	16.7	3.47	-19.78	13.53
74	old adult	female	9000-7600	2.53	28.68	16.64	2.01	-19.42	13.1
78	old adult	male	9000-7600	4.81	45.28	14.83	3.56	-19.28	15.55
78a	old adult	male	9000-7600	2.39	35.51	14.61	2.64	-18.91	14.45
79	young adult	female	9000-7600	1.5	36.26	15.96	2.83	-18.56	14.84
80	young adult	female	7600-6500	2.39	38.62	15.5	2.91	-19.51	13.14
80a	young adult	female	7600-6500	1.76	36.52	12.04	3.54	-19.45	14.81
81	old adult	female	7600-6500	1.25	30.146	13.24	2.46	-18.97	13.72
84M	adult	male	?	1.38	44.02	16.2	3.17	-19.37	14.98
n=51			mean	2.43	40.93	14.87	3.21	-19.72	14.38
			stand.dev.	1.07	6.35	2.01	0.38	0.67	1.46
			minimum	0.37	27.88	9.15	2.01	-21.93	9.87
			maximum	5	58.66	18.13	3.87	-18.39	18.31

Tab. 1b: Datings, biological and analytical data for the human skeletal finds from Vlasac.

hunting activities of their owners, or had been left to forage on their own. This food would have consisted of leftovers since interestingly, the median $\delta^{13}C$-values of dogs and pine martens (the latter sometimes occurring near human dwellings) are nearly identical, while $\delta^{13}C$-values of the wolves, animals that probably rather avoided human camp sites, are slightly more negative. Trophic relationships among the terrestrial animals are maintained, represented by the $\delta^{15}N$-ratios of the wild boars and the single brown bear whose $\delta^{15}N$ ratio of 8.22 is indicative of its largely herbivorous feeding habits. Unfortunately, no skeletal parts of herbivores were available for measurement, but the three data on Neolithic bovids from Lepenski Vir published by Bonsall and his co-workers (1997) may serve for comparison. Although separated by time, the animals had at least dwelt in the very same environment. These bovids, and hence strict herbivores, had $\delta^{13}C$-values varying from –21.8 to –20.6 ‰, and $\delta^{15}N$-values between 5.2 and 7.7 ‰ (Bonsall et al. 1997). The trophic level effect for the terrestrial animals of about 3 ‰ is therefore maintained. The same holds for the aquatic food chain where the lowest $\delta^{15}N$-ratios were detected in the detritivorous carp, higher values in the piscivorous sturgeons, and especially for the modern catfish and zander. The latter two values appear rather separated in the bivariate plot (Fig. 2) and it cannot be excluded that this is also due to their being modern, and therefore considerably separated from the other fish bones by time. The carps exhibit a very large variability in terms of $\delta^{13}C$ which would be due to their feeding behaviour (see below).

Human bones from both Vlasac and Lepenski Vir are clearly at the top of the food chain with mean $\delta^{15}N$-values around 14 ‰, being thus separated from the carnivores by a full trophic level despite an overall high variability in terms of stable nitrogen isotope ratios. Since the people at the Iron Gates were archaeologically defined as successful hunters and fishermen, this find is not surprising. On the assumption that the $\delta^{15}N$ represents the protein part of the diet only, and that any meat in the diet of an omnivore will therefore always be over-represented compared to plant food since meat contains much more protein (van Klinken et al. 2000), the regular consumption of considerable amounts of freshwater fish would be the logical explanation for these results (cf. also Bonsall et al. 1997). The fish bones recovered at Vlasac and Lepenski Vir belonged to very large fishes and many fish species become increasingly carnivorous as they increase in size. This is also supported by $\delta^{15}N$-ratios around 12 ‰ measured in the modern catfish and zander.

Comparison of the human data from the earlier and later phases at Vlasac and Lepenski Vir reveals an interesting

Vlasac	animals							
Species	no	skeletal element	% gelatine	%C	%N	molar C/N	$\delta^{13}C$	$\delta^{15}N$
Accipenser sturio	P11, comga B		2.27	44.35	15.82	3.27	-19.21	10.07
Accipenser sturio	P10, comga B		0.4	36.28	12.75	3.32	-21.26	8.32
Accipenser sturio	P2, comga B		1.12	42.01	14.31	3.42	-19.4	8.78
Canis familiaris	comga B	tibia	1.3	35.57	14.22	2.92	-19.09	10.65
Canis familiaris	comga B	humerus	2.14	43.36	15.06	3.36	-19.5	10.21
Canis familiaris	comga A		0.45	29.87	10.55	3.3	-19.56	10.97
Canis familiaris	comga B	mandible	0.68	29.78	10.82	3.21	-18.86	10.47
Canis lupus	comga B	mandible	1.31	49.09	16.86	3.43	-19.68	11.87
Canis lupus	comga B	metacarpus	2.33	44.18	15.41	3.34	-20.01	9.62
Canis lupus	comga B	calcaneus	1.33	51.17	17.67	3.38	-20.1	10.1
Cyprinus carpio	P11, comga B	vertebra	1.34	47.12	16.36	3.34	-20.34	6.59
Cyprinus carpio	P12, comga A	Os phar. inf.	0.34	29.08	12.74	2.66	-21.19	6.3
Cyprinus carpio	P25, comga A	hyomandible	0.35	23.67	8.49	3.26	-23.68	9.8
Cyprinus carpio	P10, comga B	Os phar. inf.	0.17	20.01	7.02	3.32	-20.84	7.12
Lynx lynx	comga B	ulna	0.69	45.3	15.08	3.51	-21.4	10.64
Martes martes	m9, comga B		0.68	34.89	11.63	3.5	-20.16	10.14
Martes martes	m1, comga B		0.88	35.51	12.39	3.32	-19.38	9.12
Martes martes	m7-12, comga B	pelvis	1.8	46.55	15.98	3.39	-18.94	9.24
Martes martes	m8, comga B		1.08	39.77	13.66	3.39	-19.24	10.62
Sus scrofa	comga A	tibia	1.1	43.75	15.17	3.27	-20.32	7.06
Sus scrofa	comga A	scapula, juv.	0.99	46.83	16.75	3.26	-20.66	11.94
Ursus arctos	m18, comga A		0.28	30.97	10.84	3.34	-20.54	8.22
n=22		mean	1.05	38.6	13.62	3.3		
		stand.dev.	0.63	8.4	2.76	0.18		
		minimum	0.17	20.01	7.02	2.66		
		maximum	2.33	51.17	17.67	3.51		
Sander lucioperca	modern		3.68	41.93	13.31	3.68	-23.46	12.2
Silurus glanis	modern		2.57	48.33	14.97	3.77	-23.41	11.64
n=2		mean	3.13	45.13	14.14	3.73	-23.44	11.92

Tab. 2: Species and analytical data for the animal bone finds from Vlasac.

phenomenon. Vlasac, a site that was permanently settled during its earlier phase but which later became a seasonal camp showed a variability of stable isotope ratios in the human bones that remains largely identical (Fig. 3). In contrast, the individuals from the later Neolithic phase at Lepenski Vir show a considerable broadening of the dietary spectrum in terms of $\delta^{15}N$, especially in terms of lower ratios. In this case, the site had probably not been permanently settled before the late Neolithic phase. The evolving farming lifestyle and cultivated plants would have led to a higher diversity of daily food with more emphasis on terrestrial herbivores – one must remember that certain stages of husbandry have to elapse before animal bones with the indubitable morphological features of full domestication show up in archaeozoological records. This interpretation is supported by the anthropological observation that the younger human finds at Lepenski Vir are not only morphologically totally different from the older finds, but tend, at the same time, to have the lowest $\delta^{15}N$-values.

Highest $\delta^{15}N$-ratios were measured in neonates and small infants at both sites, indicating that these children died before being weaned. Whether the fact that more

Lepenski Vir

no	age	sex	cal BC	% gelatine	%C	%N	molar C/N	$\delta^{13}C$	$\delta^{15}N$
7(I)	old adult	male	6300-6000	0.46	21.1	7.43	3.31	-19.7	11.49
11	14 y	nd	6300-6000	1.47	42.69	14.98	3.33	-18.76	15.15
14	young adult	female	7000	0.17	26.28	10.53	2.91	-21.38	9.35
17	juvenile	female	7000	0.7	40.31	14.61	3.22	-19.98	11.79
19	old adult	female	6000-5600	3.65	42.53	16.95	2.93	-18.95	15.52
26	young adult	male	6300-6000	0.98	24.55	8.2	3.49	-19.66	10.57
31/a	old adult	male	6404-5826	3.87	32.28	11.18	3.89	-18.61	16.18
32/b	old adult	female	6156-5721	1.37	35.38	14.44	2.86	-18.79	12.58
41	young adult	female	7000	3.35	45.89	17.95	2.98	-18.78	14.41
42a	adult	male	6000-5600	1.62	44.32	15.37	3.36	-20.15	11.65
42b	old adult	female	partial	2.29	41.88	16.36	2.99	-18.26	15.24
45/b	old adult	male	6300-6000	3.26	37.59	11.6	3.79	-18.15	16.82
47	old adult	female	7000	1.56	42.08	16.16	3.04	-18.54	15.02
50	old adult	male	7000	2.4	35.73	10.88	3.83	-19.46	14.2
54/b	old adult	female	6000-5600	1.64	37.61	14.74	2.98	-19.78	10.38
54/c	old adult	female	6000-5600	2.31	43.2	14.64	3.44	-20.16	14.08
54/d	old adult	female	6300-6000	0.51	29.24	10.75	3.17	-19.93	13.41
54/e	young adult	female	6000-5600	1.57	43.06	13.78	3.44	-19.65	13.87
60	young adult	male	7000	1.23	29.65	9.56	3.62	-19.02	15.32
63	neonate	female	6300-6000	1.93	45.01	15.58	3.37	-19.27	15.49
64	old adult	male	6000-5600	1.25	42.31	15.97	3.09	-19.71	14.69
68	senile	female	7000	1.15	32.2	11.92	3.15	-20.31	12.91
69	old adult	male	7000	0.56	32.43	11.76	3.22	-19.4	14.44
70	old adult	male	partial	1.18	44.83	16.18	3.23	-18.51	15.97
72	2-4 y	nd	partial	1.77	38.96	13.98	3.25	-20.49	11.32
84	3-4 y	nd	partial	2.38	39.65	14.3	3.24	-18.81	13.2
89/b	5 y	nd	6300-6000	2.16	46.19	16.26	3.32	-19.25	15.25
90	young adult	male	7000	2.15	41.27	15.95	3.02	-18.25	16.61
91	young adult	female	6000-5600	0.18	22.23	8.31	3.12	-20.42	12.98
93	old adult	female	6300-6000	0.3	34.18	12.06	3.31	-20.13	12.2
94	neonate	male	6000-5600	0.35	34.79	13.03	3.12	-19.38	15.16
99	12-13 y	nd	partial	1.37	37.82	13.69	3.22	-20.31	9.14
100	12-14 y	nd	partial	0.64	37.34	14.56	2.99	-20.43	9.79
101	neonate	nd	6000-5600	1.4	37.11	13.28	3.26	-19.37	15.25
103	neonate	male	6000-5600	1.88	57.14	19.24	3.47	-18.84	17.66
104	15-17 y	nd	6300-6000	4.22	53.43	16.83	3.38	-20.06	15.59
107	neonate	male	6300-6000	2.65	49.1	17.14	3.34	-18.41	16.79

Tab. 3a: Datings, biological and analytical data for the human skeletal finds from Lepenski Vir.

Lepenski	Vir								
109/a	neonate	female	6300-6000	3.22	44.31	15.4	3.36	-19.43	16.21
110	neonate	female	6300-6000	2.07	45.29	16.01	3.3	-20.2	16.23
111	neonate	female	6300-6000	1.12	48.25	16.48	3.42	-20.05	15.37
113	neonate	female	6000-5600	2.85	46.15	16.01	3.36	-20.24	15.02
116	neonate	female	6300-6000	2.5	43.4	15.32	3.3	-19.53	15.32
118	neonate	female	6300-6000	3.09	40.75	14.78	3.22	-18.84	15.9
125	neonate	male	6000-5600	2.13	61.7	21.56	3.34	-19.28	16.21
127	neonate	female	6000-5600	0.66	33.5	12.26	3.19	-18.67	16.6
N = 45			mean	1.77	39.66	14.18	3.27	-19.45	14.19
			stand.dev	1.03	8.24	2.87	0.23	0.73	2.16
			minimum	0.17	21.1	7.43	2.86	-21.38	9.14
			maximum	4.22	61.7	21.56	3.89	-18.15	17.66

Tab. 3b: Datings, biological and analytical data for the human skeletal finds from Lepenski Vir.

Ofnet Cave								
no	sex	age	complex	%C	% N	molar C/N	$\delta^{13}C$	$\delta^{15}N$
2484-11	male	old adult	large	39.7	14.3	3.2	-19.4	10.9
2493-21	male	adult	large	41.3	14.8	3.3	-19.6	10.6
2476-3	female	adult	large	41.2	14.9	3.2	-19.6	10.7
2481-8	female	adult	large	40.9	14.8	3.2	-19.6	10.9
2486-13	female	adult	large	39.9	14.4	3.2	-19.8	10.7
2488-15	female	young adult	large	40.1	14.6	3.2	-19.7	10.6
2490-18	female	adult	large	40.3	14.3	3.3	-20	10.5
2497-25	female	adult	large	40.7	14.8	3.2	-19.6	11.1
2504-32	female	adult	small		14.8	3.3	-19.7	10.9
2474-1	nd	5-9 years	large	41.3	15.1	3.2	-19.4	10
2478-5	nd	7 years	large		13.6	3.3	-19.9	10.7
2479-6	nd	2-3 years	large	42.2			-19.3	11.7
2482-9	nd	1.5 years	large	38.8	13.8	3.3	-19.5	11.6
2485-12	nd	7 years	large	40	14.4	3.2	-19.7	11.3
2489-17	nd	5-6 years	large	38.7	13.7	3.3	-19.9	11.1
2491-19	nd	5 years	large	40.9	14.8	3.2	-19.6	11.7
2492-20	nd	8-9 years	large	42.4	15.4	3.2	-19.6	11.1
2494-22	nd	4 years	large	39.4	14.3	3.2	-19.6	10.5
2495-23	nd	6-7 years	large	39.1	14.3	3.2	-19.6	11
2498-26	nd	7 years	large	40.2	14.6	3.2	-19.3	10.7
2499-27	nd	2-3 years	large	40.3	14.7	3.2	-19.2	12.2
2500-28	nd	7 years	small	41.3	15	3.2	-19.6	11.2
2502-30	nd	7 years	small	42.2	15.3	3.2	-19.4	10.1
2503-31	nd	7 years	small	40.8	14.1	3.4	-19.7	10.8
2505-33	nd	3-5 years	small	40.7	14.7	3.2	-19.6	10.8
n=25			mean	40.54	14.56	3.23	-19.6	10.94
			stand.dev	1.03	0.46	0.06	0.19	0.5
			minimum	38.7	13.6	3.2	-20	10
			maximum	42.4	15.4	3.4	-19.2	12.2

Tab. 4: Biological and analytical data for the human finds from the Ofnet Cave.

Pestenacker	animals							
species	no	skeletal element	% gelatine	%C	%N	molar C/N	$\delta^{13}C$	$\delta^{15}N$
Anas platyrhinchos	Pe 92 C IV Bef 4.1	humerus	2.06	45.66	16.98	3.14	-23.59	7.68
Anas platyrhinchos	Pe 91 H IV Bef 6.75	ulna	2.22	41.27	14.8	3.25	-21.23	8.32
Bos taurus	82	metacarpus	3.35	39.19	13.96	3.21	-21.43	6.32
Bos taurus	336	metacarpus	1.66	45.39	15.72	3.37	-21.39	6.12
Bos taurus	196	rib	3.8	57.69	17.76	3.79	-21.5	7.14
Bos taurus	237	ulna	4.64	43.63	13.76	3.7	-21.19	5.07
Canis familiaris	387	mandible	1.62	29.53	10.78	3.07	-20.97	8.5
Canis familiaris	389	humerus	3.31	42.01	15.19	3.23	-21.09	7.79
Canis familiaris	169	humerus	3.29	39.51	14.31	3.22	-21.19	7.19
Canis familiaris	405	ulna	3.02	41.59	15.31	3.17	-20.14	7.21
Canis familiaris	501	mandible	4.93	43.76	15.59	3.27	-20.25	7.52
Capreolus capreolus	114 non-ad	metatarsus	1.31	34.3	15.4	2.8	-22.27	5.2
Capreolus capreolus	110	metacarpus	1.69	44.41	16.67	3.11	-21.04	6.32
Capreolus capreolus	560	metacarpus	2.48	40.86	15.33	3.11	-22.09	5.64
Capreolus capreolus	435	pelvis	2.88	45.54	15.54	3.42	-22.12	4.84
Capreolus capreolus	Pe K IV Bef 6.80	humerus	3.03	46.42	14.6	3.71	-22.25	5.05
Castor fiber	Hu 517	humerus	5.02	34.68	12.13	3.33	-22.46	6.78
Castor fiber	Pe Ti 168 non-ad	tibia	0.5	24.23	13.42	3.8	-22.15	5.84
Castor fiber	Pe Mand 242	mandible	1	28.64	14.89	2.24	-22.7	4.82
Cervus elaphus	123	humerus	2.19	48.04	18.43	3.04	-21.24	5.23
Cervus elaphus	365	humerus	2.74	42.19	15.6	3.16	-22.06	5
Cervus elaphus	375	scapula	3.9	47.38	17.51	3.16	-21.61	3.94
Cervus elaphus	349	metatarsus	4.69	44.43	13.86	3.74	-21.5	4.83
cyprinide	Pe 91 K III Bef 6.19A	rib	1.26	31.82	11.17	3.28	-24.78	7.59
cyprinide	Pe 91 K IV Bef 6.151	rib	0.57	29.26	15.52	2.2	-21.75	8.5
Emys orbicularis	Pe 91 K II Bef 6.22	carapax	1.98	38.34	14.03	3.19	-23.66	7.77
Emys orbicularis	Pe 92 K IV Bef 6.155	carapax	1	39.1	15.67	2.82	-23.48	9.12
Equus ferus	296	metacarpus	4.72	45.11	15.99	3.29	-21.18	5.38
Equus ferus	301	metacarpus	3.55	46.78	16.9	3.23	-21.35	4.53
Equus ferus	305	metacarpus	1.99	50.1	18.14	3.22	-21.36	5.88
Equus ferus	Pe D IV Bef 2.2	metacarpus	3.81	43.73	15.75	3.24	-21.53	6.09
Equus ferus	264	metacarpus	1.68	46.34	14.35	3.76	-22.07	4.55
Esox lucius	Pe 89 K IV Bef 6.44	vertebra	0.22	35.08	13.02	3.15	-22.77	6.91

Tab. 5a: Species and analytical data for the animal bone finds from Pestenacker and Unfriedshausen (sheep bones only).

Pestenacker	animals							
Grus grus	Pe 90 K V Bef 6.114	ulna	0.66	36.6	12.46	3.43	-19.89	8.62
Grus grus	Pe 89 B III Bef 3.0	radius	3.25	49.31	17.48	3.29	-20.14	7.6
Grus grus	Pe 92 B III Bef 3.0	tibiotarsus	0.43	18.67	10.43	2.09	-20.42	10.04
Lutra lutra	Pe 92 CI	mandible	0.35	27.91	13.08	2.49	-23.05	8.54
Ovis aries	37	scapula	2.26	45.45	17.41	3.07	-22.28	5.24
Ovis aries	316	metatarsus	2.39	34.79	13.54	3	-21.35	6.85
Ovis aries	UN 86	radius	2.96	37.58	14.85	2.95	-21.09	6.93
Sus domesticus	558 non-ad	pelvis	0.79	31.1	12.18	2.98	-21.78	7.21
Sus domesticus	225 non-ad	tibia	2.88	43.19	15.75	3.2	-21.07	3.83
Sus domesticus	126	humerus	4.74	46.21	16.54	3.26	-21.28	4.23
Sus domesticus	369	femur	2.97	43.48	15.33	3.31	-21.49	7.03
Sus scrofa	125	tibia	1.09	38.95	14	3.3	-21.34	6.3
Sus scrofa	542	pelvis	1.78	36.7	15.08	2.84	-20.09	6.98
Sus scrofa	205	ulna	2.5	38.09	13.28	3.35	-20.95	5.83
Sus scrofa	34	tibia	2.18	53.72	17.59	3.56	-20.98	4.9
Sus scrofa	35	lumbar vertebra	3.47	48.62	15.46	3.67	-22.29	5.12
Ursus arctos	459	mandible	1.62	34.4	13.7	2.93	-20.51	6.66
Ursus arctos	Pe Hu 274	humerus	4.23	44.29	16.74	3.01	-20.08	6.13
Ursus arctos	Pe Hu 408	humerus	4.65	45.2	16.53	3.19	-20.35	5.85
Vulpes vulpes	Pe o. Nr.	maxilla	4.32	34.92	13.52	3.01	-19.74	8.26
n=53		**mean**	2.56	40.48	14.96	3.18		
		stand. dev.	1.35	7.4	1.84	0.36		
		minimum	0.22	18.67	10.43	2.09		
		maximum	5.02	57.69	18.43	3.8		

Tab. 5b: Species and analytical data for the animal bone finds from Pestenacker and Unfriedshausen (sheep bones only).

neonates who had obviously never been fed mother's milk were encountered at Mesolithic Vlasac than at Neolithic Lepenski Vir (Tables 1, 3) has any meaning in terms of the ceremonial uses of the sites or is simply due to a sampling artefact is not deducible.

The animal bones recovered from the Bavarian Neolithic site at Pestenacker are extremely well-preserved, and due to the excavation technique a considerable number of terrestrial and aquatic species were recovered. Fig. 4 shows that the trophic relationships among both terrestrial and aquatic vertebrates are maintained: carnivorous dogs and red foxes are at the top of the terrestrial food chain, and the herbivores have lowest δ^{15}N-values, while the omnivorous pigs, wild boars and brown bears fall somewhat in between. Comparable to the Iron Gates, domestic dogs are hardly distinguishable from red foxes, which seek the vicinity of human dwellings to forage for leftovers. The largely until exclusively carnivorous European pond turtle is located at the top of the aquatic food chain as expected, and the omnivorous crane is still high up, while the exclusively herbivorous beaver has lowest δ^{15}N-values. Again, aquatic vertebrates exhibit lower δ^{13}C-values in contrast to the terrestrial ones, and the cyprinids show considerably high variability. The wide range of δ^{13}C in freshwater ecosystems is due to several specifities, one of which is whether phytoplankton or attached algae are the primary introducers of carbon into the food chain (Rozanski et al. 2001). Plankton utilizes CO_2 rather than HCO_3 and is therefore depleted in terms of ^{13}C. In fact, δ^{13}C in aquatic plants may vary from –30 to –10 ‰ (Deines 1980), which accounts for the low δ^{13}C-values encountered in the aquatic vertebrate bone finds.

Metabolic specifities are also maintained in the isotopic data, i.e., the slightly higher δ^{15}N-values in the bones of ruminants (cattle, sheep) in contrast to the nonruminant wild horse (Hristov 2002). Among the terrestrial animals, forest-dwelling species have generally lower δ^{13}C-values which would be the result of the canopy-effect since Bavaria would have been much more densely wooded during the Neolithic Period. The amino acids in plants (the source of carbon and nitrogen in the consumer's collagen) as a whole tend to be rich in ^{13}C with respect to the total plant (Deines 1980) and therefore cannot account for this observation.

Just like the Iron Gates' finds, the human finds from the Bavarian sites are rich in ^{15}N, up to a full trophic level

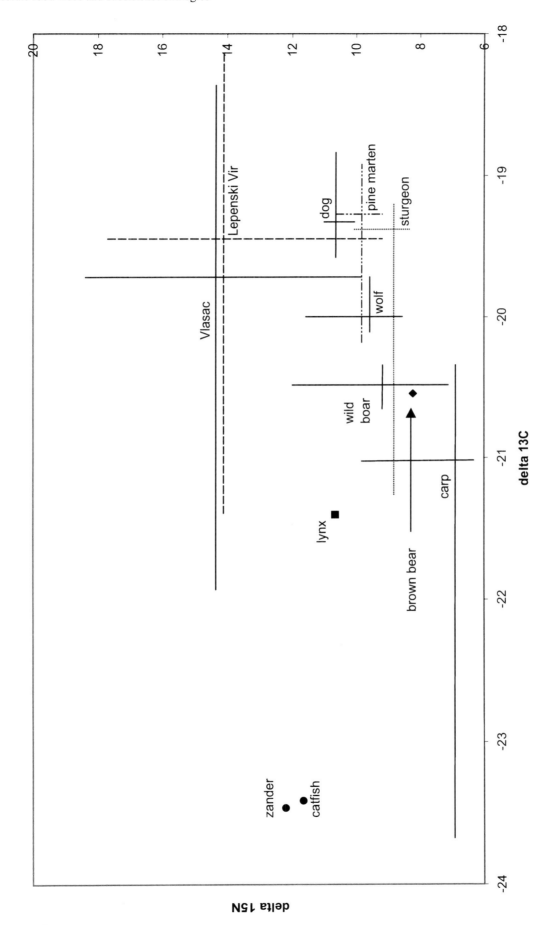

Fig. 2: Trophic web at the Iron Gates according to stable isotope analyses. Mean values and range are plotted for the human finds, median values and range for the animal bone finds.

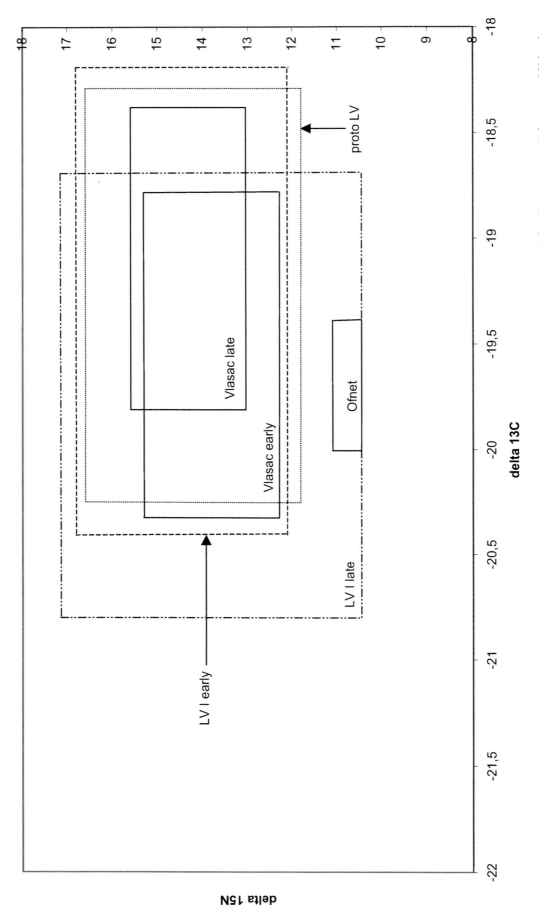

Fig. 3: Diachronic trends in stable isotopic variability for Mesolithic and Neolithic human bones, adult individuals only. One young adult female (grave no 14) from proto LV has been omitted because of her deviant isotopic signature (cf. Table 3). The particular dietary behaviour of this individual should have had special reasons, but would have obscured the general pattern of the distribution of isotopic values.

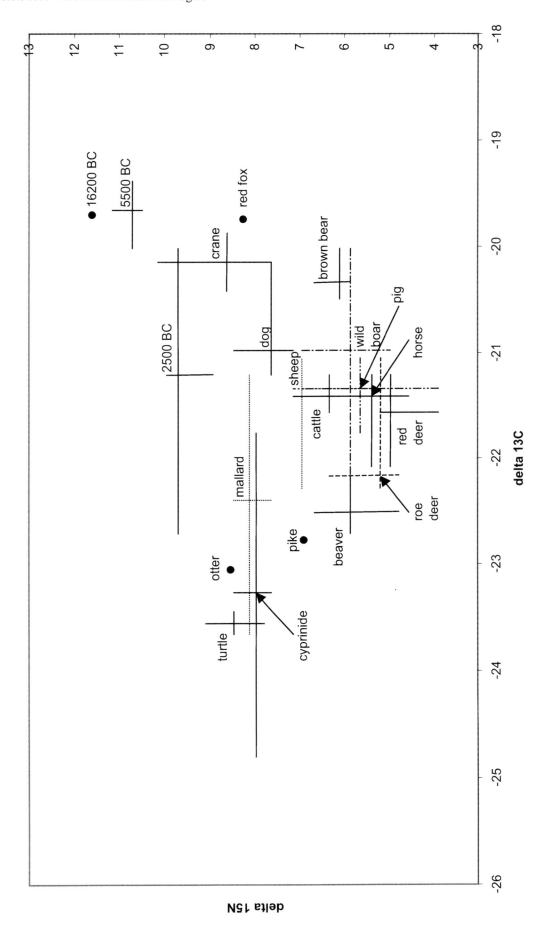

Fig. 4: Neolithic trophic web, including the human epipalaeolithic find from Neu-Essing (16 200 BC), the Ofnet Cave finds (5500 BC), and some Bell Beaker individuals from the Weichering site (2500 BC). Median values and total range are plotted, with the exception of the Ofnet finds, where the data base permitted the calculation of arithmetic means.

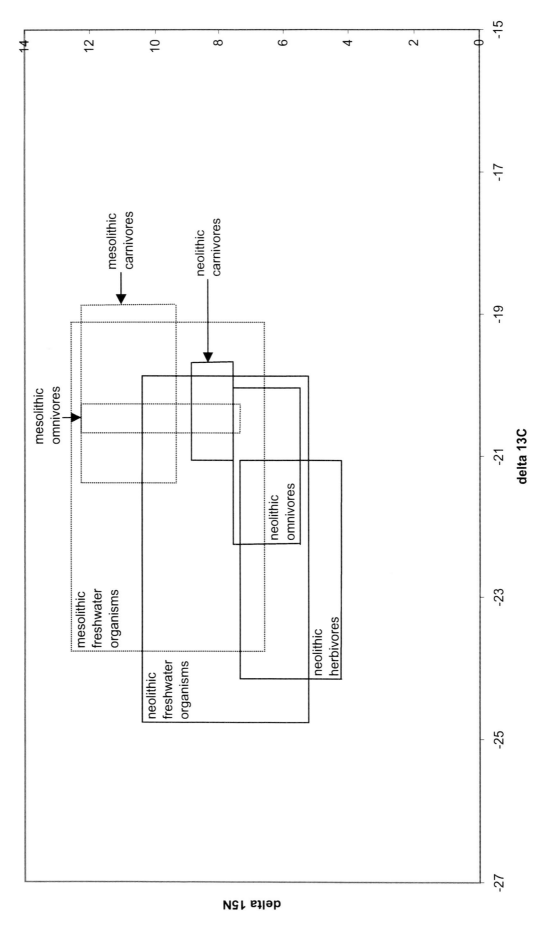

Fig. 5: Boxplots for minimal and maximal isotopic values detected in animal bones according to their feeding habits. The Mesolithic food web at Vlasac is systematically enriched with ^{15}N.

effect compared with the Neolithic carnivores. Even the late Neolithic farmers are hardly distinguishable from dogs, the red fox or the cranes in their stable nitrogen isotope ratios. This latter group also shows a tendency towards lower $\delta^{13}C$-values. The overall high $\delta^{15}N$-values in human bones are not only due to the unavoidable over-representation of animal protein in any mixed diet. All animal species in the archaeozoological record from Pestenacker exhibited butchering marks which means that these humans not only hunted game, but also consumed their dogs, water birds and turtles, thereby behaving like secondary carnivores which introduces an additional trophic level effect. Moreover, with regard to human metabolic needs, one should think of Stone Age people not only as successful hunters, but also as fishermen. This is not easily deduced from the archaeofaunal evidence since the most tiny fish bones are only recovered by sieving but are also more prone to decomposition. As a result, fish bone numbers usually do not exceed 3-5 % of the total faunal assemblage at European Neolithic inland sites and an underestimation of the contribution of freshwater resources to the daily human diet is most probable.

This hypothesis would necessitate the revision of an earlier interpretation of high $\delta^{15}N$-values in the Mesolithic human finds from Ofnet Cave in terms of a predominantly hunting subsistence (Bocherens et al. 1997). As depicted in Fig. 3, Mesolithic Ofnet finds are isotopically indistinguishable from some individuals from the early Neolithic, late LV I period at the Iron Gates. Despite the observed broadening of the dietary spectrum at this time at Lepenski Vir, according to clear archaeological evidence, people still heavily relied on the Danube as a protein source. Hence, a contribution of freshwater fish to the diet of the people buried in the Ofnet Cave can no longer be excluded. The *caveat* that reservoir effects brought about by a freshwater subsistence for the Iron Gates could be relevant for radiocarbon dating, as published by Cook et al. recently (2002), should be taken seriously. This interpretation gains considerable support from the observation that generally, stable nitrogen isotopic ratios are higher in the vertebrates from the Iron Gates than in Bavaria (Fig. 5). This overall shift may have climatic reasons, but could also be related to the environmental influence brought about by the riverine ecosystems, since the background data for such systems are heavily dependent on the characteristics of the river and its segments (such as pH, temperature, oxygen availability, type and amount of transported terrestrial organic matter). This last observation clearly demonstrates that any palaeodietary interpretation, or any reconstruction of prehistoric food webs must be based on contemporary human and animal bone finds from the same site, and at least, from the same restricted geographical, ecologically well-defined area.

Summary / Zusammenfassung

This paper presents the first results of a joint research project currently being undertaken by the State Collection of Anthropology and Palaeoanatomy of Bavaria, the Anthropological Collection at the University of Belgrade and the Geobio-Center of the University of Munich. It evaluates the effects of human subsistence strategies on the local palaeoecosystem. The still unanswered question as to whether the spread of the Neolithic culture was the result of several successive waves of emigrating farmers, or the result of adoption by neighbouring hunter-gatherer populations might probably be better solved through an understanding of the local vertebrate trophic webs first. Another aspect frequently neglected is the exploitation of freshwater food resources by humans and their omnivorous domestic animals.

This ongoing project aims at the reconstruction of vertebrate food webs by stable isotope analyses of bone collagen and bone structural carbonate of human and animal bone finds from three areas relevant for the Neolithic transition in Europe. The first region is Anatolia, place of origin of the new lifestyle. Bone analyses are underway but are yet to be completed. Next, finds from the Iron Gates sites on the Balkan, a very important transition zone and also early centre of animal domestication, together with some Mesolithic and Neolithic finds from southern Germany have already been completely investigated in terms of stable isotopes from bone collagen.

The data set from the Iron Gates serves as a very good model for trophic relationships and in particular, the human exploitation of freshwater food resources, which was clearly evident from the archaeological finds of fish bones and fishing devices. All terrestrial carnivores cluster around $\delta^{15}N$-values of about 10 ‰, but do differ slightly in $\delta^{13}C$ values. It is noteworthy that the domestic dogs at Vlasac are hardly, if ever, distinguishable from pine martens and wolves. In terms of diet, the dogs have most probably not been fed by their owners deliberately, but rather, been left to find food on their own. This food would have consisted of leftovers.

Human bones from both Vlasac and Lepenski Vir are clearly at the top of the food chain with mean $\delta^{15}N$-values around 14 ‰, being thus separated from the carnivores by a full trophic level despite an overall high variability in terms of stable nitrogen isotope ratios. Since the people at the Iron Gates were archaeologically defined as successful hunters and fishermen, this result is not surprising and can largely be attributed to the regular consumption of considerable amounts of freshwater fish.

Comparison of the human data from the earlier and later phases at Vlasac and Lepenski Vir reveals an interesting phenomenon. Vlasac was permanently settled during its earlier phase and became a seasonal camp later. However, variability of stable isotope ratios in the human bones remains largely identical. In contrast, the individuals from the later Neolithic phase at Lepenski Vir show a considerable broadening of the dietary spectrum in terms of $\delta^{15}N$, especially in terms of lower ratios. In this case, the site had probably not been permanently settled before the late Neolithic phase. The evolving producing lifestyle would have led to a larger diversity of daily food with more emphasis on terrestrial herbivores. This interpretation is supported by the anthropological observation that the younger human finds at Lepenski Vir are not only morphologically totally different from the older finds, but at the same time tend to have the lowest $\delta^{15}N$-values.

The animal bones recovered from the Bavarian Neolithic site at Pestenacker are extremely well-preserved and due to the excavation technique, a considerable number of terrestrial and aquatic species were recovered. Again, the trophic relationships among both terrestrial and aquatic vertebrates are maintained. The carnivorous dogs and red foxes are at the top of the terrestrial food chain, and the herbivores have lowest $\delta^{15}N$-values while the omnivorous pigs, wild boars and brown bears fall somewhat in between. Comparable to the Iron Gates, domestic dogs are hardly distinguishable from red foxes which seek the vicinity of human dwellings to live off leftovers. The carnivorous European pond turtle is located at the top of the aquatic food chain as expected, and the omnivorous crane is still high up while the exclusively herbivorous beaver has lowest $\delta^{15}N$-values. Again, aquatic vertebrates exhibit lower $\delta^{13}C$-values in contrast to the terrestrial ones.

Just as in the Iron Gates finds, the human finds from the Bavarian sites are rich in ^{15}N, up to a full trophic level in comparison with the Neolithic carnivores. Even the late Neolithic farmers are hardly distinguishable from dogs, red foxes or cranes in stable nitrogen isotope ratios. One should therefore think of Stone Age people not only as successful hunters, but also as fishermen. This is not easily deduced from the archaeofaunal evidence since the tiny fish bones are only recovered by sifting and are also more prone to decomposition. As a result, fish bone numbers usually do not exceed 3-5 % of the total faunal assemblage at European Neolithic inland sites, and an underestimation of the contribution of freshwater resources to the daily human diet is most probable. This hypothesis would necessitate the revision of an earlier interpretation of high $\delta^{15}N$-values in the Mesolithic human finds from Ofnet Cave in terms of a predominantly hunting subsistence.

It is noteworthy that in general, stable nitrogen isotopic ratios are higher in the vertebrates from the Iron Gates than in Bavaria. This overall shift may have climatic reasons, but should also be related to the environmental influence brought about by the riverine ecosystems since the background data for such systems are heavily dependent on the characteristics of the river and its segments (such as pH, temperature, oxygen availability, type and amount of transported terrestrial organic matter).

Im Rahmen eines gemeinsamen Projektes der Staatssammlung für Anthropologie und Paläoanatomie in München und der Anthropologischen Sammlung der Philosophischen Fakultät der Universität Belgrad, durchgeführt in Kooperation mit dem GeoBio-Center der Universität München, werden die Auswirkungen menschlicher Subsistenzstrategien auf das jeweilige Paläoökosystem untersucht. Erste Ergebnisse dieses Projektes werden in diesem Beitrag vorgestellt. Die noch immer kontrovers diskutierte Frage, ob die Neolithisierung Europas auf mehrfachen Einwanderungswellen früher Ackerbauern oder auf der schrittweisen Adaptation der produzierenden Lebensweise durch jeweils benachbarte Bevölkerungen beruhte, könnte durch ein vertieftes Verständnis lokaler Nahrungsnetze beantwortet werden. Ein weiterer, häufig vernachlässigter Aspekt betrifft die Nutzung von Wirbeltieren aus Süßgewässern als Nahrung für Menschen und deren omnivore Haustiere.

Ziel unseres Projektes ist die Rekonstruktion von Nahrungsnetzen bei Wirbeltieren mittels Analyse stabiler Isotope aus dem Kollagen und strukturellem Karbonat von menschlichen und tierischen Skelettfunden aus drei Regionen, welche in bezug auf die neolithische Transition in Europa eine herausragende Rolle spielten. Die erste dieser Regionen ist Anatolien, Ursprungsgebiet der produzierenden Lebensweise. Die entsprechenden Analysen werden derzeit durchgeführt, sind aber noch nicht abgeschlossen. Zum zweiten wird die Balkanregion des Eisernen Tores untersucht, welche gleichermaßen eine wichtige Transitionszone darstellte, als auch ein frühes Zentrum der Tierzucht. Die Untersuchung ist ebenso abgeschlossen wie jene einiger mesolithischer und neolithischer Funde in Süddeutschland.

Aufgrund der zahlreichen archäologischen Funde von Fischknochen und Fischereigeräten, können die für das Eiserne Tor erarbeiteten Isotopendaten als Modell für Nahrungsnetze insbesondere in bezug auf die Erschließung von Süßgewässerressourcen durch den Menschen herangezogen werden. Die terrestrischen Carnivoren weisen $\delta^{15}N$-Werte um 10 ‰ auf, unterscheiden sich jedoch geringfügig in bezug auf $\delta^{13}C$. Auffällig ist, dass mesolithische Haushunde von Vlasac sich in ihrer Isotopie praktisch nicht von Baummardern und Wölfen unterscheiden. Die Hunde sind höchstwahrscheinlich nicht gezielt von ihren Besitzern gefüttert worden, sondern mussten sich ihre Nahrung selber suchen. Diese dürfte auch Überreste der menschlichen Mahlzeiten beinhaltet haben.

Mit $\delta^{15}N$-Werten um 14 ‰ liegen die Isotopien der Menschenknochen aus Vlasac und Lepenski Vir eindeutig an der Spitze der Nahrungskette und sind trotz insgesamt hoher Variabilität im Mittel um eine volle Trophiestufe von den entsprechenden Werten der Carnivoren verschieden. Da die Menschen vom Eisernen Tor archäologisch als hervorragende Jäger und Fischer ausgewiesen sind, ist dieses Ergebnis nicht überraschend und sollte auf den regelmäßigen Konsum von Flussfischen zurückzuführen sein.

Im Vergleich der menschlichen Isotopien aus den jeweils frühen und späten Phasen von Vlasac und dem frühneolithischen Lepenski Vir zeigt sich ein interessantes Muster. Vlasac war während der frühen Phasen ständig bewohnt und wandelte sich in der Spätphase zu einem saisonalen Camp. Die Variabilität der stabilen Isotope aus menschlichem Knochenkollagen blieb jedoch weitestgehend unverändert. Im Gegensatz hierzu weisen die Individuen aus der Spätphase des frühneolithischen Lepenski Vir eine Ausweitung des Nahrungsspektrums in bezug auf $\delta^{15}N$ auf, insbesondere im Hinblick auf niedrigere Werte. Lepenski Vir war vermutlich nicht vor dem Spätneolithikum permanent bewohnt. Die aufkommende produzierende Lebensweise sollte zu dieser höheren Variabilität des Nahrungsspektrums geführt haben, welche durch eine zunehmend stärkeren Nutzung terrestrischer Herbivorer gekennzeichnet war. Diese Interpretation wird durch den anthropologischen Befund unterstützt, demzufolge die zeitlich jüngeren menschlichen Skelettfunde von Lepenski nicht nur morphologisch von den älteren grundverschieden sind, sondern zugleich auch zu den niedrigsten $\delta^{15}N$-Werten tendieren.

Die Tierknochenfunde des neolithischen Fundplatzes von Pestenacker in Bayern sind außerordentlich gut konserviert und enthalten Dank der angewandten Ausgrabungstechnik auch überdurchschnittlich viele terrestrische und aquatische Wirbeltierspezies. Auch hier sind die Nahrungsketten sowohl innerhalb der terrestrischen als auch der aquatisch lebenden Vertebraten im Hinblick auf deren Isotopien gewahrt. Haushunde und der Rotfuchs befinden sich an der Spitze der terrestrischen Nahrungskette, Herbivore weisen die niedristen $\delta^{15}N$-Werte auf, und die omnivoren Hausschweine, Wildschweine und Braunbären zeigen mittlere Isotopenverhältnisse. Wie erwartet, liegt die carnivore Europäische Sumpfschildkröte an der Spitze der aquatischen Nahrungskette, und auch der omnivore Kranich ist noch durch hohe Stickstoffisotopien ausgezeichnet, wohingegen der ausschließlich herbivore Biber die niedrigsten $\delta^{15}N$-Werte liefert. Wie am Eisernen Tor sind die aquatischen Wirbeltiere durch niedrigere $\delta^{13}C$-Werte im Vergleich zu den terrestrischen ausgewiesen.

Entsprechend der Befunde an den Skeletten des Eisernen Tores sind auch die menschlichen Skelette von den bayerischen Fundplätzen mit ^{15}N im Sinne einer vollen Trophiestufe im Vergleich zu den terrestrischen Carnivoren angereichert. Selbst die spätneolithischen Ackerbauern unterscheiden sich kaum von Haushunden und Rotfuchs, und sind in bezug auf ihre Isotopie nicht vom Kranich zu unterscheiden. Man sollte daher in Betracht ziehen, neolithische menschliche Bevölkerungen nicht nur als Wildbeuter im Sinne von erfolgreichen Jägern zu bezeichnen, sondern ebenso als Fischer. Eine solche Sichtweise ergibt sich nicht ohne Weiteres allein aufgrund der archäologischen Funde, da die zierlichen Fischknochen nicht nur bevorzugt der Dekomposition anheim fallen, sondern nur durch Einsatz der Siebtechnik überhaupt geborgen werden können. Entsprechend stellen Fischknochen in der Regel lediglich 3-5 % des gesamten Tierknochenkonvolutes aus neolithischen Fundplätzen des Binnenlandes, was zu einer Unterschätzung der Bedeutung von aquatischen Ressourcen für die menschliche Ernährung führt. Diese Hypothese würde allerdings eine Revision der früheren Interpretation einer Isotopenanalyse der mesolithischen Schädelfunde aus der Ofnet-Höhle notwendig machen, welche hohe $\delta^{15}N$-Werte als Resultat einer hauptsächlich auf der Jagd beruhenden Subsistenz bewertet hatte.

Ein wichtiger Befund ist die generelle Anreicherung der Wirbeltiere einschließlich des Menschen vom Eisernen Tor mit ^{15}N im Vergleich zu bayerischen Funden. Diese Verschiebung zu höheren $\delta^{15}N$-Werten dürfte klimabedingt sein, könnte aber ebenso auch eine Folge der speziellen Ökologie von Süßgewässersystemen sein, da die Basisdaten dieser Systeme in hohem Maße von den charakteristischen Parametern des Flusses beziehungsweise seiner Abschnitte abhängen (pH, Temperatur, Sauerstoffverfügbarkeit, Art und Menge mitgeführten terrestrischen Materiales, etc.).

Acknowledgements

This project is financially supported by the Deutsche Forschungsgemeinschaft. Special thanks are due to Dr. U. Struck, Geobio-Center of the University of München, for the mass spectrometry.

References

Ambrose S.H., 1986.
Stable carbon and nitrogen isotope analysis of human and animal diet in Africa. *Journal of Human Evolution* **15**: 707-731.

Ambrose S.H., 1993.
Isotopic analysis of paleodiets: Methodological and interpretive considerations. In: Sandford M.K. (ed). *Investigations of Ancient Human Tissue: Chemical Analyses in Anthropology*: 59-130. Longhorne: Gordon & Breach.

Ambrose S.H. & Norr L., 1993.
Experimental evidence for the relationship of the carbon isotope ratios of whole diet and dietary protein to those of bone collagen and carbonate. In: Lambert J.B. & Grupe G. (eds). *Prehistoric Human Bone. Archaeology at the Molecular Level*: 1-37. Heidelberg: Springer.

Balasse M., Bocherens H. & Mariotti A., 1999.
Intra-bone variability of collagen and apatite isotopic composition used as evidence of a change of diet. *Journal of Archaeological Science* **26**: 593-598.

Bocherens H., 1997.
Isotopic biogeochemistry as a marker of Neandertal diet. *Anthropologischer Anzeiger* **55**: 101-120.

Bocherens H., Billiou D., Mariotti A., Patou-Mathis M., Otte M., Bonjean D. & Toussaint M., 1999.
Palaeoenvironmental and palaeodietary implications of isotopic biogeochemistry of last interglacian Neanderthal and mammal bones in Scladina Cave (Belgium). *Journal of Archaeological Science* **26**: 599-607.

Bocherens H., Grupe G., Mariotti A. & Turban-Just S., 1997.
Molecular preservation and isotopy of Mesolithic human finds from the Ofnet Cave (Bavaria, Germany). *Anthropologischer Anzeiger* **55**: 121-129.

Bökönyi S., 1970.
Animal remains from Lepenski Vir. *Nature* **167**: 1702-1704.

Bökönyi S., 1978.
The vertebrate fauna of Vlasac. In: *Vlasac. A mesolithic settlement in the Iron Gates. Volume II: Geology – Biology – Anthropology. Serbian Academy of Sciences and Arts Monographies Vol. DXII, Department of Historical Sciences Vol. 5*: 35-65. Beograd.

Bonsall C., Cook G., Lennon R., Harkness D., Scott M., Bartosiewicz L. & McSweeney K., 2000.
Stable isotopes, radiocarbon and the Mesolithic-Neolithic transition in the Iron Gates. *Documenta Praehistorica* **27**:119-132.

Bonsall C., Lennon R., McSweeney K., Stewart K., Harkness D., Borooneant V., Barosiewicz L., Payton R. & Chapman J., 1997.
Mesolithic and early Neolithic in the Iron Gates: A palaeodietary perspective. *Journal of European Archaeology* **5**: 50-92.

Cook G.T., Bonsall C., Hedges R.E.M., McSweeny K., Boroneant V., Bartosiewicz L. & Pettitt P.B., 2002.
Problems of dating human bones from the Iron Gates. *Antiquity* **76**: 77-85.

De Laet S.J., 1994.
Europe during the Neolithic. In: De Laet S.J. (ed). *History of Humanity, Vol. I: Prehistory and the Beginnings of Civilization*: 490-500. London: Routledge.

Dittmann K. & Grupe G., 2000.
Biochemical and palaeopathological investigations on weaning and infant mortality in the early Middle Ages. *Anthropologischer Anzeiger* **58**: 345-355.

Deines P., 1980.
The isotopic composition of reduced organic carbon. In: Fritz P. & Fontes J. (eds). *Handbook of Environmental Isotope Geochemistry. Volume 1: The Terrestrial Environment*: 329-406. Amsterdam: Elsevier.

Dufour E., Bocherens H. & Mariotti A., 1999.
Palaeodietary implications of isotopic variabiality in Eurasian lacustrine fish. *Journal of Archaeological Science* **26**: 617-627.

Gieseler W., 1977.
Das jungpaläolithische Skelett von Neuessing. In: Schröter P. (ed). *75 Jahre Anthropologische Staatssammlung München 1902-1977*: 39-52. Selbstverlag der Anthropologischen Staatssammlung München.

Herrmann B., Grupe G., Hummel S., Piepenbrink H. & Schutkowski H., 1990.
Prähistorische Anthropologie. Leitfaden der Feld- und Labormethoden. Heidelberg: Springer.

Hristov A.N., 2002.
Fractionation of ammonia nitrogen isotopes by ruminal bacteria in vitro. *Animal Feed Science and Technology* **100**: 71-77.

Katzenberg M.A. & Weber A., 1999.
Stable isotope ecology and palaeodiet in the Lake Baikal region of Siberia. *Journal of Archaeological Science* **26**: 651-659.

van Klinken G.J., Richards M.P. & Hedges R.E.M., 2000.
An overview of causes for stable isotopic variations in past European human populations: Environmental, ecophysiological, and cultural effects. In: Ambrose S.H. & Katzenberg M.A. (eds). *Biogeochemical Approaches to Paleodietary Analysis*: 39-63. New York: Kluwer Academic/Plenum Publishers.

Lubell D., Jackes M., Schwarcz H., Knyf M. & Meiklejohn C., 1994.
The Mesolithic-Neolithic transition in Portugal: Isotopic and dental evidence of diet. *Journal of Archaeological Science* **21**: 201-216.

Mikic Z., 1992.
The Mesolithic population of the Iron Gates region. *Balcanica* **23**: 33-45.

Nemeskeri J., 1978.
Demographic structure of the Vlasac epipaleolithic population. In: *Vlasac. A mesolithic settlement in the Iron Gates. Volume II: Geology – Biology – Anthropology. Serbian Academiy of Sciences and Arts Monographies Vol. DXII, Department of Historical Sciences Vol. 5*: 231-247. Beograd.

Richards M.P. & Hedges R.E.M., 1999a.
A Neolithic revolution? New evidence of diet in the British Neolithic. *Antiquity* **73**: 891-897.

Richards M.P. & Hedges R.E.M., 1999b.
Stable isotope evidence for similarities in the types of marine foods used by late Mesolithic humans at sites along the Atlantic coast of Europe. *Journal of Archaeological Science* **26**: 717-722.

Rozanski K., Fröhlich K., Mook W.G. & Stichler W. (eds), 2001.
Environmental Isotopes in the Hydrological Cycle. Principles and Applications. Vol. III: Surface Water. International Atomic Energy Agency, Vienna. (www.iaea.or.at/programmes/ripc/ih/volumes/ volume3.htm, last update June 2001).

Schmidt R.R., 1913.
Die altsteinzeitlichen Schädelgräber der Ofnet und der Bestattungsritus der Diluvialzeit. Stuttgart: E. Schweizerbart.

Schönfeld G., 1991.
Die Ausgrabung in der jungneolithischen Talbodensiedlung von Pestenacker, Ldkr. Landsberg am Lech, und ihre siedlungsarchäologischen Aspekte. *Bericht der Römisch-Germanischen Kommission* **71**: 355-380.

Schulting R.J. & Richards M.P., 2002.
Dogs, ducks, deer and diet: New stable isotope evidence on early Mesolithic dogs from the Vale of Pickering, North-east England. *Journal of Archaeological Science* **29**: 327-333.

Schwarcz H.P. & Schoeninger M.J., 1992.
Stable isotope analyses in human nutritional ecology. *Yearbook of Physical Anthropology* **34**: 283-321.

Srejovic D., 1975.
Lepenski Vir. Eine vorgeschichtliche Geburtsstätte europäischer Kultur. Bergisch Gladbach: Lübbe. 2nd edition.

Tauber H., 1981.
^{13}C evidence for dietary habits of prehistoric man in Denmark. *Nature* **292**: 323-333.

Vagedes K., 1998.
Haus- und Wildtiere im Umfeld jungneolithischer Siedlungen bei Landsberg am Lech. *Documenta Naturae* **118**. München.

Weber A.W., Link D.W. & Katzenberg M.A., 2002.
Hunter-gatherer culture change and continuity in the Middle Holocene of the Cis-Baikal, Siberia. *Journal of Anthropological Archaeology* **21**: 230-299.

Weinig J., 1992.
Ein neues Gräberfeld der Kupfer- und Frühbronzezeit bei Weichering, Landkreis Neuburg-Schrobenhausen, Oberbayern. *Das archäologische Jahr in Bayern* **12**: 64-67.

Histomorphometric analysis of primate and domesticated animal long bone microstructure

Karola Dittmann, Department Biologie I,
Bereich Biodiversitätsforschung/Anthropologie, Universität München

Abstract / Zusammenfassung

The microstructure of limb bones of animals and humans is determined by species-specific biomechanical demands such as locomotion and weight. Histomorphometry and statistics were used to identify various primate species *(Hylobates moloch, Pongo satyrus borneensis, Pan tr. troglodytes, Gorilla g. gorilla, Homo sapiens)*, equid species *(Equus caballus, Equus asinus, mule, Equus hemionus kulan, Equus ferus Przewalskii)* and also pre- and early historic horses e.g., Iron Age, medieval and Neolithic forms on the microstructural level. Furthermore, bones from domesticated cattle and their Neolithic forms as well as pigs, sheep and goats *(Bos taurus, Sus scrofa, Ovis aries, Capra hircus)* were examined. Thin sections from the proximal metacarpi or radii from each species of the domesticated animals and from the distal humeri of the primates were taken. Areas, perimeters, minimal and maximal axes of Haversian canals and secondary osteons were measured on digital images.

Canonical discriminant analyses of these bone microstructure parameters permit a differentiation of the species. It makes it possible to distinguish between the different primate species, between sheep and goats, between modern and extinct horses, as well as between donkeys, mules and kulans, but not between cattle and pig, or between *Equus caballus* and *Equus ferus przewalskii*, or between medieval and Iron Age *Equus caballus*.

Die Knochenmikrostruktur der Extremitätenknochen ist das Resultat von speziesspezifischen biomechanischen Ansprüchen, wie Lokomotionsweisen und Gewicht. Mit Hilfe histomorphometrischer und statistischer Verfahren wurden verschiedene Primatenspezies *(Hylobates moloch, Pongo satyrus borneensis, Pan tr. troglodytes, Gorilla g. gorilla, Homo sapiens)*, Equidenspezies *(Equus caballus, Equus asinus, Maultiere, Equus hemionus kulan, Equus ferus przewalskii)*, sowie vor- und frühgeschichtliche Pferde aus Eisenzeit, Mittelalter und Steinzeit, domestizierte Rinder, deren neolithische Formen, Schweine, Schafe und Ziegen *(Sus scrofa, Ovis aries, Capra hircus)* untersucht.

Auf digitalen Aufnahmen von Knochendünnschnitten aus proximalen Metacarpi oder Radii (bei Haustieren) und distalen Humeri (bei Primaten) wurden histomorphometrische Parameter wie Flächen, Umfänge, maximale und minimale Durchmesser von sekundären Osteonen und Havers'schen Kanälen erhoben. Die statistische Methode der Diskriminanzanalyse erlaubt die Unterscheidung der Arten mittels linearer Diskriminanzfunktionen. Es ist möglich, sowohl die Primaten untereinander, als auch Rinder, Schweine und nah verwandte Spezies wie Ziege und Schafe, Pferde, historische Pferde, Esel, Maultiere, Halbesel, zu unterscheiden. Jedoch gelingt keine Trennung von Rindern und Schweinen, Hauspferden und Przewalskipferden, sowie von mittelalterlichen und eisenzeitlichen Pferden.

Keywords: Bone microstructure, histomorphometry, discriminant analysis, species identification, domestication
Knochenmikrostruktur, Histomorphometrie, Diskriminanzanalyse, Speziesidentifikation, Domestikation

Introduction

Bone is living, connective tissue that is capable of lifelong remodeling. Possessing great plasticity, it can adapt to meet different biomechanical demands such as activity and weight by changing its mineral density, shape and microstructure.

The three main principal cell types that mediate the structural and functional properties of bone are: *osteoblasts*, which synthesize the bone matrix, *osteoclasts* which enable resorption of bone matrix and *osteocytes*, which support the bone structure and are organized throughout the mineral bone matrix. The ability of bone to adapt can be seen in different types of bone. Higher primates, including man, for example, mostly show a type of bone called *secondary osteonal* or *Haversian* bone, which shows good elastic properties (Fig. 1,2,3).

Secondary osteons are a result of remodeling. This is achieved when *osteoclasts* resorb existing bone by drilling a tunnel within the existing bone matrix. *Osteo-*

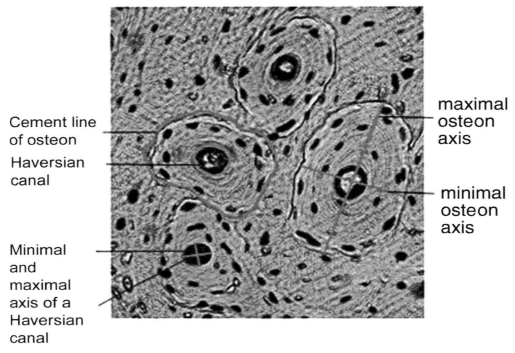

Fig. 1: Histomorphometric analysis at 160 x, thin section from the humerus of a chimpanzee *(Pan tr. troglodytes)*. Digital image of bone microstructure showing outlined osteons and Haversian canals and the measured maximal/minimal axis.

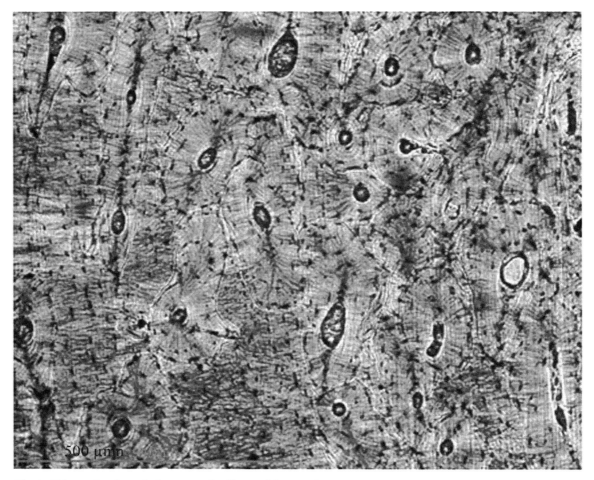

Fig. 2: Thin section from the humerus of a silvery gibbon *(Hylobates moloch)* at 160 x. Secondary osteons are shown.

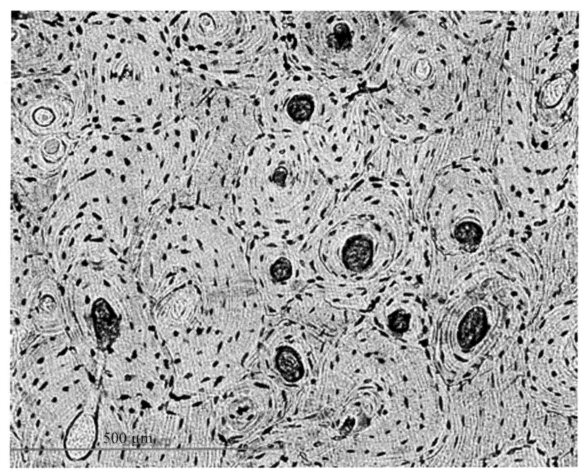

Fig. 3: Thin section from the humerus of a lowland gorilla *(Gorilla g. gorilla)* at 160 x. Secondary osteons are shown.

blasts then fill the lacuna with concentric lamellae of bone and the process is completed by a subsequent invasion of blood vessels and finally, in the formation of a new osteon. The boundary that separates primary (formed *de novo*) from secondary osteons (result of remodeling) is known as the *cement line*.

Reorganization of bone continues throughout life to mend bone damage, such as microfractures, as well as to repair wear and tear. It is likely that secondary osteons whose cement lines conteract the spread of cracks, confer an evolutionary advantage (Martin & Burr 1982).

Another type of bone, the *plexiform* bone, is found when strength and fast formation rate is needed, such as in cattle, sheep, elephants – but also primary and secondary osteons are found within *plexiform* bone.

The basic idea for our study was actually laid down as early as 1870, when Julius Wolff in his visionary treatise on the architecture of bones and its relationship to growth, formulated what is known today as "Wolff's Law." Conceptualized in an age without the aid of microtechnology and when the concept of micromechanical principles was not part of the scientific vocabulary, Wolff was nevertheless able to state that, "Bone will adapt its structure to meet the functional demands placed upon it, and will therefore alter its mass and morphology so that it can withstand the extremes of functional loading" (Wolff 1870). His theories have been confirmed because in accordance with his law, different animal species with specific phylogenetic and life histories exhibit distinctive bone microstructures as a result of adaptation to species-specific demands. In the higher primates, for example, different forms of locomotion, like knuckle walking in the case of gorillas and chimpanzees, suspensory behavior in orangutans, brachiatory behaviour in gibbons and bipedalism in man, have lead to distinctive microstructures in the humeri. Where recently and historically domesticated animals are concerned, the different bone microstructures are the result of breeding techniques, farming conditions, work, restriction of mobility and heavier weight and body size. This study presents a method which permits the identification of distinct species as well as differentiation between closely related species.

Material and Methods

The research substrate consisted of macerated bones from a total of 75 individuals (Tab. 1), all of which were provided by the State Collection of Anthropology and Palaeoanatomy of Munich. The sample primates (*Hylobates, Pongo, Pan, Gorilla*), all wild animals from Borneo, south Cameroon and south Congo, were shot at the end of the 19th century and the beginning of the 20th century. Our research includes all primate humeri that are still available in Munich. However, due to the fact that part of the collection was lost in the Second World War, its samples are of most but not all primates. In addition, bone microstructure of our living next kin is hardly examined because of the difficulty in obtaining material. If and when samples were available, they were usually from dead zoo animals (Schaffler & Burr 1984).

The representatives of the species Homo are from the medieval Zeholfing, Bavaria (A. D. 1200 – 1500, unpublished material) and the Peruvian desert zone, Pacatnamu (A. D. 115 – 1400, Ubbelohde-Doering 1958, 1959, 1960). All the historic animal bones come from Bavarian excavation sites, namely, the Neolithic Pestenacker (3496-3517 B.C., Vagedes 1998, von den Driesch & Gerstner 1993), the Roman Period Oberstimm (A.D. 14-70, Stettmer 1997), the Iron Age Celt-Oppidum Manching (450 B.C., Maier 1992), as well as from the medieval Karlburg (A.D. 741 – 1236, Vagedes 1994). Recent material (pigs, cattle) was sourced from several abattoirs and zoos (Przewalski horses, kulans) and the mule samples were provided by the Berchtesgaden Mountain Regiment whose stables include a large number of pack mules.

To exclude age-related differences like fewer number of secondary osteons in young individuals and osteoporotic bones in older samples, only adult animals were selected for study. The bones showed no pathological variations on their surfaces.

Cores of bone were drilled from adult and healthy metacarpi (cattle, equids), radii (pigs, sheep, goats) and humeri (primates), at 80% distal (primates) or proximal (domesticated animals) diaphyseal length on the front side and then embedded in epoxy resin Biodur (Gunther von Hagens). The selection of bone types under study was based on their frequency in archaeological finds.

No muscle marks, which alter the microstructure (Currey 1984) were found at the named drilling locations.

Thin, transverse sections (70 – 100 µm) were prepared with a sawing microtome (Leitz 1600) and stuck on glass slides with Eukitt (Kindler). Examination was done by transmission light microscopy at 80 x and 160 x and digital images were taken (microscope: Axioskop 2, digital camera: SV Sound Micro, software: Axio Vision; all Zeiss). The sections were photographed starting with the periosteal side (outer layer) and ending with the endosteal side (inner layer) of bone. Measurements were taken of all mature *osteons* that were surrounded by concentric lamellae of bone, were bounded by *cement lines,* were not in a resorption phase and could clearly be outlined by the Adobe Photoshop 5.0 computer program. The software SCION Image allows measuring of the following parameters: maximum/minimum osteon axis, osteon perimeter, osteon area, maximum/minimum Haversian canal axis, Haversian canal perimeter, Haversian canal area (see Fig. 1). The results were expressed in µm and µm^2.

Statistical analyses were performed in the computer program SPSS 10.0 using canonical discriminant analysis.

Results

Tab. 1 shows all examined species as well as the averages and standard deviations of the analyzed parameters. The last columns provide information about the amount of Haversian canals and osteons measured per species. The smaller number of osteons is due to two reasons.

Osteons are often found truncated in digital pictures and therefore not measurable in contrast to the accompanying Haversian canals. Thus, only secondary osteons that were clearly outlined on their cement lines were used for the analysis. However, since numerous species showed many primary osteons, only the central blood canals could be evaluated in these cases.

A look at the averages of each single species shows that sheep, goats and pigs have smaller Haversian canals in comparison with all other examined species. Within the closely related equids, kulans show the largest area of Haversian canals. Donkeys and mules, on the other hand, show similar Haversian canal areas but different osteon areas while Neolithic horses have the smallest osteon areas. Within the primates, gorillas show the largest Haversian canals, followed by orang-utans and chimpanzees. Compared to chimpanzees, humans show slightly smaller Haversian canals, but differ in osteon size.

Since different species thus exhibit different microstructural parameters, a mathematical combination of all parameters should permit a separation of the species. This combination is done via discriminant analysis by calculating variables for each parameter (canonical variables), which are then multiplied with the parameter. The aim is to maximize the differences between the groups' means. A linear combination of these factors plus a constant leads to a typical discriminant value for

	Species	Number	Area Hc	Perimeter Hc	Max. axis Hc	Min. axis Hc	Area Ost.	Perimeter Ost.	Max. axis Ost.	Min. axis Ost	Number HK	Number Ost.
Mean values	Modern cattle	4	368,1	69,5	23,5	18,4	15601,1	474,2	157,5	121,7	437	217
	Neolith. cattle	4	769,2	104,9	34,9	26,9	10762,7	394,9	130,8	97,3	322	303
	Modern pigs	2	325,5	65,0	21,7	17,5	13701,5	436,9	142,5	114,9	50	38
	Roman period pigs	3	397,4	71,2	24,3	18,5	15626,2	460,7	154,9	115,7	90	72
	Przewalski horses	4	968,5	108,4	36,4	29,1	28997,5	642,6	211,9	164,1	377	214
	Modern horses	5	786,7	100,3	33,7	26,9	27294,3	619,6	205,8	158,2	591	460
	Modern donkeys	3	691,6	92,4	30,9	24,8	29220,8	647,8	213,5	162,6	292	129
	Modern mules	3	690,7	95,5	31,3	26,2	26385,1	616,9	200,9	155,7	299	216
	Modern kulans	5	1073,2	120,9	40,6	31,8	16359,3	477,3	161,1	119,7	161	157
	Neolith. horses	5	903,2	111,8	37,2	28,7	14428,9	461,3	151,9	112,2	220	220
	Iron age horses	3	935,0	111,3	37,2	29,6	7708,9	312,7	103,2	82,9	200	200
	Medieval horses	3	755,5	101,3	34,0	26,7	11366,2	373,5	122,6	98,6	140	140
	Modern sheep	4	396,5	74,1	25,4	19,2	10568,1	372,9	130,7	86,4	87	80
	Modern goats	5	233,4	55,1	18,8	14,4	17880,7	513,8	176,8	123,9	319	130
	Gibbons	5	1256,1	132,7	45,4	33,4	23471,0	597,2	203,4	139,1	595,0	280,0
	Orang-Utans	4	1992,5	164,2	54,5	43,0	34640,5	719,3	236,7	177,5	240,0	190,0
	Chimpanzees	5	1547,5	143,1	49,8	36,1	27858,0	633,3	215,6	150,8	287,0	270,0
	Gorillas	3	2356,2	176,8	59,9	47,4	38018,0	746,0	251,0	186,2	211,0	184,0
	Humans	4	1428,6	140,2	50,2	33,3	39823,3	805,3	281,1	174,0	223,0	170,0
Standard-deviation	Modern cattle	4	87,67	8,02	2,79	1,85	1799,29	27,96	10,32	6,72		
	Neolith. cattle	4	90,34	6,14	2,97	1,67	447,03	8,32	3,63	3,61		
	Modern pigs	2	96,56	8,75	3,09	2,29	593,29	19,66	4,95	8,88		
	Roman period pigs	3	95,46	7,89	2,31	2,39	7085,51	100,96	30,68	25,82		
	Przewalski horses	4	205,69	10,75	3,80	2,62	4979,88	48,83	16,02	15,69		
	Modern horses	5	200,43	12,09	4,03	3,20	4781,25	54,12	18,42	14,03		
	Modern donkeys	3	98,91	5,93	2,16	1,26	3331,51	41,09	15,92	12,88		
	Modern mules	3	86,82	6,36	2,18	1,61	2396,66	40,72	9,50	11,60		
	Modern kulans	5	166,91	9,30	3,15	2,47	3310,16	48,50	15,86	12,61		
	Neolith. horses	5	316,29	21,51	7,02	5,52	3899,56	67,94	16,65	21,07		
	Iron age horses	3	194,86	10,73	3,50	3,08	3036,80	64,11	20,99	17,85		
	Medieval horses	3	82,49	5,86	2,22	1,52	9156,10	167,30	54,39	43,74		
	Modern sheep	4	83,50	7,24	1,75	2,69	5436,51	106,65	33,79	28,83		
	Modern goats	5	102,33	10,18	3,25	2,88	3310,18	43,64	14,48	15,04		
	Gibbons	5	254,97	12,67	3,94	3,71	4367,31	44,60	12,10	14,12		
	Orang-Utans	4	490,01	16,41	5,96	5,54	3447,90	30,53	12,34	3,88		
	Chimpanzees	5	347,56	13,06	4,60	2,90	4855,76	70,35	27,24	9,89		
	Gorillas	3	619,10	20,04	6,24	9,08	1809,35	29,50	17,50	4,50		
	Humans	4	139,80	9,81	4,61	1,50	5884,62	65,80	33,75	8,26		

Tab: 1: Number of examined animals, means of measured parameters in µm (perimeter, minimal, maximal axis) and µm² (area) and standard deviation. Number of counted H*aversian canals* (Hc) and osteons (Ost).

each species. Group members will scatter around this discriminant value (Bortz 1999). In the case of the equids (without Przewalski horses) a 100% separation of the species was achieved with the significant discriminant functions (P<0.000; Wilks-Lambda = 0.000, Chi-square = 137,079). The Haversian canal minimal axis parameter was excluded because it does not have significant influence on the differentiation of the species. The other seven variables permit the application of linear discriminant functions that may be used to determine whether the osteons (Ost) and Haversian canals (Hc) are of modern, Neolithic or Przewalski horse, mule, kulan or donkey origin.

SPSS calculates n-1 discriminant functions (n = number of groups). In case of the equids, six groups, this means five functions were calculated. The first two, at the same time the best functions are shown.

Function 1 = − 0,0966845 + 0.0259231*AreaHc − 0,5655926*PerimeterHc + 0,3275892*Max.axisHc −0,0006901*AreaOst − 0,0389993*PerimeterOst + 0,2079689*Max.axisOst +0,1751808*Min.axisOst

Function 2 = − 9,052214 + 0,0104503*AreaHc − 0,7528728*PerimeterHc + 1,849940*Max.axisHc − 0,0010132*AreaOst − 0,2273484*PerimeterOst + 0,4632554*Max.axisOst + 0,5575127*Min.axisOst

Fig. 6 illustrates the first two discriminant functions in a two-dimensional diagram for the equids. For each group a mean discriminant value, the "group centroid," is evaluated and the case classifications are distributed around it.

Discriminant analysis does not allow distinguishing between modern horses and Przewalski horses (Fig. 8). Also medieval and Iron Age horses form an overlap, but from the others distinct group (without Fig.).

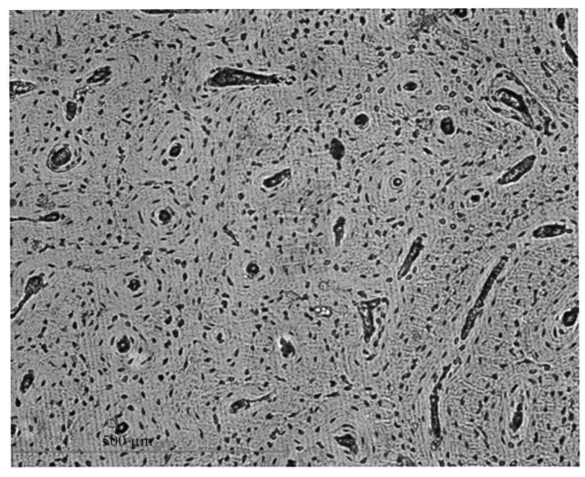

Fig. 4: Thin section from the metacarpus of a modern cattle *(Bos taurus)* at 160 x. Primary osteons and osteonbanding (on the left) are shown.

In the case of the primates, a separation of 90.9% was achieved (P<0.002; Wilks-Lambda = 0.016 Chi-square = 59.718) (Fig. 7).

For the domesticated animals, an 83.3% correct classification was achieved. However, it was not possible to distinguish cattle from pigs and modern horses could not be distinguished from Przewalski horses. The separation succeeds very well between goats and sheep (Fig. 8). It was also possible to distinguish cattle from horses. Test of the functions yielded significance in P<0.000; Wilks-Lambda = 0.003, Chi-square = 93.388.

Roman period pigs and modern pigs form an overlapping group and so they are not distinguishable from one another (without figures).

Discussion

Apart from the functional aspects of bone microstructure for all the species examined here, the principle observed already by Jowsey (1966) is fully demonstrated in this study, namely, that the size of osteons and Haversian canals increases with the size of the animal. Accordingly to Jowsey's findings, the largest primates, gorillas, show the largest Haversian canals; orang-utans and chimpanzees do exceed humans in the size of the Haversian canals, but not in osteon size. The sexual dimorphism, particularly found in gorillas and orang-utans provides no influence on the size of the microstructural units.

The groups of gorillas and orang-utans as well as the groups of chimpanzees and gibbons overlap in the two-dimensional representation (fig. 7). Because SPSS calculates four discriminant functions, i.e., always n – 1 function, the separation of the species succeeds up to 90.9 %. Humans with non-locomotory upper limbs are distinct from the other species. The great apes, gorillas and orang-utans are grouped closely together by discriminant

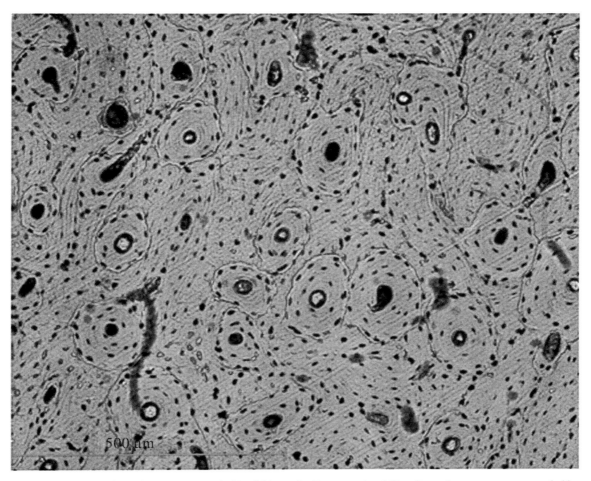

Fig. 5: Thin section from the metacarpus of a Neolithic cattle *(Bos taurus)* at 160 x. Secondary osteons, surrounded by the cement line, are shown.

analysis, just as the relatively smaller species, chimpanzees and gibbons are. Two animals were not classified correctly. One gibbon was wrongly classified as a chimpanzee and one chimpanzee was wrongly classified as a human. The body size of an animal and /or genetic relationships seem to have a greater influence on the size of the measured microstructural parameters than locomotory aspects. Qualitatively, the bone microstructure of all primates may be described as "mostly secondary Haversian bone" (Fig. 2, 3).

The species identification in the case of the recent equids succeeds, with exception of the Przewalski horses, 100%. Since wild Przewalski horses are extinct, a zoo population was examined. Life circumstances of these Przewalski horses are similar to that of domesticated horses and are probably the reason why they show similar microstructure.

The research potential of the field of palaeoanatomy can be expanded when more cases can be cleared, for instance, to distinguish donkeys, mules and horses in archaeological finds that contain bone fragments, under the prerequisite of a reproducible research location on the bones. A good example of its application is during the discovery of an equid mass grave of a Roman castle near Weißenburg, Bavaria. Using the histomorphometric metacarpal bone microstructure analysis, it should be possible to distinguish the species of the 35 mules and horses excavated. The findings could establish the important role mules played as pack and draught animals in the Roman army (Peters 1998). Until then, because of identification difficulties, the number of mules used by the Roman army was probably underestimated. Another advantage is that the analysis used metacarpi, a bone that is often well preserved in archaeological finds, just as it was in this case.

Histomorphometry, in addition to being a tool for species identification can also offer morphological information. Just as it is possible to distinguish between sheep and goats using the microstructure of the radius, it is also possible to see morphological sense because domestication leaves its mark on the size of the microstructural units. During the Roman period, for example, when increased animal breeding led to a

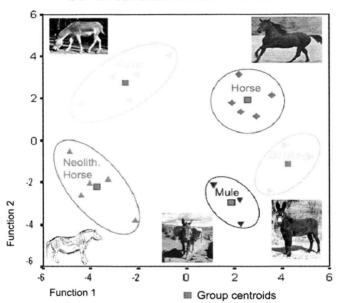

Fig. 6: Graphical representation of the canonical discriminant analysis for equids. Group centroids, dispersion of individuals are shown. 100% correct classification of group cases.

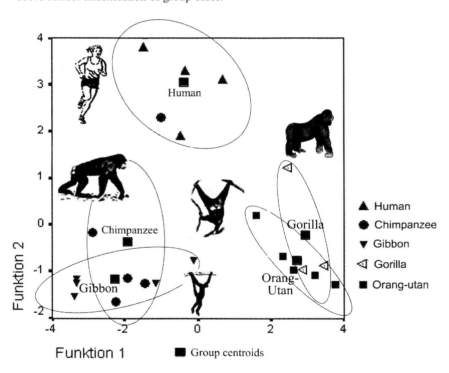

Fig. 7: Graphical representation of the canonical discriminant analysis for primates. Group centroids, dispersion of individuals are shown. 90.9% correct classification of group cases.

marked increase in the size of the animals (Benecke 1994), it was reflected in a corresponding increase in the bone microstructure of the animals. Hence, Roman period pigs show microstructural units even larger than those of recent pigs examined (Tab. 1), though they are not separable from each other by discriminant analysis. In contrast to the Roman period (Oberstimm: A.D.14 – 70) when increased animal farming led to an increase in animal size, its decline in the Middle Ages is reflected in the microstructural units of horses. The microstructures of the horses from the medieval valley village of Karlburg (A. D. 741 – 1236) are clearly smaller than those of the Iron Age (Manching: 450 – 12 B.C.) and Neolithic horses (Pestenacker: 3496 B.C.). Neolithic horses can

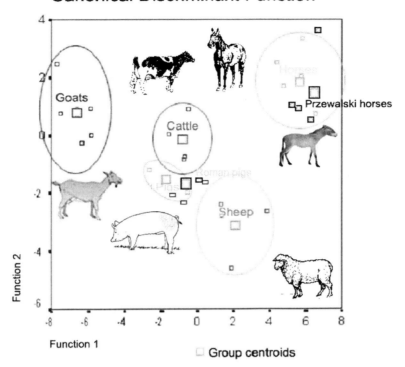

Fig. 8: Graphical representation of the canonical discriminant analysis for domesticated animals. Group centroids, dispersion of individuals are shown. Cattle and pigs, like modern horses and Przewalski horses are not distinguishable from one another. Sheep and goats can be distinguished.

be separated clearly from all others while medieval and Iron Age horses overlap.

Apart from the temporal influence of domestication, differences or similarities in the size of microstructural units can also be due to different animal farming habits. Modern cattle, for example, were kept for milk and meat supply, while recent horses were used mostly for leisure and sport. Furthermore, microstructural units from modern cows are very similar to those of modern and Roman period pigs, but different from Neolithic cattle. The Neolithic forms probably were more mobile.

Qualitatively, the microstructure of Neolithic cattle also differs from that of recent cattle (Fig. 4, 5). Neolithic animals tended to show more secondary osteons. This confirms the studies done by Lasota-Moskalewska & Moskalewski (1980). They found that medieval pigs also showed more secondary osteons in comparison to recent mast pigs. It is probable that diminished movement limits the development of secondary osteons. This thesis is supported by the experiments of Owerkowicz & Crompton (1995), who found a reinforced remodeling, and therewith connected, more secondary osteons, in bones of animals which had been forced to run in an impeller. Larger Haversian canals permit a better blood supply (Martinko et al. 1989) and could therefore be expected in more active types, such as the Neolithic animals. This was also true in kulans (they show big Haversian canals), the Asian desert donkeys, some of which were captured wild.

Osteonbanding, or the stringing of osteons like pearls on a string, and which can be used to distinguish human from nonhuman bone (Mulhern & Ubelaker 2001), also appeared reinforced in the remains of domesticated animals (Fig. 4). It is therefore possible, that modern domestication with its selective breeding systems and emphasis on diminished movement and nourishment supplementation with hormone and antibiotics could have an influence on the bone microstructure of recent animals. Prerequisite for all encountered statements of this investigation was however the strict retention of a defined research location on selected long bones.

Conclusions

Discriminant analytical combination of microstructural parameters (area, circumferences, minimal and maximal axes of osteons and Haversian canals) can identify a species in most cases. In future, increase in the number of species studied will lead to better models of species differentiation, and eventually, better differentiation within closely related species will be possible.

Summary / Zusammenfassung

Microstructural parameters (areas, circumferences, maximal and minimal diameters of osteons and Haversian canals) of long bones, such as the humeri of higher primates, the metacarpi of equids and cattle, the radii of pigs, sheep and goats were measured and species-specific bone microstructures were proven in most cases. Within the equid group, modern horses, donkeys, mules, kulans and Neolithic horses can be separated up to 100% from one another. Modern horses and Przewalski horses, however, cannot be distinguished via discriminant analysis from one another, probably due to what could be similar farming conditions. Since wild Przewalskii horses are extinct, a zoo population was examined. Also medieval and Iron Age horses, both of which show relatively small osteonal structures, cannot be separated from one another.

Higher primates such as gorillas, chimpanzees, gibbons, orang-utans and humans differ on the basis of their locomoto adaptations, namely, knuckle walking in gorillas, knuckle walking and acrobatic climbing in chimpanzees, brachiatory behaviour in gibbons, suspensory behaviour in orangutans and bipedal locomotion in humans. However, only humans with their non-locomotory upper limbs are separated from the other primates by discriminant analysis. The values of the examined microstructural parameters seem to depend to a higher degree on the body size of the species than on locomotory factors. So, gorilla and orang-utan are grouped together same as chimpanzee and gibbon (fig. 7). Yet, the separation of the species succeeds up to 90.9 % applying discriminant analysis.

The sexual dimorphism found particular in gorilla and orang-utan with big males and smaller females has no influence on the values of the microstructural parameters. Therefore, apart from the relationship between body size and bone microstructure, one should also consider genetic factors.

Interestingly, where the domestic animals are concerned, sheep and goats can be distinguished based on their bone microstructure, although they share similar bone morphology. Modern horses can be separated well from modern pigs and cattle. However, modern cattle are very similar to modern and Roman period pigs and cannot be separated with security from them. Recent and Roman period pigs form an overstocking, not distinguishable group.

Mikrostruktureinheiten (Flächen, Umfänge, maximale und minimale Durchmesser von Osteonen und Havers'schen Kanälen) von Langknochen wie Oberarmknochen von höheren Primaten, Mittelhandknochen von Pferdeartigen und Rindern, sowie Speichen von Schweinen, Schafen und Ziegen wurden vermessen und eine speziesspezifische Knochenmikrostruktur konnte in den meisten Fällen nachgewiesen werden.

So lassen sich innerhalb der Equiden-Gruppe Hauspferde, Esel, Maultiere, Halbesel und neolithische Pferde zu 100% voneinander trennen. Haus- und Przewalskipferde sind diskriminanzanalytisch nicht voneinander zu unterscheiden, was auf ähnliche Haltebedingungen zurückgeführt werden könnte. Da Przewalskipferde in der Natur ausgestorben sind, wurde eine Zoopopulation untersucht. Auch mittelalterliche und eisenzeitliche Pferde, die beide relativ kleine Osteostrukturen aufweisen, lassen sich nicht voneinander trennen.

Höhere Primaten wie Gorillas, Schimpansen, Gibbons, Orang-Utans und Menschen unterscheiden sich auf Grund ihrer Lokomotionsweisen wie Knöchelgang (Gorillas), Knöchelgang und akrobatisches Klettern (Schimpansen), Schwinghangeln (Gibbons), suspensorisch (Orang-Utans) und biped (Menschen). Allerdings grenzt sich nur der Mensch mit seinen freien Vorderextremitäten diskriminanzanalytisch von den anderen Primaten ab. Die Werte der mikrostrukturellen Parameter sind stärker von der Körpergröße der jeweiligen Spezies abhängig als von lokomotorischen Faktoren, so dass die Gruppen der Gorillas und Orang-Utans ebenso wie die der Schimpansen und Gibbons zusammenfallen (Fig. 7).

Der besonders bei Gorilla und Orang-Utan ausgeprägte Sexualdimorphismus mit großen Männchen und kleinen Weibchen wirkt sich dagegen nicht auf die Werte der Mikrostrukturparameter aus. Daher muss neben einem Zusammenhang von Individuengröße und Knochenmikrostruktur auch an genetische und verwandtschaftliche Faktoren gedacht werden.

Interessant bei den Haustieren ist, dass sich Ziegen und Schafe anhand ihrer Knochenmikrostruktur unterscheiden lassen, obwohl die Morphologie der Knochen beider Spezies recht ähnlich ist. Rezente Pferde lassen sich gut von rezenten Schweinen und Rindern trennen. Rezente Rinder sind modernen Schweinen und römerzeitlichen Schweinen (beide letztgenannten bilden eine überlagernde Gruppe) sehr ähnlich und nicht mit Sicherheit zu trennen.

Acknowledgements

The author thanks the Ludwig-Maximilian-University of Munich for providing a scholarship, the Deutsche Forschungsgemeinschaft for kindly sponsoring the research project and the State Collection of Anthropology and Palaeoanatomy of Munich for free access to the samples.

References

Benecke N., 1994.
  Der Mensch und seine Haustiere. Die Geschichte einer Jahrtausende alten Beziehung. Stuttgart: Theiss.

Bortz J., 1999.
  Statistik für Sozialwissenschaftler. Berlin, Heidelberg, New York: Springer-Verlag.

Cattaneo C., DiMartino S., Scali S., Craig O.E., Grandt M. & Sokol R.J., 1999.
  Determining the human origin of fragments of burnt bone: a comparative study of histological, immunological and DNA techniques. *Forensic Science International* **102**: 181-191.

Currey J., 1984.
  The Mechanical Adaptations of Bones. New Jersey: Princeton University Press

von den Driesch A. & Gerstner H., 1993.
  Tierreste aus der jungneolithischen Siedlung von Mamming, Ldkr. Dingolfing-Landau. Festschrift Torbrügge: 48-55.

Jowsey J., 1966.
  Studies of Haversian systems in man and some animals. *Journal of Anatomy* **100/4**: 857-864.

Lasota-Moskalewska A. & Moskalewski S., 1980.
  Microscopic comparison of bones from medieval domestic and wild pigs. *Ossa* **7**: 173-178.

Maier F., 1992.
  Ergebnisse der Ausgrabungen 1984 – 1987 in Manching. Stuttgart: Steiner

Martin R.B. & Burr D.B., 1982.
  A hypothetical mechanism for the stimulation of osteonal remodelling by fatigue damage. *Journal of Biomechanics* **15**: 137-139.

Martinko V., Belay M., Machay S. & Jelinek L., 1989.
  Biomechanics of the bones and skeleton. III. Microstructure. *Acta Chir. Orthop. Traumatol. Cech.* **56/2**: 160-168.

Mulhern D.M. & Ubelaker D.H., 2001.
  Differences in osteon banding between human and nonhuman bone. *Journal of Forensic Science* **46/2**: 220-222.

Owerkowicz T. & Crompton A., 1995.
  Bone of contention in the evolution of endothermy. *Journal of Vertebrate Palaeontology* **15/3**: 47 Abstract.

Peters J., 1998.
  Römische Tierhaltung und Tierzucht. Eine Synthese aus archäozoologischer Untersuchung und schriftlich-bildliche Überlieferung Passauer. Universitätsschriften zur Archäologie **5**. Rahden/Westfalen: Verlag Marie Leidorf GmbH.

Schaffler M.B. & Burr D.B., 1984.
  Primate Cortical Bone Microstructure: Relationship to Locomotion. American Journal of Physical Anthropology **65**: 191-197.

Stettmer A., 1997.
  Die Tierknochenfunde aus dem römischen Kastell Oberstimm, Ldrk. Ingolstadt/Bayern (Grabung 1994). Dissertation der Ludwig-Maximilians-Universität München.

Ubbelohde-Doering H., 1958.
  Berichte über archäologische Feldarbeiten in Peru 1. *Ethnos* **23**: 67-99.

Ubbelohde-Doering H., 1959.
  Berichte über archäologische Feldarbeiten in Peru 2. *Ethnos* **24**: 1-32.

Ubbelohde-Doering, H., 1960.
  Berichte über archäologische Feldarbeiten in Peru 3. *Ethnos* **25**: 153-182.

Vagedes K., 1994.
  Die Tierknochenfunde aus Karlburg – ein osteoarchäologischer Vergleich zwischen mittelalterlicher Burg und Talsiedlung. Dissertation der Ludwig-Maximilians-Universität München.

Vagedes K., 1998.
  Haus- und Wildtiere im Umfeld jungneolithischer Siedlungen bei Landsberg am Lech. Documenta naturae: **118**. München

Wolff J., 1870.
  Über die innere Architektur der Knochen und ihre Bedeutung für die Frage vom Knochenwachstum. *Virchows Archiv für Pathologische Anatomie und Physiologie* **50**: 389-453.

Variations in dental microwear and abrasion in ancient human groups of southern Germany: 7500 BP to the Early Middle Ages

Irene Luise Gügel, Department Biologie I,
Bereich Biodiversitätsforschung/Anthropologie, Universität München

Abstract / Zusammenfassung

Dental wear and microwear is of special interest when reconstructing patterns of dietary behaviour in the human past. We examined the wear and microwear patterns on the molars of three south German human groups, namely, a Mesolithic hunter-gatherer population, a Bell Beaker population from the end of the Neolithic and three populations from the Early Middle Ages. The study also includes examinations of the Kaufertsberg individual (Mesolithic) and the first Hominid of Oldoway OH1 (Paleolithic). The rate of wear was calculated by scoring the four quadrants of the Lower First and Lower Second Molars of the individuals. Analyses of the principal axis provided distinct slopes and scores for M_1 at eruption point M_2 as well as scores for M_2 at saturation score of M_1 for each of the groups. The angles and degrees of wear were taken and they showed that the regression slopes separate the hunter-gatherers from the agriculturalists. The dental microwear analyses show significant differences between the human groups in terms of the enamel regions analysed, i.e., in feature densities, proportions, sizes and orientation. In particular, a gradual change of microwear pattern from Mesolithic (Ofnet Cave) to end Neolithic (Weichering) to the Early Middle Ages (Bittenbrunn, Waging and Brombach) is indicated. In addition, the analyses done in this study further established the Kaufertsberg individual as, in all likelihood, to have been a carnivorous hunter-gatherer, like the Ofnet-Cave individuals. This supports the anthropological hypothesis that the Kaufertsberg Skull belongs to the series from the Mesolithic. It also established the Oldoway Hominid OH 1 as a hunter-gatherer as well, but one with lower feature density, slightly higher pit size and scratch length than those of the Bavarian populations.

Die Zahnabtragung und dentale Microwear ist zur Rekonstruktion von Ernährungs- und Nahrungsverhalten in der menschlichen Vergangenheit von besonderer Bedeutung. Muster der Abtragung und der dentalen Microwear wurden an menschlichen Gruppen aus dem süddeutschen Raum untersucht. Eine Jäger-Sammler Gruppe aus dem Mesolithikum, eine Gruppe der Glockenbecherkultur des Endneolithikums, und drei Gruppen aus dem frühen Mittelalter. Die Kopfbestattung vom Kaufertsberg, die in das Mesolithikum datiert wird, als auch der erste Hominide Oldoway OH 1 wurden in die Untersuchung aufgenommen. Die Geschwindigkeit der Zahnabtragung wurde ermittelt, indem jeder der vier Quadranten des unteren ersten und zweiten Molaren der Individuen bewertet (score) wurde. Die Ermittlung der Hauptachsengleichung ergab Unterschiede in der Steigung, den Abtragungswerten von M_1 zum Zeitpunkt der Eruption von M_2 und den Abtragungswerten von M_2 am Sättigungswert von M_1. Die ermittelte Steigung aus der Regression zwischen Abtragungswinkel- und grad ergab eine Trennung zwischen Jäger-Sammler Gruppen und Ackerbauern. Die Analyse der dentalen Microwear macht signifikante Unterschiede zwischen den menschlichen Gruppen deutlich. Diese betreffen im besonderen die untersuchte Zahnschmelzregion, die Merkmalsdichten, ihre Anteile, Größen und Orientierung. Eine stufenweise Änderung des Microwear-Musters vom Mesolithikum (Ofnet-Höhle), über das Endneolithikum (Weichering) in das frühe Mittelalter (Bittenbrunn, Waging und Brombach) ist zu erkennen. Die Untersuchungen ergaben einen weiteren Beleg dafür das das Individuum vom Kaufertsberg ebenfalls als vorwiegend carnivorer Jäger-Sammler gelebt haben dürfte wie die Individuen aus der Ofnet-Höhle. Dies unterstützt die Hypothese der Anthropologen, dass der Schädel vom Kaufertsberg zur Serie von Kopfbestattungen aus dem Mesolithikum gehört. Die Untersuchungen gaben auch Hinweise dafür, dass der Oldoway Hominide OH 1 als Jäger-Sammler gelebt haben könnte, seine Zahnoberflächen zeigen jedoch eine niedrigere Merkmalsdichte, geringfügig größere Gruben und längere Kratzspuren als diejenigen der Bayerischen Populationen.

Keywords: Dental abrasion and microwear, SEM, dietary behavior, food preparation
Abtragung, dentale Microwear, SEM, Ernährungsverhalten, Nahrungsmittelzubereitung

Introduction

The reconstruction of paleodiets is of major concern to physical anthropology because the living conditions and health of human populations, their growth and developmental power, are all linked to food supply. Changes in subsistence strategies produce a variety of environmental and cultural consequences for humans and the human societies that employ them. Therefore, jaws from individuals living in different historical eras will indicate different degrees of intensity, patterns of wear and physiological and pathological biting positions. Teeth, in particular, are exposed to continuous mechanical, chemical and pathogenic stress from the time they erupt through the alveolar bone and are exposed to the oral environment. The resultant wear of dental enamel and dentine is irreparable, and because of the extraordinary physicochemical composition of hydroxylapatite, macroscopically and microscopically visible traces left by corrosion from acidic liquids (King et al. 1999a), abrasion (tooth-diet-tooth contact) and attrition (tooth-tooth contact) will remain on the surfaces as long as the teeth are not damaged posthumously (Gordon 1984, Gordon & Walker 1983, King et al. 1999a, Puech et al. 1985) or destroyed.

The advantages of using dentition in anthropological research are further enhanced by the fact that for much of human history, there have been no or almost no orthodontic preservative or therapeutic measures available. Therefore, well-preserved alterations on fossil teeth and jaws will reflect masticatory stress and dietary behaviour including aspects of food preparation in the human past (Rose & Ungar 1998, Teaford 1991, Walker et al. 1978).

Characteristic trace patterns and wear on functional teeth develop primarily during chewing (Fine & Craig 1981, Gordon 1982, Grine 1986, Grine & Kay 1988, Gügel 2001, Gügel et al. 2001, King et al. 1999b, Lalueza et al. 1993, Molleson et al. 1993, Puech 1992, Schmidt 2001, Ungar & Spencer 1999). They also result from distinctive dietary components, such as the toughness and hardness of foods consumed (Peters 1987), the quality and quantity of abrasive particles inherent in the foods, like siliceous opal phytoliths or exogenous grit (Danielson & Reinhard 1998, Lalueza & Pérez-Pérez 1994, Lalueza et al. 1996, Lucas & Teaford 1995, Teaford & Runestad 1992, Ungar et al. 1995), and the technological changes involved in food preparation (Molleson et al. 1993). The intensity of wear also increases with age (Bullington 1991), masticatory strength for breaking and grinding food (Gordon & Walker 1983, Ungar & Spencer 1999) and increasing content of abrasive particles in the diet (Gügel et al. 2001, Leek 1972, Teaford & Lytle 1996, Teaford & Oyen 1989a). The microscopic features of dental wear are mainly scratches, pits, gouges, furrows and the extent of polished area (Gordon 1982, 1984). To quantify dental microwear, scratches and pits on selected functional areas of enamel surfaces were measured and analysed by Dental Microwear Analysis (DMA).

The main purpose of the present study was to systematically investigate the wear and dental microwear of mandibular molars, to quantify their patterns and to compare the various samples taken from human groups that lived in different parts of southern Germany and on different dietary regimens.

Material and Methods

This study uses remains from several well-documented south German human populations from the Mesolithic to the Early Middle Ages. It also includes work on the Oldoway Hominid OH 1 (Paleolithic) and the remains of the individual excavated from Kaufertsberg at Lierheim (Donau-Ries; Mesolithic). All the specimens are housed in the State Collection of Anthropology and Palaeoanatomy in Munich.

An in-depth comparative study was made of the jaws from juvenile and adult human remains excavated from burial sites in Waging am See (n=10, Traunstein, late 6th to early 8th century), Bittenbrunn (n=8, Neuburg-Schrobenhausen, middle 5th to middle 7th century) and Brombach (n=4, Gunzenhausen-Weißenburg, late 7th to early 8th century), and Weichering (n=6, Neuburg-Schrobenhausen, Late Neoliithic), all of which are sites in Bavaria. In addition, adult individuals of the Ofnet Cave (n=12, Holheim near Nördlingen, Late Mesolithic, 7500 BP) were studied. The samples were selected on the basis of their all having nearly complete postcanine dentition. The Ofnet-Cave samples represented a unique dietary situation because they were victims of violence who died while still quite young and within a short time span (Mollison 1936, Orschiedt 1998) and were therefore expected to show a homogenous microwear pattern.

To analyse the patterns of dental wear, the following procedure was applied. First, the degree of abrasion on the Lower First (M_1) and Lower Second (M_2) Molars was determined by visually dividing the occlusal surfaces, which are clearly defined by their fissures, into four quadrants. Each quadrant was then scored on a scale from 1 to 10 according to Scott (1979a). The resulting composite score for each tooth was then registered on a 4-to-40 scale based on the quantity of enamel still present on the tooth.

Secondly, since pairs of adjacent molars erupt at six-yearly intervals, it is possible to calculate the degree of wear, which is independent from the age of the individual, from the time the second molar occludes with the first. The slope of the resulting equation – the principal axis analysis (Scott 1979b, Sokal & Rohlf 1969) – represents the rate of wear. None of the score values of the selected individuals studied exceeded the critical score of 36 (which would indicate severely worn teeth) and the pairs that had a first molar score lower than the second were excluded.

In the third phase of our tests the occlusal plane of M_2 was taken using a special instrument that had a long arm centred at the highest homologous points of the Lower Second Right and Left Molars. A second rotating arm was brought down across the occlusal surface and the pointer indicated the angle of deviation of crown surface from horizontal defined by the base of the instrument (Smith 1984). Positive values indicate tilts to buccal, negative values to lingual. The newly erupted molars showed physiological occlusal planes with negative tilt values.

DMA was performed on matched pairs of the Lower Second Molars which were selected according to the criteria mentioned by Gügel et al. (2001). All teeth were carefully cleaned and dehydrated by an increasing alcohol series, treated with acetone and vacuum-dried. The early medieval and late Neolithic tooth samples were used as originals. The Mesolithic and Paleolithic samples were prepared as casts according to Beynon (1987). The carefully prepared originals were covered with high-resolving plastic (President plus jet light body, Coltène Whaledent) to make negative casts which were in turn filled with epoxy resin (Araldite XW 396 and Araldite XW 397, 10:3, Ciba Geigy) to make positive casts. Several studies (Teaford & Lytle 1996, Teaford & Oyen 1989b, Ungar 1996) as well as internal tests show that this material reproduces features with resolutions of only 1 µm (approximately two-pixel distance) between the original and the casts. The originals and the positive casts were then sputtered with a thin, soft layer of gold (EMITECH K 550) and analysed using Scanning Electron Microscopy (SEM, LEO 1430VP and Philips XL 20 Series). The enamel surfaces were adjusted perpendicular to the electron beam to ensure symmetrical distribution of contrast and brightness and two-dimensional sharpness of picture in order to avoid any distortion of linear features. The total surfaces of regions of facets (which are created during the chewing cycle) on the second mandibular molars (region of facet 6: phase I – shearing, and region of facet 9: phase II – crushing and grinding, [Kay & Hiiemae 1974]) were inspected, as prescribed by Gordon (1982), in a standardised orientation from buccal to lingual and left to right on every surface (figure 1). At least two images (1024 x 725 pixels, resolution 260 dpi, LEO) were taken and stored as a – tiff-

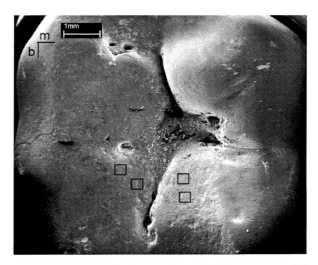

Fig. 1: SEM-image 10x, Lower Second Molar, areas sampled in this study.

file. The area of every image was approximately 0.500 x 0.350 mm^2 (0.176 mm^2). The analysis of the microwear features was done with the aid of the semi-automatic software, Microware 3.0, developed by Ungar (Ungar et al. 1991, Ungar 1995).

Every feature is identified by four coordinates. One pair describes the length and orientation of the major axis of the feature and the second pair, its lateral extension. The following data were analysed 1) total feature density (feature counts per image), 2) pit density, 3) scratch density, 4) mean length of pits, 5) mean width of pits, 6) mean length of scratches, 7) orientation of scratches and 8) distribution of scratch orientation (mean striation vector length = R [Ungar 1994]). The accumulation of artefacts, posthumous destruction of the surfaces and asymmetrical distributions of features led to an exclusion of the surfaces or the individual.

Statistical analysis of data was done on SPSS 11.0 (SPSS Inc., Chicago, IL.). Multivariate Analysis of Variance (MANOVA) was done to determine an overall difference in microwear patterns. The Analysis of Variance for Each Variable (ANOVA) was done with an F-test. Statistically significant variables were further analysed by applying a pairwise comparison with TukeyHSD (Honestly Significant Difference) to establish patterns of microwear variation. This was then followed by a canonical discriminant analysis to classify the human groups. The parameters which were considered useful for describing and discriminating samples of different origins were nine in total: mean total number of features, number of pits and scratches, proportion of pits, the length and width of pits, the length and standard deviation of length of scratches and the mean striation vector length.

Results

Rate of wear Although the range of the mean degree of wear on the mandibular First and Second Molars is small (Table 1: Bittenbrunn M_1/M_2 21.1/14.8; Brombach M_1/M_2 23.7/14.2; Waging M_1/M_2 18.6/14.7; Weichering M_1/M_2 22.7/16.4; Ofnet Cave M_1/M_2 20.6/13.6), it nevertheless demonstrates differential rates of wear between the groups. A slope of wear of approximately 1.0 would indicate that the wear rates at M_1 and M_2 are nearly equal, whereas a slope-of-wear value smaller than 1.0 would indicate higher rates for M_2 or lower rates for M_1. Conversely, a slope-of-wear value that is greater than 1.0 would indicate lower rates for M_2 or higher rates for M_1. Bittenbrunn shows the lowest slope (s = 0.589) with no overlap compared to all the other human groups and starts with the highest score (14.7) at the point of eruption of M_2 (Table 1, fig. 2). Thus, Bittenbrunn would reach its saturation score of 40 only when M_2 reaches an estimated saturation score of about 47.0. Among the other groups, the 95% Regions of Confidence overlap, but the mean slopes tend to 1.0 in Weichering (1.073) and Waging (1.083). It is a little higher in Brombach (1.156) and is highest in the Ofnet-Cave samples (1.394). This indicates that their molars were worn at an equal or at a slightly faster rate than molars which had earlier occlusal contact. The values for M_1 at the moment of eruption of M_2 (score = 4) are: Brombach (11.9), Weichering (9.3), Waging (7.0) and Ofnet Cave (6.4). All groups, except Bittenbrunn, reach the saturation score of 40 on M_1 at a score approximately between 28.1 and 34.5 on M_2.

Wear angle Due to the low integrity of the alveolar bones for an analysis of the occlusal tilt of the molars, the pattern of wear angles and linear regressions were calculated only for the groups from Waging, Bittenbrunn and Ofnet Cave. The data for Kaufertsberg and Oldoway were included as isolated cases (fig. 3). The slopes of linear regression for Bittenbrunn (1.341) and Waging (1.534) are distinctly higher than that of the Ofnet Cave (0.534). The angles of the occlusal planes of the individuals from Kaufertsberg and Oldoway OH 1 are flat, and their patterns of wear on all the posterior teeth are similar to those of the Ofnet Cave. Figures 4 and 5 show regions of facets 6 and 9 of the enamel surfaces of every human group and the remains from Kaufertsberg and the Oldoway Hominid OH 1.

Dental microwear The results of dental microwear are summarized in Table 2 and include the mean values, standard deviation and numbers of areas. The total number of areas analysed appear low because half of the SEM analyses was done at a lower resolution on a Philips XL 20 Series SEM (177 dpi), and the other half on a LEO 1430 VP SEM (260 dpi). Although the surface areas photographed were the same, the change of picture resolution from 177 to 260 dpi also changed the absolute pixel size from 0.717 µm to 0.4885 µm. This has influenced the detection of dimensions and numbers of the parameters statistically significantly, so that the datafiles could not be taken together. The main differences are in the number and dimensions of short and thin scratches. The resulting tendencies of the patterns are similar, but the two sets of data needed separate analyses. While this lowers the statistical validity, it does, at the same time, reiterate the need for standardised procedures for dental microwear analyses as well as precise comparison of absolute values.

The results presented here show the data collected on 260 dpi. The MANOVA (Table 3a) indicates statistically significant differences in the microwear patterns among the groups in all three tests, Pillai-trace, Wilks Lambda and Hotelling traces. The ANOVA F-test indicates significant differences of some individual parameters among the groups (Table 3b). These variables include the density of features (N), scratches (NS), proportion of pits (%), length (SL) and standard deviation of length (SL SD) of scratches, and mean striation vector length (R) for the region of facet 6. Region facet 9 shows similar results without density (NP) and proportion (%) of pits which are not statistically significant. Multiple pairwise comparisons of these variables by TukeyHSD are presented in Table 4. There are some differences in the results of the different regions of facets (Table 2, 4). The Ofnet-Cave results indicate statistically significant lower feature, higher pit and lower scratch densities, higher length of scratches and higher mean striation vector length than Bittenbrunn on facet 6. This holds only for feature and scratch densities and mean striation vector length on the region of facet 9. Also, Ofnet Cave is separated from Waging by statistically significant higher scratch and mean striation vector length, and from Brombach by density and proportion of pits and scratch length deviation, but these differences are only detected on the region of facet 6. No significant differences were observed between Waging and Bittenbrunn, or between Ofnet Cave and Weichering, except for a significantly higher mean striation vector length observed in Ofnet Cave on the region of facet 6. Compared to Waging, Weichering shows a statistically significant lower scratch length deviation, a lower feature and scratch density in comparison to Bittenbrunn, and a lower scratch length in comparison to Brombach. This result is reversed, but not significantly, on the region of facet 9.

The same results are only found for Bittenbrunn on the region of facet 9. Brombach shows statistically significant lower feature and pit densities and proportion, and higher scratch length and standard deviation of scratch length than Waging, but lower feature and scratch densities, as well as higher scratch and standard deviation

Table 1 Principal axis analysis based on the Scott Wear-Scoring System (1979a, b)

	*M₁	*M₂	n of pairs	r	Equation of principal axis	95% region of convidence	*M₁ at M₂ = 4	*M₂ at M₁ = 40
Bittenbrunn	21.1	14.8	22	0.783	12.32 + 0.589y	0.387<b<0.834	14.7	47.0
Brombach	23.7	14.2	15	0.720	7.31 + 1.156y	0.655<b<2.150	11.9	28.3
Waging	18.6	14.7	20	0.910	2.63 + 1.083y	0.836<b<1.410	7.0	34.5
Weichering	22.7	16.4	9	0.560	5.03 + 1.073y	0.226<b<6.519	9.3	32.6
Ofnet-Cave	20.6	13.6	21	0.640	0.87 + 1.394y	0.963<b<2.123	6.4	28.1

Annotation: *mean wear scores on four quadrants

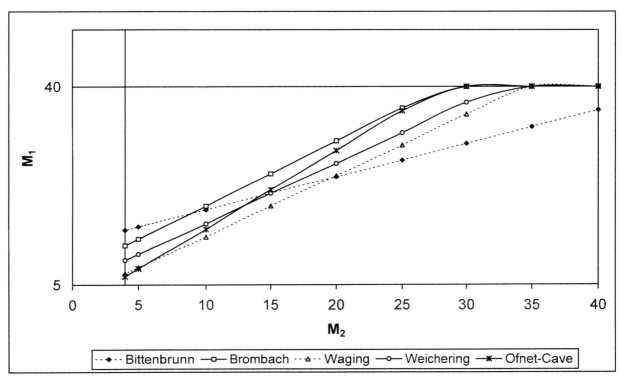

Fig. 2: Regression of rate of wear according to the Equation of Principal Axis, saturation at a score of 40.

of scratch length than Bittenbrunn on the region of facet 6. On the region of facet 9 there are only statistically significant lower feature and scratch densities than Bittenbrunn. Between Brombach and Waging there are no significant differences.

As mentioned, the mean striation vector length describes the distribution of the orientation of scratches on the occlusal surface. This means that high values indicate the existence of a preferred direction of scratch orientation while low values indicate an increasing degree of random distribution. The differences in distribution which are statistically significant are illustrated in histograms (figure 6a – d). The histograms show the orientation of scratches classified in 18 steps of 10°. An increase in preferred orientation is visible in the samples from Weichering which has the lowest correlation (0.286/0.338), Waging, Brombach and Ofnet Cave (0.587/0.506) with preferred orientations from 160 to 60°. This is parallel to the buccolingual and slightly between distobuccal to mesiolingual orientation on the tooth surface. Bittenbrunn, which is not illustrated, is similar to Waging. The Mesolithic Kaufertsberg individual and the Oldoway Hominid OH 1 show even higher correlations to normal distribution, but they represent only single individuals without the statistical variation of a group. The Mesolithic Kaufertsberg individual differs from the Ofnet-Cave Mesolithic individuals in low-

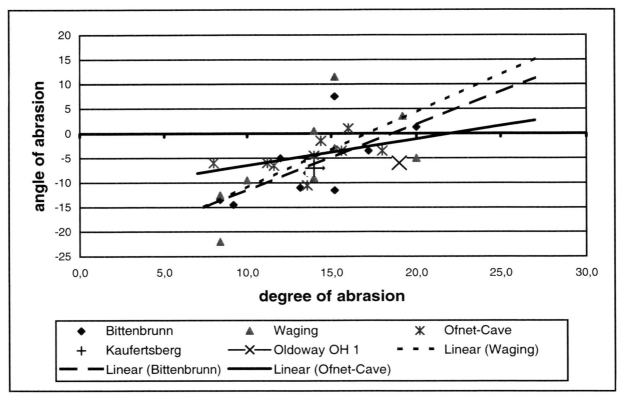

Fig. 3: Raw data for human M_2 wear plane angle (ordinate) at measured stage of wear (abscissa). Included is the Least Squares Line from regression for the human groups of Waging, Bittenbrunn and the Ofnet Cave. The physiological lingual tilt of occlusal plane at eruption of lower molars is changed to buccal with increasing wear. The slope of the line indicates the degree of change from flat (hunter-gatherers) to steep (agriculturalists).

er feature, pit and scratch densities, lower proportion of pits and higher pit length, but they are similar in scratch length, pit width and the tendency to high mean striation vector length. The Oldoway Hominid OH 1 is more similar to Weichering than to the Ofnet-Cave individuals in terms of lower feature and pit densities and higher pit length, but it is more similar to Ofnet Cave than to Weichering in a higher scratch and mean striation vector length and higher pit width. The two functions that best describe dental microwear patterns of human groups are illustrated in Figures 7a and 7b. The results of the two different mandibular molar regions are shown. 92% of the original grouped cases for the region of facet 6 and 78.7% for the region of facet 9 were correctly classified by applying functions one and two with the standardised coefficients of canonical discriminance.

Discussion

The results presented show significant differences in dental wear and microwear patterns for human groups of southern Germany from the Mesolithic to the Early Middle Ages. In accordance with data collected by Smith (1984) and Baum (1991), our results confirm the observation that hunter-gatherers develop flat occlusal angles on their mandibular molars whereas those of agriculturalists show steep tilt of wear. The functional implications of this observation is that it is a result of demands made on the posterior teeth during mastication of different foods. Tough, fibrous, fleshy and dried foods need to be ground several times with expansive movements of the jaws before being swallowed, leading to tooth wear which concerns all the cusps. The early wear of cusps results in a loss of guided and limited movement of the mandibula because lateral anchoring is lost. It is not really clear as to whether this cause may be valid for the individuals of the Ofnet Cave: the group contained no juvenile individuals whose stages of dentition would allow for a study of the cusps, but the deciduous teeth of the children in the samples were relatively worn, implying that the cusps were worn rapidly after eruption and occlusal contact. The dietary shift to soft, crushed, ground or cooked food with only marginal portions of fibre resulted in the wear mainly of the buccal area of the tooth crown. The lingual cusps maintain and provide the leading edge for limited biting movement. The age-independent method of comparing dental wear across samples prescribed by Scott (1979b) was applied in a modified version. First, the wear stages were reversed in calculating the principal axis. This is feasible because deviations are measured perpendicular to the direction of principal and minor axes rather than perpendicular to the abscissa, as was done in the regression

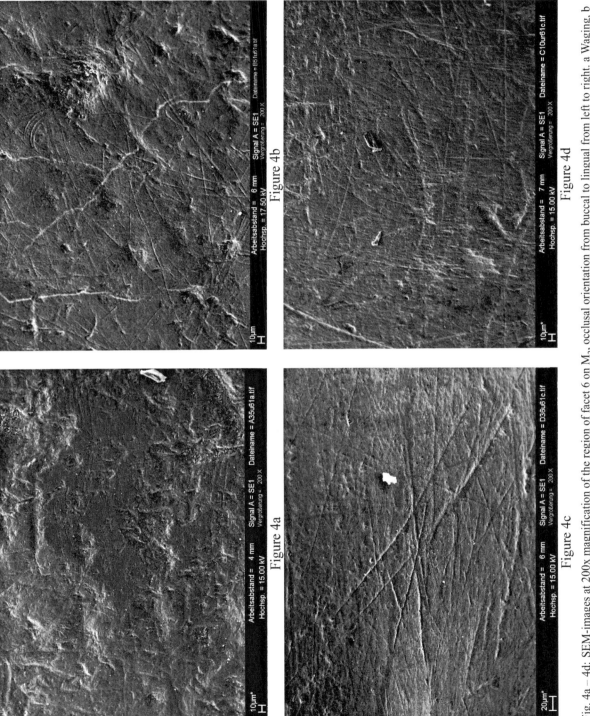

Fig. 4a – 4d: SEM-images at 200x magnification of the region of facet 6 on M_2, occlusal orientation from buccal to lingual from left to right. a Waging, b Bittenbrunn, c Brombach, d Weichering, e Ofnet Cave, f Kaufertsberg, g Oldoway OH 1.

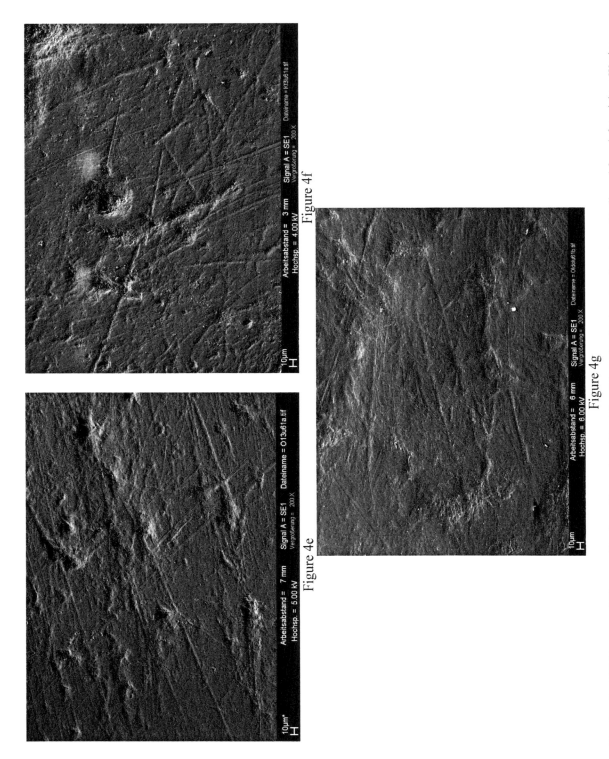

Fig. 4e – 4g: SEM-images at 200x magnification of the region of facet 6 on M_2, occlusal orientation from buccal to lingual from left to right. a Waging, b Bittenbrunn, c Brombach, d Weichering, e Ofnet Cave, f Kaufertsberg, g Oldoway OH 1.

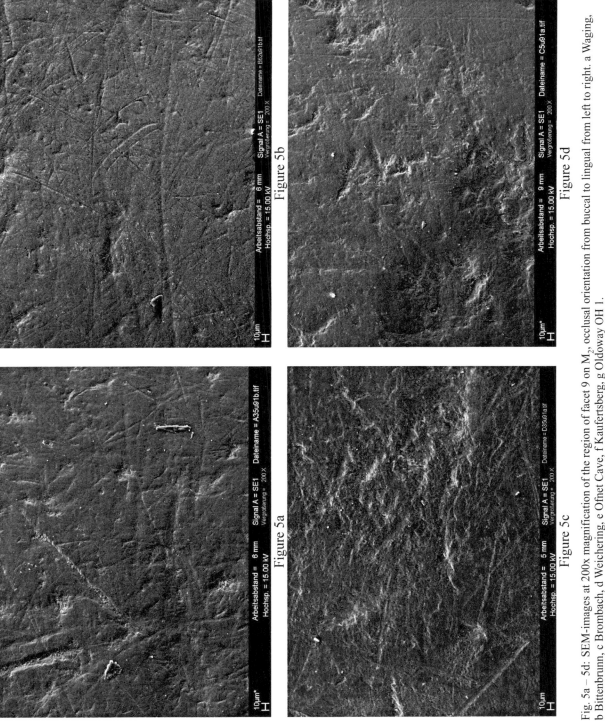

Fig. 5a – 5d: SEM-images at 200x magnification of the region of facet 9 on M_2, occlusal orientation from buccal to lingual from left to right. a Waging, b Bittenbrunn, c Brombach, d Weichering, e Ofnet Cave, f Kaufertsberg, g Oldoway OH 1.

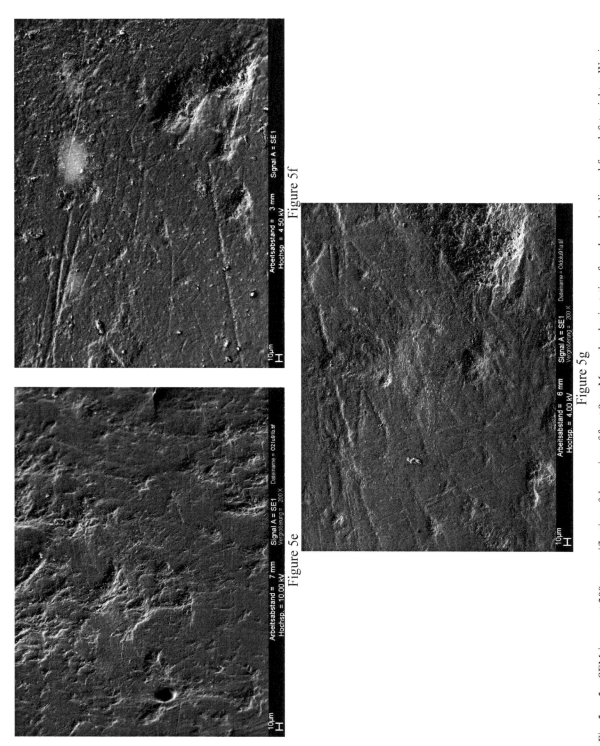

Fig. 5e – 5g: SEM-images at 200x magnification of the region of facet 9 on M_2, occlusal orientation from buccal to lingual from left to right. a Waging, b Bittenbrunn, c Brombach, d Weichering, e Ofnet Cave, f Kaufertsberg, g Oldoway OH 1.

Table 2 Statistical parameters of dental microwear on mandibular human molars

					M_2 region of facet 6										M_2 region of facet 9					
	N_{total}	N	NP	% of pits	PL (µm)	PW (µm)	NS	SL	SL SD	R		N	NP	% of pits	PL (µm)	PW (µm)	NS	SL	SL SD	R
Waging																				
mean	169.9	184.0	37.1	22.8	30.2	19.2	146.9	34.7	19.7	0.363		158.4	24.1	14.9	26.2	17.3	134.3	46.3	32.8	0.324
s +/-	71.2	79.9	13.3	8.3	5.4	4.6	77.2	7.6	5.9	0.128		40.7	15.2	8.5	4.0	3.0	34.4	14.5	12.1	0.186
n	44	7	7	7	7	7	7	7	7.0	7		7	7	7	7	7	7	7	7.0	7
Bittenbrunn																				
mean	182.7	194.4	16.8	9.1	28.2	19.5	177.7	41.1	29.9	0.412		202.3	23.0	12.7	34.9	18.3	179.3	43.1	34.2	0.256
s +/-	49.1	53.6	15.4	8.5	7.1	7.9	55.0	6.6	8.0	0.161		34.1	25.0	14.3	26.8	4.4	53.0	5.9	8.3	0.094
n	44	12	12	12	12	12	12	12	12.0	12		6	6	6	6	6	6	6	6.0	6
Brombach																				
mean	123.7	88.0	4.6	5.4	32.0	17.8	83.4	74.0	56.8	0.496		126.3	23.0	17.8	34.3	21.9	103.3	53.7	32.7	0.357
s +/-	39.8	22.8	3.1	3.7	15.1	8.2	22.8	14.7	10.9	0.188		21.2	8.7	5.2	9.2	7.1	15.2	9.9	7.0	0.182
n	25	5	5	5	5	5	5	5	5.0	5		6	6	6	6	6	6	6	6.0	6
Weichering																				
mean	122.9	130.7	18.1	15.3	25.3	14.5	112.6	50.4	38.8	0.286		118.8	16.2	14.3	31.5	20.5	102.6	61.6	44.8	0.338
s +/-	29.4	41.0	10.5	9.5	3.8	3.2	44.3	7.2	9.9	0.135		22.3	18.1	16.9	8.4	4.7	30.5	12.8	14.3	0.209
n	38	9	9	9	9	9	9	9	9.0	9		12	12	12	12	12	12	12	12.0	12
Ofnet-Cave																				
mean	143.1	135.4	33.1	26.0	28.6	17.7	102.2	56.4	35.3	0.587		145.7	26.9	19.3	28.3	18.3	118.8	57.2	37.4	0.506
s +/-	50.5	52.2	20.3	12.7	5.8	3.8	47.9	19.8	18.0	0.146		43.6	13.2	8.0	4.8	2.9	38.9	12.3	11.2	0.166
n	78	17	17	17	17	17	17	17	17.0	17		17	17	17	17	17	17	17	17.0	17
Kaufertsberg																				
mean	99.2	100.4	17.8	18.3	31.5	17.8	82.6	57.5	43.1	0.613		91.6	23.2	27.2	28.5	16.9	68.4	65.7	51.5	0.447
s +/-	26.0	23.8	7.9	8.1	7.5	3.1	22.7	4.7	6.0	0.097		20.3	16.3	19.5	5.3	4.8	30.5	4.4	7.0	0.217
n	39	5	5	5	5	5	5	5	9.0	5		5	5	5	5	5	5	5	10.0	5
Olduvay OH 1																				
mean	120.3	129.4	16.9	13.2	31.4	20.7	112.6	62.1	41.6	0.640		109.3	15.9	15.4	31.7	20.0	93.4	59.9	41.3	0.609
s +/-	24.5	17.3	6.5	5.2	7.1	5.4	18.5	8.5	6.6	0.086		26.6	5.1	5.8	6.3	5.6	28.1	7.6	9.7	0.062
n	28	9	9	9	9	9	9	9	9.0	9		10	10	10	10	10	10	10	10.0	10

Annotation: all parameters are mean values of the features on an area of 0.176 mm^2
Abbreviations: N_{total} = Number of features on all analysed areas (0.176 mm^2); NP = density of pits; PL = length of pits; PW = width of pits; NS = number of scratches; SL = length of scratches; SL SD = standard deviation of scratch length; R = mean striation vector

Table 3a Multivariate analysis of variance test results

		value	F-value	df	df error	P
M_2 facet 6	Pillai-trace	1,941	3,526	48,000	354,000	,000
M_2 facet 6	Pillai-trace	1,941	3,526	48,000	354,000	,000
	Wilks-Lambda	,048	4,814	48,000	269,765	,000
	Wilks-Lambda	,048	4,814	48,000	269,765	,000
	Hotelling-traces	5,497	5,994	48,000	314,000	,000
	Hotelling-traces	5,497	5,994	48,000	314,000	,000

		value	F-value	df	df error	P
M_2 facet 9	Pillai-trace	1,669	2,841	48,000	354,000	,000
M_2 facet 9	Pillai-trace	1,669	2,841	48,000	354,000	,000
	Wilks-Lambda	,099	3,377	48,000	269,765	,000
	Wilks-Lambda	,099	3,377	48,000	269,765	,000
	Hotelling-traces	3,445	3,756	48,000	314,000	,000
	Hotelling-traces	3,445	3,756	48,000	314,000	,000

Table 3b Univariate F-tests (all human groups)

M_2 facet 6	SS	df	MS	F	P
N	82420,508	6	13736,751	6,157	,000
NP	5927,033	6	987,839	4,954	,000
% of pits	3370,267	6	561,711	6,202	,000
PL	248,490	6	41,415	,850	,536
PW	231,779	6	38,630	1,388	,234
NS	75423,714	6	12570,619	5,869	,000
SL	7698,362	6	1283,060	8,531	,000
SL SD.	5253,764	6	875,627	6,689	,000
R	,939	6	,156	8,083	,000

M_2 facet 9	SS	df	MS	F	P
N	61290,531	6	10215,088	9,988	,000
NP	1269,797	6	211,633	,930	,480
% of pits	1325,579	6	220,930	1,505	,192
PL	556,219	6	92,703	,967	,455
PW	177,027	6	29,504	1,469	,204
NS	55695,335	6	9282,556	8,179	,000
SL	2597,960	6	432,993	3,790	,003
SL SD.	2599,628	6	433,271	3,773	,003
R	,815	6	,136	5,102	,000

Annotation: SS = sum square; MS = mean square; P = Significance.
Abbreviations: N_{total} = number of features on all analysed areas (0,176 mm^2); NP = density of pits;
PL = Length of pits; PW = Width of pits; NS = number of scratches; SL = Length of scratches;
SL SD = standard deviation of Length of scratches; R = mean striation vector

analysis (Sokal & Rohlf 1969). Secondly, this procedure enables the direct evaluation of M_1 at the moment of eruption and first occlusal contact with M_2. Our results prove that the rate of wear varies with the different subsistence strategies as well as among the human groups living contemporarily but in different locations. The amount of wear on the first molar at the moment when the second molar erupts gives an idea of how intensively the first molar was used and worn during a span of approximately six years. As a consequence, the rate of wear on the teeth of the Ofnet Cave individuals was not rapid as is perhaps indicated by the low score of 6.4 (M_1, Table 1). This means that while some signs of wear are visible on all four quadrants, the cusps are still present. Therefore, the differences in slope and score in comparison to the other groups and the lower rate of wear of M_2 (indicated by the higher slope and the score of 28.1 at the time of saturation of M_1) may reflect the supporting and protecting function of a flat plane. The latter probably acts as a distributor of mastication stress by increasing the chewing area. If we postulate a shortened time span of eruption between the First and Second Molars, it may explain the low score of M_1 at eruption of M_2, but not the low score of M_2 at the saturation point of M_1, or the comparable similar mean scores of M_1 and M_2 to the other groups. As the angles of occlusal planes among the early medieval groups are similarly steep, the differences in the rate of wear cannot be explained by differences in the supporting function of the occlusal areas. The macroscopically visible traces on the sam-

Variations in dental microwear and abrasion

Table 4 Pairwise statistical comparison (Tukey HSD) of mean dental microwear parameters

region facet 6		N MD	N P	NP MD	NP P	% of pits MD	% of pits P	NS MD	NS P	SL MD	SL P	SL SD MD	SL SD P	scratch R MD	scratch R P
Waging	Bittenbrunn	-10,417	0,999	20,393	0,052	13,728	0,800	-30,810	0,052	-6,325	0,930	-10,146	0,511	-0,049	0,989
	Brombach	96,000	0,016	32,543	0,004	17,458	0,242	63,457	0,040	-39,310	0,000	-37,082	0,000	-0,133	0,661
	Weichering	53,333	0,290	19,032	0,123	7,582	0,761	34,302	0,695	-15,677	0,165	-19,102	0,025	0,077	0,928
	Ofnet-Cave	48,647	0,264	4,025	0,995	-3,113	0,339	44,622	0,990	-21,661	0,004	-15,552	0,053	-0,224	0,012
Bittenbrunn	Brombach	106,417	0,001	12,150	0,672	3,730	0,005	94,267	0,990	-32,985	0,000	-26,936	0,001	-0,084	0,915
	Weichering	63,750	0,048	-1,361	1,000	-6,146	0,034	65,111	0,764	-9,352	0,600	-8,956	0,570	0,126	0,395
	Ofnet-Cave	59,064	0,024	-16,368	0,047	-16,841	0,001	75,431	0,000	-15,336	0,024	-5,407	0,870	-0,174	0,024
Brombach	Weichering	-42,667	0,670	-13,511	0,609	-9,876	0,916	-29,156	0,514	23,633	0,017	17,980	0,088	0,210	0,115
	Ofnet-Cave	-47,353	0,443	-28,518	0,003	-20,571	0,984	-18,835	0,001	17,649	0,086	21,530	0,008	-0,090	0,859
Weichering	Ofnet-Cave	-4,686	1,000	-15,007	0,151	-10,695	0,998	10,320	0,109	-5,984	0,898	3,549	0,988	-0,300	0,000
region facet 9		MD	P					MD	P	MD	P	MD	P	MD	P
Waging	Bittenbrunn	-43,905	0,189					-45,048	0,215	3,203	0,998	-1,409	1,000	0,068	0,989
	Brombach	32,095	0,551					30,952	0,650	-7,362	0,876	0,097	1,000	-0,033	1,000
	Weichering	39,679	0,141					31,702	0,438	-15,280	0,056	-11,934	0,242	-0,014	1,000
	Ofnet-Cave	12,723	0,973					15,462	0,947	-10,931	0,272	-4,527	0,964	-0,183	0,180
Bittenbrunn	Brombach	76,000	0,002					76,000	0,004	-10,565	0,611	1,506	1,000	-0,101	0,935
	Weichering	83,583	0,000					76,750	0,000	-18,483	0,016	-10,525	0,447	-0,082	0,951
	Ofnet-Cave	56,627	0,007					60,510	0,006	-14,134	0,095	-3,118	0,996	-0,250	0,031
Brombach	Weichering	7,583	0,999					0,750	1,000	-7,918	0,754	-12,031	0,287	0,018	1,000
	Ofnet-Cave	-19,373	0,860					-15,490	0,959	-3,569	0,992	-4,624	0,970	-0,150	0,467
Weichering	Ofnet-Cave	-26,956	0,292					-16,240	0,859	4,348	0,932	7,407	0,531	-0,168	0,107

Abbreviations: N_{total} = number of features on all analysed areas (0,176 mm^2); NP = density of pits; PL = Length of pits; PW = Width of pits; NS = number of scratches; SL = Length of scratches; SL SD = meand standard deviation of scratch length; R = mean striation vector; MD = mean differences; P = probability.

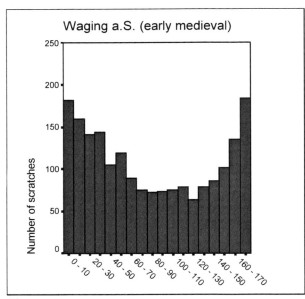
Figure 6a

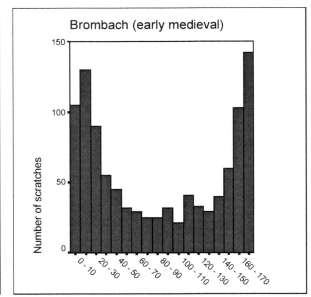
Figure 6b

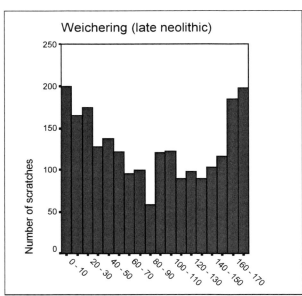
Figure 6c

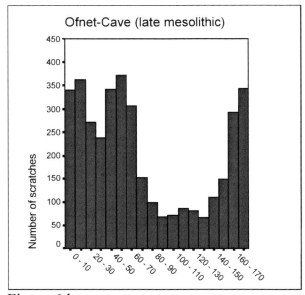
Figure 6d

Fig. 6a – 6d: Orientation of the scratches on the molar enamel surfaces; 0 – 45° and 135.1 – 180° parallel to the bucco-lingual orientation, 45.1 – 90° distobuccal to mesiolingual, 90.1 – 135° mesiobuccal to distolingual. The symmetry of left and right molars is reflected.

ples of the group from Weichering (end Neolithic) indicate an intermediate pattern of occlusal plane.

The results of the dental microwear analysis presented in this study show signficant differences between the groups in terms of feature density, proportion, size and orientation. They indicate a change of microwear pattern from Ofnet Cave to Weichering, Bittenbrunn, Waging and to Brombach. The Ofnet-Cave and Kaufertsberg samples show medium and low densities of features, higher proportions of pits and long scratches with preferences in orientation. Waging and Bittenbrunn indicate high densities of features, variable proportions of pits, short scratches with little preferential orientation. Weichering also shows medium feature and scratch densities, intermediate proportion of pits, medium scratch length with nearly no preference of orientation. Brombach, although a medieval human group, stands out with low density of features, an almost complete absence of pits and long scratches with preference in orientation. The Paleolithic Oldoway Hominid OH 1 also shows medium feature and scratch densities, intermediate proportion of pits (similar to Weichering), but high scratch length and a high

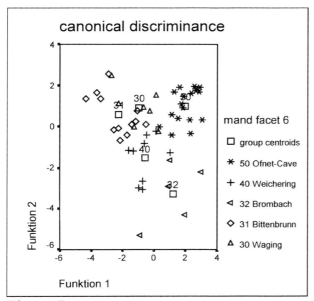
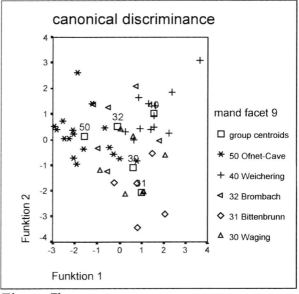

Figure 7a   Figure 7b

Fig. 7a – 7b: Canonical discriminance: a) region of facet 6 on M_2, b) region of facet 9 on M_2.

preference of orientation. The study conducted by Lalueza et al. (1993) on buccal molar surfaces of six modern human groups that included both agriculturalists and hunter-gatherers (subdivided in three categories) concluded that more carnivorous hunter-gatherers have fewer, but frequently, more vertical and longer striations than vegetarian or mixed diet groups. The composition of the microwear patterns of this study supports that idea. As meat does not contain endogenous abrasive particles, an increase of feature density and variability of scratch orientation would indicate the proportion of collected wild plants in the diet. The dietary shift from meat to plants implies lower masticatory stress, but higher amounts of abrasive particles in the diet originating from exogenous grit or silicious opal phytoliths from plants. These particles cause small pits, higher densities of features and scratches that tend to be short (Gügel 2001, Lucas & Teaford 1995, Ungar et al. 1995). As a consequence of a high rate of wear, microwear features undergo change. Constant turnover during mastication (Teaford & Lytle 1996, Teaford & Oyen 1989a Friedel 2000), leaves behind interrupted and shortened older scratches, less frequent but flattened pits as well as smaller, newly produced features. This shift of microwear pattern is seen as a gradient from Ofnet Cave to Weichering, Waging and Bittenbrunn. Bittenbrunn indicates a complete pattern of rapid dental wear and microwear, the result of a diet high in abrasive particle content. There is some evidence (Christlein 1971) that the population of Bittenbrunn, especially the individuals found at the end of the burial site, must have been very poor, whereas the population at Waging is described as having been economically balanced (Knöchlein 1995).

The pattern of Brombach is different. Low feature and pit densities, coupled with high scratch lengths and preferred orientation of scratches, imply that the population of Brombach lived as carnivorous hunter-gatherers like the Ofnet-Cave people. However, historical sources offer evidence of a well-developed agricultural practice that is characteristic of the Middle Ages (Wurm 1982), with a local but highly advanced milling technology existing along the former Brombach River shortly after the burial of the individuals used in this study (Lidl & Hahn 1989). These people probably processed their plant food, especially corn, and removed the husks that contained the phytoliths, and thereby eliminated a large part of the abrasive content from their mixed diet. That may explain the similarity to a carnivorous pattern of microwear in comparison to Waging and Bittenbrunn. The pattern of microwear of Kaufertsberg identifies this fossil individual as a mainly carnivorous hunter-gatherer, like the Ofnet-Cave people. These results support the anthropological theory that the Kaufertsberg Skull belongs to the series of skulls from the Mesolithic (Schröter 1983). The Oldoway Hominid OH 1 originates from the Oldoway site in Tanzania and has been dated to the Paleolithic (Oakley et al. 1977). The pattern of microwear identifies the teeth as originating from a meat-eating individual with lower feature density and little higher pit size and scratch length. Differences in the power and turn of mandibular movement during the mastication of hard and solid dietary components must be the cause of the different lengths and orientation of scratches and, may be one cause of scratch length variation and preferred orientation of scratches in diets with low amounts of abrasive particles.

Conclusions

The presented results prove that significantly different dental wear and microwear patterns appear in the dentition of human groups of southern Germany in relation to their dietary behaviour and/or their cultural development. The individuals of the Ofnet Cave show a carnivorous hunter-gatherer lifestyle, similar to that of Kaufertsberg. A slightly different pattern is seen on the Oldoway Hominid OH 1. The individuals of Weichering appear to have adopted a mixed diet, with a higher tendency towards meat. The groups of Waging and Bittenbrunn, with different amounts of abrasive particles in their diet, show a complete pattern of agricultural behaviour. A distinct pattern of wear and microwear is detectable in Brombach which is in concurrence with current literature that describes these people as probably having refined methods for cleaning and preparing food, especially grain.

Tough foods, and both exogenous or endogenous hard, abrasive particles in the diet, cause a higher rate of wear and significantly different patterns of microwear, which separate the human groups that have them in their diets from the human groups that live off soft, well-prepared diets. The analysis of these patterns on surfaces of human teeth has the potential to decode cultural developments in the human past. Further analyses will be necessary to create a basic database of these variables and the variety of dental wear and microwear patterns. In particular, attention has to be paid to the technical aspects of SEM and standardised methods, including the selection of analysed areas on the human teeth.

Summary / Zusammenfassung

The aim of the study was to investigate patterns of dental wear and microwear of human groups of southern Germany from the Mesolithic to the Early Middle Ages. The fossil from Kaufertsberg, dated to the Mesolithic, and the first Hominid of Oldoway OH 1, dated to the Paleolithic were included. All human groups are housed in the State Collection of Anthropology and Palaeoanatomy in Munich.

The patterns of wear differed in the tilt of the occlusal abraded planes, (hunter-gatherers exhibited flat planes on the mandibular molars while the agriculturalists had steep planes), as well as in the rate of wear, the degree of abrasion of M_1 at the moment of eruption of M_2, and the degree of abrasion of M_2 at the moment of saturation of M_1. Differences in patterns were not only detected between the human groups from the Mesolithic, Neolithic and the Early Middle Ages, but also between human groups of the Early Middle Ages.

The analysis of dental microwear on well-defined regions of the Lower Second Molars resulted in statistically significant differences between the groups in terms of dental microwear feature density, proportion, size and orientation. These differences indicate a gradual change of microwear pattern from Ofnet Cave (Mesolithic) to Weichering (end Neolithic), Bittenbrunn and Waging. The more carnivorous groups tended to show fewer but longer scratches and these scratches run from buccal to lingual more frequently. As meat does not contain endogenous abrasive particles, an increase of feature density and variability of scratch orientation would indicate the proportion of wild plants in the diet. The dietary shift from meat to plants implies lower masticatory stress, but higher amounts of abrasive particles in the diet originating from exogenous grit or silicious opal phytoliths from wild or cultivated plants These particles cause small pits, higher densities of features and scratches with the tendency of being short. As a consequence of the high rate of wear brought about by frequent turnover during mastication, rapid changes in the features occur. Characteristics of this change are interrupted and shortened older scratches, less but more flattened pits and smaller, newly produced features.

The results also showed a different dental microwear pattern for the Brombach population. Low feature and pit densities in combination with a high scratch length and preferred orientation of scratches imply that the population lived as carnivorous hunter-gatherers, similar to the Ofnet-Cave population. However, there is historical evidence that the Brombach individuals had well-developed agricultural practices, which is characteristic of the Middle Ages. There is also evidence that milling technology was developed locally to clean and prepare plant food, especially corn. The removal of the abrasive phytoliths together with the husks and the subsequent milling of the corn, has given this population a relatively low rate of wear and the low microwear feature density.

The pattern of wear and microwear of the fossil of the Kaufertsberg identifies it as belonging to a mainly carnivorous hunter-gatherer group, similar to the Ofnet-Cave population. It supports the anthropological theory that the skull of Kaufertsberg belongs to the Mesolithic. The Oldoway Hominid OH 1, originating from the Oldoway site in Tanzania, shows a pattern of microwear which identifies it as originating from hunter-gatherer with lower feature density and little higher pit size and scratch length.

The results show that the analysis of microwear patterns on human occlusal tooth surfaces bears the potential to further decode the causes of microwear features and create more and new opportunities for describing cultural developments. Further analyses are necessary to create a basic database about the spectrum of informative variables and the spectrum of variability of dental wear and microwear patterns. It is worth mentioning that more attention has to be paid to the technical aspects of SEM and standardised methods, including the selection of analysed areas on human teeth.

Das Ziel der vorliegenden Studie war, dentale Zahnabtragungs- und Microwear Muster von menschlichen Gruppen aus dem süddeutschen Raum vom Mesolithikum bis in das frühe Mittelalter zu untersuchen und zu vergleichen. Das Individuum vom Kaufertsberg, welches in das Mesolithikum datiert wird und der erste Hominide aus der Oldoway Schlucht OH 1, der in das Paläolithikum datiert wird, konnten in die Untersuchung aufgenommen werden. Alle menschlichen Gruppen sind in der Staatssammlung für Anthropologie und Paläoanatomie in München beherbergt.

Das Abtragungsmuster unterschied sich im Neigungswinkel der okklusal abgetragenen Flächen, die Jäger-Sammler entwickelten flachere Winkel als die Ackerbauern, die steile Neigungen aufzeigen, in der Rate der Zahnabtragung, im Grad der Abtragung vom M_1 zum Zeitpunkt des Durchbruches von M_2 und im Grad der Abtragung von M_2 zum Sättigungswert von M_1. Musterunterschiede wurden nicht nur im Vergleich der menschlichen Gruppen vom Mesolithikum zum frühen Mittelalter erfasst, sondern auch zwischen den frühmittelalterlichen Gruppen.

Die Analyse der dentalen Microwear an definierten Regionen der unteren zweiten Molaren erbrachten statistisch signifikante Unterschiede zwischen den Gruppen die vor allem die Dichte der Microwearmerkmale, deren Anteil, ihre Dimensionen und ihre Orientierung betrafen. Diese Unterschiede zeigen eine fortschreitende Veränderung des Microwear-Musters von der Ofnet-Höhle (Mesolithikum) nach Weichering (Endneolithikum), Bittenbrunn und Waging an. Die vorwiegend carnivor lebenden Gruppen tendieren zu weniger, aber längeren Kratzspuren und diese Kratzspuren verlaufen gehäuft parallel zu einer bukko-lingualen Orientierung. Da Fleisch keine endogenen abrasiven Partikel enthält sollte eine Zunahme in der Merkmalsdichte und der Variabilität der Orientierung von Kratzspuren den Anteil an verzehrten Wildfrüchten in der Ernährung widerspiegeln. Die Verschiebung von einer vorwiegend fleischhaltigen zu einer überwiegend pflanzlichen Ernährungsweise impliziert ebenfalls niedrigere Kaubelastungen, aber auch einen höheren Anteil abrasiver Partikel in der Nahrung, deren Ursprung exogener Verunreinigung oder silikathaltiger Opal-Phytolithe sind, die sowohl in Wild- als auch in Kulturpflanzen enthalten sind. Diese Partikel verursachen höhere Merkmalsdichten, tendenziell vorwiegend kurze Kratzspuren und kleine Gruben. Als Konsequenz einer hohen Abtragungsrate werden die Merkmale durch den Turnover während des Kauens rasch verändert, es bleiben unterbrochene und verkürzte Kratzspuren zurück, weniger häufig Gruben, die abgeflacht und kleiner sind, und neu produzierte Merkmale.

Das dentale Microwear-Muster von Brombach zeigt nicht die fortschreitende Änderung. Die niedrige Merkmals- und Grubendichte in Verbindung mit langen Kratzspuren, die bevorzugt parallel zur bukko-lingualen Orientierung verlaufen, ließe annehmen, dass diese Individuen eine vorwiegend carnivore Ernährungsweise hatten, ähnlich der Individuen aus der Ofnet-Höhle. Es ist allerdings historisch belegt, dass diese Individuen eine gut entwickelte Ackerbauerkultur charakteristisch für das frühe Mittelalter hatten. Es gibt weiterhin Hinweise aus etwas späterer Zeit dafür, dass diese Region eine hoch entwickelte Mühlentechnologie besaß, die es ermöglichte pflanzliche Nahrung im besonderen Getreidekörner, deren Spelzen Phytolithe beinhalten, zu reinigen und zur Ernährung zuzubereiten. Die relativ niedrige Abtragungsrate und die niedrige Dichte dentaler Microwear-Merkmale läßt eine effektive Entfernung abrasiver Partikel aus der Ernährung annehmen.

Das Zahnabtragungs- und dentale Microwear-Muster des Individuums vom Kaufertsberg identifiziert dieses Fossil als zu einer vorwiegend carnivor lebenden Jäger-Sammler Gruppe gehörig ähnlich wie die Individuen der Ofnet-Höhle. Dies unterstützt die Theorie der Anthropologen, dass der Schädel vom Kaufertsberg zu der Serie von Kopfbestattungen aus dem Mesolithikum gehört. Der Oldoway Hominide OH 1, der seinen Ursprung in der Oldoway Schlucht in Tanzania hat, zeigt ein Microwear-Muster welches die Oberfläche als ursprünglich für einen Jäger-Sammler mit niedrigerer Merkmalsdichte und geringfügig größeren Gruben und Kratzspuren identifiziert.

Die Ergebnisse zeigen, dass die Analyse von Microwear-Mustern auf menschlichen okklusalen Zahnschmelzoberflächen ein großes Potential birgt, um noch feiner die Ursachen für Microwear-Merkmale zu entziffern und sie zeigen auch, dass diese weitere und neue Möglichkeiten eröffnen wird, kulturelle Entwicklungen zu detektieren. Weitere Untersuchungen sind notwendig, um eine grundlegende Datenbasis entstehen zu lassen, die das Spektrum an informativen Variablen und das Spektrum der Variabilität der dentalen Zahnabtragungs- und Microwear-Muster erfasst. Es ist wichtig zu bemerken, dass eine genaue Beachtung der technischen Aspekte der SEM und der standardisierten Vorgehensweise, eingeschlossen der Auswahl der zu analysierenden Oberflächen von menschlichen Zähnen notwendig ist.

Acknowledgements

I am most indebted to Gisela Grupe (Department I, Biodiversity research/Anthropology, Ludwig-Maximilian-University, Munich), for her advice and encouragement. Many thanks are due to her for making it possible to collect data from the specimens (State Collection of Anthropology and Palaeoanatomy of Munich) and for the provision of facilities for the microscopy and data analysis. I am also much indebted to Peter Ungar (University of Arkansas) for his advice on the statistics and on working with Microware 3.0, as well as his inspiration for the present research. I am grateful to the Deutsche Forschungsgemeinschaft for providing grants for parts of the program and technical requirements.

Bibliography

Baum N., 1991.
  Sammler/Jäger oder Ackerbauern? Eine Paläoodontologische Untersuchung zur kulturhistorischen Stellung der Kopfbestattungen aus der grossen Ofnet-Höhle in Schwaben. *Archäologisches Korrespondenzblatt* **21**: 469-474.

Beynon A.D., 1987.
  Replication technique for studying microstructure in fossil enamel. *Scanning Microscopy* **1**: 663-669.

Bullington J., 1991.
  Deciduous dental microwear of prehistoric juveniles from the lower Illinois River valley. *American Journal of Physical Anthropology* **84**: 59-73.

Christlein R., 1971.
  Ausgrabung eines Gräberfeldes des 5.-7. Jahrhunderts bei Bittenbrunn, Ldkr. Neuburg a.d.Donau. Jahresbericht der Bayerischen Bodendenkmalpflege, Habelt Bonn **8/9**: 87-103.

Danielson D.R. & Reinhard K.J., 1998.
  Human dental microwear caused by calcium oxalate phytoliths in prehistoic diet of the lower pecos region, Texas. *American Journal of Physical Anthropology* **107**: 297-304.

Fine D. & Craig G.T., 1981.
  Buccal surface wear of human premolar an molar teeth: A potential indicator of dietary and social differentiation. *Journal of Human Evolution* **10**: 335-344.

Friedel S., 2000.
  Die Entwicklung einer Kausimulation zur experimentellen Erzeugung dentaler Microwear. Diplomarbeit Institut für Anthropologie und Humangenetik der LMU München.

Gordon K.D., 1982.
  A study of microwear on chimpanzee molars: Implications for dental microwear analysis. *American Journal of Physical Anthropology* **59**: 195-215.

Gordon K.D., 1984.
  The assessment of jaw movement direction from dental microwear. *American Journal of Physical Anthropology* **63**: 77-84.

Gordon K.D. & Walker A.C., 1983.
  Playing Possum: A microwear experiment. *American Journal of Physical Anthropology* **60**: 109-112.

Grine F.E., 1986.
  Dental evidence for dietary differences in Australopithecus and Paranthropus: a quantitative analysis of permanent molar microwear. *Journal of Human Evolution* **15**: 783-822.

Grine F.E. & Kay R.F., 1988.
  Early hominid diets from quantitative image analysis of dental microwear. *Nature* **333**: 765-768

Gügel I.L., 2001.
  Microwear und Abrasion: Untersuchungen an Zahnschmelz historischer Populationen und von Simulationsexperimenten. In: Schultz et al. (eds) Homo – *Unsere Herkunft und Zukunft*. Proceedings – 4. Kongress der Gesellschaft für Anthropologie. Cuvillier, Göttingen, 384-390.

Gügel I.L., Grupe G. & Kunzelmann K.-H., 2001.
  Simulation of dental microwear: Characteristic traces by opal phytoliths give clues to ancient human dietary behavior. *American Journal of Physical Anthropology* **114**: 124-138.

Kay R.F. & Hiiemae K.M., 1974.
  Jaw movement and tooth use in recent and fossil primates. *American Journal of Physical Anthropology* **40**: 227-256.

King T., Andrews P. & Boz B., 1999a.
  Effect of Taphonomic Processes on Dental Microwear. *American Journal of Physical Anthropology* **108**: 359-373.

King T., Aiello L.C. & Andrews P., 1999b.
  Dental microwear of *Griphopithecus alpani*. *Journal of Human Evolution* **36**: 3-31

Knöchlein R., 1995.
  Das Reihengräberfeld von Waging am See. Schriftenreihe des Bajuwarenmuseums Nr. 1, Erdl Trostberg.

Lalueza C., Pérez-Pérez A. & Turbon D., 1993.
Microscopic study of the Banyoles mandible (Girona, Spain): diet, cultural activity and toothpick use. *Journal of Human Evolution* **24**: 281-300.

Lalueza-Fox C., Juan J. & Albert R.M., 1996.
Phytolith analysis on dental calculus, enamel surface, and burial soil: information about diet and palaeoenvironment. *American Journal of Physical Anthropology* **101**: 101-113.

Lalueza-Fox C. & Pérez-Pérez A., 1994.
Dietary information through the examination of plant phytoliths on the enamel surface of human dentition. *Journal of Archaeological Science* **21**: 29-34.

Leek F., 1972.
Teeth and bread in ancient Egypt. *Journal of Egyptian Archaeology* **58**: 126-132.

Lidl J. & Hahn W., 1989.
An der Mühlstrasse: mit dem Zeichenstift durchs Brombachtal. Zweckverband Brombachsee, Pleinfeld. Keller, Treuchtlingen.

Lucas P.W. & Teaford M.F., 1995.
Significance of silica in leaves to long-tailed Macaques (*Macaca fascicularis*). *Folia Primatology* **64**: 30-36.

Molleson T., Jones K. and Jones S., 1993.
Dietary change and the effects of food preparation on microwear patterns in the late Neolithic of abu Hureyra, northern Syria. *Journal of Human Evolution* **24**: 455-468.

Mollison T., 1936.
Zeichen gewaltsamer Verletzungen an den Ofnet-Schädeln. *Anthropologischer Anzeiger* **13**: 79-88.

Oakley K.P., Campbell B.G. & Molleson T.I., 1977.
Catalogue of fossil Hominids. Part I: Africa. Trustees of the British Museum (Natural History) London: 184-185.

Orschiedt J., 1998.
Ergebnisse einer neuen Untersuchung der spätmesolithischen Kopfbestattungen aus Süddeutschland. *Urgeschichtliche Materialhefte* **12**: 147-160.

Peters C.R., 1987.
Nut-like oil seeds: Food for monkeys, chimpanzees, humans, and probably ape-men. *American Journal of Physical Anthropology* **73**: 333-363.

Puech P-F, 1992.
Microwear studies of early African hominid teeth. *Scanning Microscopy* **6**: 1083-1088.

Puech P.-F., Prone A., Roth H. & Cianfarani F., 1985.
Reproduction expérimentale des processus d'usure des surfaces dentaires des Hominidés fossiles: conséquences morphoscopiques et exoscopiques avec application à l'Hominidé I de Garusi. *Comptes rendus Académie des Sciences* **301**, *Série II*, 59-64.

Rose J.C. & Ungar P.S., 1998.
Gross dental wear and dental microwear in historical perspective. In: Alt K.W., Rösing F.W. & Teschler-Nicola M. (eds) *Dental Anthropology, Fundamentals, Limits and Prospects*: 349-386. Springer, New York

Schmidt C.W., 2001.
Dental microwear evidence for a dietary shift between two non-maize-reliant prehistoric human populations from Indiana. *American Journal of Physical Anthropology* **114**: 139-145.

Schröter P., 1983.
Zum Schädel vom Kaufertsberg bei Lierheim (Gem.Appetshofen, Ldkr. Donau-Ries). *Quartär* **33/34**: 99-107.

Scott E.C., 1979a.
Dental scoring technique. *American Journal of Physical Anthropology* **51**: 213-218.

Scott E.C., 1979b.
Principal axis analysis of dental attrition data. *American Journal of Physical Anthropology* **51**: 203-212.

Smith B.H., 1984.
Patterns of molar wear in hunter-gatherers and agriculturalists. *American Journal of Physical Anthropology* **63**: 39-56.

Sokal R. & Rohlf F.J., 1969.
Biometry: The principles and practice of statistics in biological research: 526-532. Freeman and Company, San Francisco

Teaford M.F., 1991.
Dental microwear: What can it tell us about diet and dental function? In: Kelley L. (eds) *Advances in dental Anthropology*: 341-356. Wiley-Liss, New York

Teaford M.F. & Lytle J.D., 1996.
Brief communication: Diet-induced changes in rates of human tooth microwear: a case study involving stone-ground maize. *American Journal of Physical Anthropology* **100**: 143-147.

Teaford M.F. & Oyen O.J., 1989a.
In vivo and in vitro turnover in dental microwear. *American Journal of Physical Anthropology* **80**: 447-460.

Teaford M.F. & Oyen O.J., 1989b.
Live primates and dental replication: New problems and new techniques. *American Journal of Physical Anthropology* **80**: 73-81.

Teaford M.F. & Runestad JA., 1992.
Dental microwear and diet in Venezuelan primates. *American Journal of Physical Anthropology* **88**: 347-364.

Ungar P.S., 1994.
Incisor Microwear of Sumatran Anthropoid Primates. *American Journal of Physical Anthropology* **94**: 339-363.

Ungar P.S., 1995.
A semiautomated image analysis procedure for the quantification of dental microwear II. *Scanning* **17**: 57-59.

Ungar P.S., 1996.
Dental microwear of European Miocene catarrhines: evidence for diets and tooth use. *Journal of Human Evolution* **31**: 335-366.

Ungar P.S., Simon J.C. & Cooper J.W., 1991.
A semiautomated image analysis procedure for the quantification of dental microwear. *Scanning* **13**: 31-36.

Ungar P.S. & Spencer M.A., 1999.
Incisor microwear, diet, and tooth use in three Amerindian populations. *American Journal of Physical Anthropology* **109**: 387-396.

Ungar P.S., Teaford M.F., Glander K.E. & Pastor R.F., 1995.
Dust accumulation in the canopy: A potential cause of dental microwear in primates. *American Journal of Physical Anthropology* **97**: 93-99.

Walker A., Hoeck H.N. & Perez L., 1978.
Microwear of mammalian teeth as an indicator of diet. *Science* **201**: 908-910.

Wurm H., 1982.
About the variation of the average height in the course of the german history and the influence of dietary protein. *Homo* **33**: 21-42.

Detection of *Yersinia pestis* DNA in early and late Medieval Bavarian burials

Christina Garrelt, Ingrid Wiechmann, Department Biologie I,
Bereich Biodiversitätsforschung/Anthropologie, Universität München

Abstract / Zusammenfassung

Biomolecular tools enable the detection of ancient microbial DNA in human remains and thereby the study of infectious diseases (for instance, plague) that do not leave macroscopic marks within the bones. As far as the history of plague is concerned, many questions regarding its etiology and epidemiology are still unanswered. In this study we examined skeletal finds from two Bavarian excavation sites that date back to the 6th and 14th centuries.

Human remains that were archaeologically dated to the 14th century and suspected to be those of plague victims were excavated from a mass grave located below the vestry of the Church of St. Leonhard in Pichl, near Manching. DNA samples extracted from the teeth of 33 individuals were tested for *Yersinia pestis*-specific DNA by means of a primer pair that is able to amplify a part of the *Y. pestis* plasmid pPCP1 *pla* sequence. Amplification products of the expected fragment size were obtained from 10 out of the 33 individuals studied.

In addition to the amplification products from this site, amplification products of equal fragment size were obtained from the remains of two other individuals excavated from an early medieval burial site at Aschheim (6th century). Using BLASTN (Basic Local Alignment Search Tool Nucleotides) 2.2.2, it was determined that the sequences derived from the amplification products and the modern *Y. pestis pla* sequence in GenBank were 94 to 100 % homologous. The observed deviations were presumably due to miscoding lesions in the template DNA.

Throughout the study, no modern *Y. pestis* DNA was introduced into the work environment and all accompanying extraction controls and PCR controls remained negative. Nevertheless, to fully exclude the possibility of contamination by amplicons derived from previous PCRs, a so-called "suicide PCR" (Raoult et al. 2000) was performed in addition. The results confirmed the independent detection of *Y. pestis*-specific DNA in different skeletons under study.

Molekularbiologische Methoden ermöglichen den Nachweis alter mikrobieller DNA in menschlichen Skelettfunden und bieten somit Zugang zum Studium von Infektionskrankheiten wie Pest, welche keine makroskopisch sichtbaren Zeichen am Knochen hinterlassen. Im Hinblick auf die Geschichte der Pest sind noch viele Fragen bezüglich der Ätiologie und Epidemiologie dieser Krankheit unbeantwortet geblieben.

In dieser Studie untersuchten wir Skelettfunde zweier Fundorte in Bayern, die in das 6. bzw. in das 14. Jahrhundert datiert wurden, und bei welchen vermutet wurde, dass es sich um Pestopfer handeln könnte. Die DNA wurde aus Zähnen der Individuen extrahiert und hinsichtlich *Yersinia pestis* spezifischer DNA untersucht, wobei ein Primerpaar verwendet wurde, mit welchem sich ein Abschnitt der auf dem *Y. pestis* Plasmid pPCP1 befindlichen *pla*-Sequenz amplifizieren lässt.

Amplifikationsprodukte der erwarteten Fragmentgröße wurden bei 10 von 33 untersuchten Individuen erhalten, welche aus einem Massengrab unterhalb der Sakristei der Kirche St. Leonhard in Manching-Pichl geborgen worden waren und archäologisch in das 14. Jahrhundert datiert wurden. Des Weiteren wurden Amplifikationsprodukte der entsprechenden Fragmentgröße bei zwei Individuen erhalten, welche aus einer frühmittelalterlichen Begräbnisstätte in Aschheim geborgen worden waren. Unter Verwendung von BLASTN 2.2.2 ergab sich für die Sequenz der PCR-Produkte eine Übereinstimmung von 94-100 % mit der modernen *Y. pestis* pla Sequenz in GenBank. Die beobachteten Abweichungen sind vermutlich auf Basensubstitutionen in der *template* DNA zurückzuführen.

Es wurde keine moderne *Y. pestis* DNA in das Institut eingeführt. Alle mitgeführten Extraktions- und PCR-Kontrollen blieben negativ. Um jedoch vollständig ausschließen zu können, dass die Ergebnisse lediglich auf Kontamination durch Amplifikationsprodukte aus vorherigen PCRs beruhen, wurde zusätzlich eine sogenannte "Suizid-PCR" (Raoult et al. 2000) durchgeführt. Mit den vorliegenden Ergebnissen ist der unabhängige Nachweis *Y. pestis*-spezifischer DNA in verschiedenen Skelettfunden gegeben.

Keywords: Ancient DNA, *Y. pestis*, plague epidemics, St. Leonhard Church (Pichl, Manching) excavation, Aschheim excavation
Alte DNA, *Y. pestis*, Pestepidemien, Kirche St. Leonhard (Manching-Pichl), Aschheim

Introduction

Yersinia pestis, the causative agent of plague, is known to have been responsible for three human pandemics: the Justinian plague, the Black Death and the modern plague (Perry & Fetherston 1997). However, in historical texts, the term "plague" or pestilence, was generally used to describe all kinds of lethal epidemics (Winkle 1997). We therefore have reason to assume that perhaps not all plagues were caused by *Y. pestis*, the bacterium first identified by Alexandre Yersin in 1894 (Yersin 1894).

Today, modern biomolecular methods like the Polymerase Chain Reaction (PCR) enable the detection of microbial DNA fragments in ancient human remains (Salo et al. 1994; Kolman et al. 1999; Haas et al. 2000). As recently as 2000, Raoult et al. provided a molecular insight into the history of plague when he and his group examined the skeletons of three suspected Black Death victims excavated from a 14th century grave in Montpellier, France, for presence of *Y. pestis-specific* DNA sequences.

The genome of the gram-negative bacterium consists of the chromosome and three different plasmids (Parkhill et al. 2001), two of which are unique to *Y. pestis*. The *Y. pestis*-specific plasmid pPCP1 encodes the plasminogen activator Pla, a putative invasin that is essential for virulence by the subcutaneous route (Sodeinde et al. 1992; Cowan et al. 2000). Different studies have demonstrated that the *pla* gene sequence is specific to *Y. pestis*. In particular, this sequence is absent in the closely related species *Yersinia pseudotuberculosis* (Hinnebusch & Schwan 1993; Campbell et al. 1993; Achtman et al. 1999). With the detection of *pla* DNA sequences in the three skeletons under study, Raoult et al. (2000) verified *Y. pestis* as the agent of the medieval Black Death.

This particular study presents the research done on skeletal finds from two southern German excavation sites which have been dated to the 6th and 14th centuries.

Human remains, which were suspected to have been of plague victims, were excavated from a mass grave that was discovered in 1984 during the renovation of the Church of St. Leonhard in Pichl, a suburb of Manching, near Ingolstadt in Bavaria. Located below the church vestry, the grave contained at least 75 individuals, buried tightly in four consecutive layers. Its skeletons were archaeologically dated to the 14th century. This unusual find of a medieval grave dating back to a period when the Black Death was sweeping through western Europe motivated our molecular genetic study of the skeletons for *Y. pestis*-specific DNA sequences.

In addition, our research included work on two other female skeletons that have been dated to the second half of the 6th century. Recovered in 1998 during an archaeological excavation of an early medieval burial site at Aschheim, near Munich (Reimann et al. 2000), they were the remains of an older adult (individual 166) and a juvenile (individual 167). These skeletons were found together with some precious grave goods of which the most remarkable was a pair of bow fibulae shared by the two females (Reimann et al. 2000). Since the two individuals were buried during the first recorded pandemic, i.e., between AD 541 and 750 when several waves of plague hit Europe (Biraben 1975; Perry & Fetherston 1997), they were ideal subjects for testing for *Y.pestis*-specific DNA as the possible cause of death.

Material and methods

Following Raoult et al. (2000), we used tooth samples for our study. The choice of teeth for sample extraction is twofold: the first is that in human plague victims, death usually occurs during the septicemic phase of the disease. Therefore, all tissues supplied with blood, including the dental pulp, would have traces of *Y. pestis*. The second aspect is that the quality of DNA extracted from teeth is usually higher than that extracted from bone (Zierdt et al. 1996).

DNA was extracted from 33 skeletons from the Church of St. Leonhard mass grave, using one tooth sample per individual. Two ancient "negative" teeth from two other individuals who were buried in separate single graves in the church nave and dated to the 12th/13th century and 14th century respectively were used as control teeth. The location of the two skeletons provided no evidence that these individuals died of plague. DNA was also extracted from two teeth of Aschheim individual 166 and four teeth of Aschheim individual 167.

Sample preparation

The tooth specimens were thoroughly washed and subsequently ultrasonically cleaned (35000 Hz, 5-10 min) with distilled water. After pretreatment, based on acid cleaning (Lalueza Fox 1996) and ultraviolet irradiation (302 nm, 15 min), the teeth were ground to a fine powder in a ZrO_2-coated mill (Retsch Type MM2000). The tooth powder was then transferred to sterile 14 ml tubes (Greiner) and stored at $-20°C$.

DNA extraction

DNA extraction was performed according to the silica-based extraction protocol C described by Yang et al. (1998). 300 mg tooth powder were transferred into a 2 ml safe-lock tube (biopur quality, Eppendorf), subsequently dissolved in 1.5 ml extraction buffer (0.5 M EDTA pH 8.0, 0.5% sodium dodecyl sulfate and 100 µg/ml proteinase K) and then incubated in a TB1 thermoblock (Biometra) at 55°C overnight, followed by a second incubation at 37°C for 24 hours.

In each extraction procedure an extraction control containing only reagents was included. After incubation, the extraction solution was centrifuged at 5000 rpm for 5 min, followed by another centrifugation at 12000 rpm for 5 min. The supernatant was then transferred to a disposable Centricon YM-30 centrifugal filter device (Millipore) and centrifuged at 5000xg until the retentate was reduced to the desired volume of approximately 100 µl. The retentate was mixed with 5 volumes of QIAquick™ PB buffer (QIAquick™ PCR purification kit, Qiagen), loaded onto a QIAquick™ column and centrifuged at 13000 rpm for 1 min. After discarding the flowthrough, the DNA was then washed by adding 750 µl of QIAquick™ PE buffer, incubated for 5 min and centrifuged at 13000 rpm for 1 min. The flowthrough was again discarded and an additional centrifugation at 13000 rpm for 1 min was performed. Finally, the DNA was eluted from the column to a 1.5 ml safe-lock tube (biopur quality, Eppendorf) by loading 50 µl QIAquick™ EB buffer and centrifuging at 13000 rpm for 1 min. The extracted DNA was stored at $-20°C$.

DNA extractions and the setting up of PCRs were carried out in a separate laboratory dedicated to ancient DNA analysis. No modern *Y. pestis* DNA was introduced into the work environment and thus no positive controls were carried out. Amplification and the analysis of amplification products were performed in a separate laboratory.

DNA amplification

The amplification reactions were performed in a Personal Cycler PC 48 (Biometra) and were accompanied by extraction controls and PCR controls containing all reagents except for DNA. Two sets of primers, YP12D/YP11R, corresponding to nucleotides 728-875, and YP11D/YP10R, corresponding to nucleotides 854-1001 of the *pla* gene (GenBank accession no. M27820) encoded by the 9.6 kb *Y. pestis*-specific plasmid pPCP1 (Sodeinde & Goguen 1989; Hu et al. 1998), were used. According to Raoult et al. (2000) the primer designations were as follows: Primer set YP12D/YP11R (YP12D: 5'-CAGCAGGATATCAGGAAACA-3', YP11R: 5'-GCAAGTCCAATATATGGCATAG-3'); primer set YP11D/YP10R (YP11D: 5'-CTATGCCATATATTG-GACTTGC-3', YP10R: 5'-GAGCCGGATGTCTTCT-CACG-3').

The PCR was performed in a total volume of 20 µl reaction mixture containing 50 mM KCl, 15 mM Tris-HCl (pH 8.0), 1.5 mM $MgCl_2$, 0.175 mM each dNTP, 1 M Betaine, 0.15 µM each primer (primer pair YP12D/YP11R or YP11D/YP10R), 1 unit AmpliTaq Gold DNA polymerase (Applied Biosystems) and 2 µl ancient DNA extract. 40 amplification cycles were started by an initial denaturation step (95°C for 10 min). Amplification cycles were as follows: 94°C for 45 sec, 55°C for 45 sec, 72°C for 45 sec. PCR was finished by final extension (72°C for 10 min).

With regard to the amplification reactions, mainly the primer pair YP12D/YP11R was used. The use of the primer set YP11D/YP10R was restricted to a so-called "suicide PCR" (Raoult et al. 2000). The suicide PCR method permits primer pairs, targeting *Y. pestis* sequences not previously targeted in the laboratory, to be used only once. This strategy prevents any possibility of molecular contamination of the ancient DNA samples by amplicons from previous investigations.

Electrophoretic separation

PCR was followed by electrophoretic separation in a 0.5 mm polyacrylamide gel (10%T/3%C) with subsequent silver staining (Schumacher 1985). These electrophoreses served as a check for amplification products from the samples as well as the extraction and PCR controls. A 'Low Range' 20 bp DNA ladder (peqlab) and a 100 bp DNA ladder (peqlab) were included into each electrophoresis.

Sequencing of *Yersinia pestis pla* DNA

Amplification products were purified using the QIAquick™ PCR Purification Kit (Qiagen). Subsequently, a second PCR was performed to increase amplicon yields to a level adequate for direct sequence determination. The PCR was performed in a total volume of 50 µl reaction mixture containing 50 mM KCl, 10 mM Tris-HCl (pH 9.0), 1.5 mM $MgCl_2$, 0.175 mM each dNTP, 1 M Betaine, 0.15 µM each primer, 1.5 units Taq DNA polymerase (Amersham Pharmacia Biotech) and 4 µl purified amplification product. 30 amplification cycles were run, starting with an initial denaturation step (95°C for 2 min). Amplification cycles were as follows: 94°C for 45 sec, 56°C for 45 sec, 72°C for 45 sec. PCR was finished by final extension (72°C for 10 min).

Reamplification products were purified with the QIAquick™ PCR Purification Kit and then directly sequenced by using the BigDye Ready Reaction Terminator Cycle Sequencing Kit and an automatic DNA Sequencer Modell 373 (Applied Biosystems).

Obtained sequences were aligned with that of modern *pla* by using BLASTN 2.2.2 [Dec-14-2001] (Altschul et al. 1997). The *pla* sequence has been demonstrated to be specific to *Y. pestis* (Hinnebusch & Schwan 1993; Campbell et al. 1993; Achtman et al. 1999).

Results and Discussion

Amplification products of the expected 148-bp fragment size, corresponding to positions 728-875 of the *pla* sequence, were obtained from 10 out of the 33 Church of St. Leonhard mass grave skeletons studied (examples in Fig.1). No amplification products were achieved from the negative control teeth or any extraction and PCR controls.

Amplification products were directly sequenced and the obtained sequences were aligned with that of modern *Y. pestis pla* in GenBank. Using BLASTN, the sequences derived from the amplification products had an identity that was 94 to 100 percent similar to that of the modern *Y. pestis pla* sequence in GenBank. The observed deviations (without exception G/C→A/T changes) were presumably due to miscoding lesions in the template DNA.

This deviation may be explained by the fact that upon the death of an organism, hydrolytic and oxidative damages starts to accumulate in its DNA (Lindahl 1993). One of the most common forms of hydrolytic damage is the deamination of cytosine which results in the conversion of cytosine to uracil in the DNA. When ancient DNA with such damage is used as a template, incorrect bases will be inserted during the PCR. Such modifications in the template DNA mean that adenosine instead of guanosine will be incorporated during the amplification and will result in G/C→A/T substitutions (Hofreiter et al. 2001).

The observed sequence heterogeneity of the PCR products achieved by the DNA samples from the mass grave individuals is in all probability due to such deamination of cytosine in these ancient DNA templates. Despite this, the overall recovery of *Y. pestis*-specific DNA sequences confirms the hypothesis that the mass grave below the vestry of the St. Leonhard church contained remains of human plague victims.

With regard to the two female skeletons excavated from an early medieval burial site at Aschheim (6th century AD), amplification products of the expected 148-bp fragment size (Fig.2) corresponding to positions 728-875 of the *pla* sequence were detected in one DNA extract from individual 166 as well as in two DNA extracts from individual 167. The sequence of the obtained PCR products and that of the modern *Y. pestis pla* sequence in GenBank were 100 percent identical, with the exception of one amplification product which revealed a single base substitution (G→A change). The detection of *Y. pestis pla* DNA sequences in these skeletons, buried in the second half of the 6th century, is nevertheless, a molecularly supported confirmation of the presence of *Y. pestis* in southern Germany during the first plague pandemic recorded.

During our research, extraction and PCR controls all remained negative. However, to exclude the possibility of contamination by amplicons from previous PCRs, a so-called "suicide PCR" (Raoult et al. 2000) was performed as well. With regard to this method, DNA extracts from two skeletons, one from Aschheim (6th century) and one from Pichl (14th century), were examined. The obtained amplification products are presented in Fig. 3. Using BLASTN, the obtained PCR products corresponding to positions 854-1001 of the *pla* sequence were confirmed as *Y. pestis*-specific. Again, all accompanying extraction and PCR controls remained negative.

Conclusions

Biomolecular tools enable the retrospective diagnosis of ancient diseases that do not cause macroscopically visible lesions within the bones. The detection of ancient microbial DNA in skeletal material thus offers the possibility of studying infectious diseases like plague and their epidemiology in historic populations. Together with the recently completed sequencing of the genome of the *Y. pestis* strain CO92 (Parkhill et al. 2001), it will also enable studies of the evolutionary aspects of plague. For example, it may be possible to test the hypothesis that the three distinct strains (biovars) of *Y. pestis* (Antiqua, Medievalis, and Orientalis), which are still present in the world today, were associated with the three known major pandemics. The comparison of the sequence data generated from ancient material with contemporary strains of *Y. pestis* will allow the reconstruction of the bacterium's evolution.

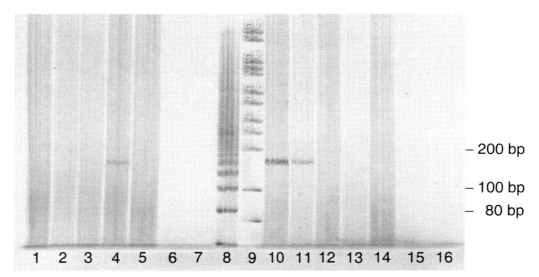

Fig. 1: Detection of *Y. pestis* DNA in skeletons from Pichl, Manching (14th century AD). Silver-stained polyacrylamide gel showing the 148-bp amplified *Y. pestis pla* fragment (YP12D/YP11R) obtained from ancient DNA.
Lanes 1-5: Individuals 1-I, 2, 5-I, 34-I, and 40. Lanes 6 and 15: Extraction controls. Lanes 7 and 16: PCR controls. Lane 8: 20 bp DNA ladder. Lane 9: 100 bp DNA ladder. Lanes 10-14: Individuals 17-I, 22, 26-I, 49, 50.

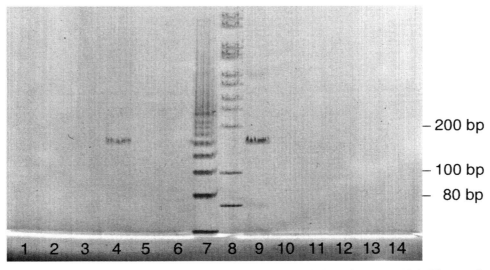

Fig. 2: Detection of *Y. pestis* DNA in two skeletons from Aschheim (6th century AD). Silver- stained polyacrylamide gel showing the 148-bp amplified *Y. pestis pla* fragment (YP12D/YP11R) obtained from ancient DNA samples.
Lanes 1, 2 and 9: Individual 166. Lanes 3, 4, 11 and 12: Individual 167. Lanes 5, 10 and 13: Extraction controls. Lanes 6 and 14: PCR controls. Lane 7: 20 bp DNA ladder. Lane 8: 100 bp DNA ladder.

Summary / Zusammenfassung

Since manifestations of bone lesions do not occur in septicemic plague, a molecular investigation in ancient human skeletal remains offers the only approach to proving an infection like plague. In this study we examined skeletal finds, dated to the 6th and 14th centuries, from two Bavarian excavation sites, for the presence of *Yersinia pestis*-specific DNA sequences.

Our investigations focused on 33 skeletons, archaeologically dated to the 14th century, from a mass grave below the vestry of the Church of St. Leonhard in Pichl, a suburb of Manching near Ingolstadt, and two female skeletons, dated to the second half of the 6th century, from a double inhumation from an early medieval cemetery at Aschheim near Munich.

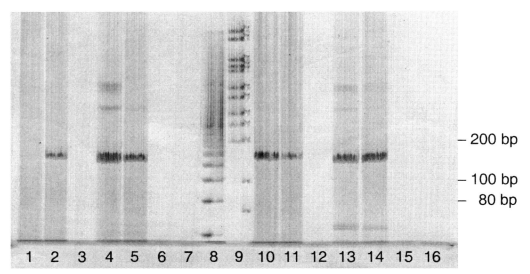

Fig. 3: Application of a "Suicide PCR" on ancient DNA samples. Silver-stained polyacrylamide gel showing the 148-bp amplified *Y. pestis pla* fragment (YP11D/YP10R) obtained from ancient DNA samples. Lanes 1, 2, 10 and 11: Individual 167 (Aschheim, 6th century AD). Lanes 3, 6, 12 and 15: Extraction controls. Lanes 4, 5, 13 and 14: Individual S1-I (Pichl, Manching, 14th century AD). Lanes 7 and 16: PCR controls. Lane 8: 20 bp DNA ladder. Lane 9: 100 bp DNA ladder

DNA was extracted from teeth, and a primer pair, YP12D/YP11R (Raoult et al. 2000), able to amplify a part of the *Y. pestis*-specific plasmid pPCP1 *pla* sequence was used. Amplification products of the expected fragment size were obtained from 10 out of 33 mass grave individuals (14th century) under study and from the two skeletons excavated from the early medieval burial site at Aschheim (6th century).

Using BLASTN, the sequences derived from the obtained amplification products shared 94 to 100 % homology with that of the modern *Y. pestis pla* sequence in GenBank. The observed deviations are attributed to miscoding lesions due to deamination of cytosine in the template DNA (Hofreiter et al. 2001).

By the overall recovery of *Y. pestis*-specific DNA sequences it was possible to confirm the hypothesis that the 14th century mass grave under study contained remains from humans who died of plague. Furthermore, the detection of *Y. pestis pla* DNA in two early medieval skeletal finds from Aschheim (6th century AD) constitutes a molecularly supported evidence of the presence of *Y. pestis* in southern Germany during the first plague pandemic recorded.

Modern *Y. pestis* DNA was not introduced into the work environment. Accompanying extraction and PCR controls all remained negative.

With regard to two skeletal finds, one from Aschheim (6th century) and one from Pichl, Manching, (14th century), a "suicide PCR" (Raoult et al. 2000) was performed in addition. The suicide PCR protocol permits primer pairs to be used only once in a laboratory, so that any amplification products obtained must result from the ongoing experiment and cannot be due to contamination by amplicons resulting from previous investigations. After application of the suicide PCR the obtained amplification products confirmed the independent detection of *Y. pestis*-specific DNA in different skeletons under study.

The recent publication of the complete genomic sequence of *Y. pestis* (Parkhill et al. 2001) provides an insight into the genetic events associated with the emergence of this pathogenic species. The inclusion of *Y. pestis* DNA sequences obtained from ancient human skeletal remains may contribute to the understanding of its evolution.

Molekulargenetische Untersuchungsmethoden stellen die einzige Möglichkeit dar, Infektionskrankheiten wie die Pest in historischen menschlichen Skelettfunden nachzuweisen, da sich die septikämische Pest nicht in Form von Knochenläsionen am Skelett manifestiert.

In dieser Studie untersuchten wir Skelettfunde zweier Fundorte in Bayern, datiert auf das 6. und 14. Jahrhundert, auf das Vorliegen *Yersinia pestis* spezifischer DNA-Sequenzen. Unsere Untersuchungen konzentrierten sich auf

33 in das 14. Jh. datierte Skelette aus einem Massengrab unterhalb der Sakristei der Kirche St. Leonhard in Manching-Pichl nahe Ingolstadt, sowie auf zwei weibliche Skelette, datiert auf die zweite Hälfte des 6. Jahrhunderts, aus einem Doppelgrab eines frühmittelalterlichen Friedhofes bei Aschheim nahe München.

Die DNA wurde aus Zähnen extrahiert, und es wurde ein Primerpaar, YP12D/YP11R (Raoult et al. 2000), verwendet, mit dem ein Teilstück der *Y. pestis* Plasmid pPCP1 *pla*-Sequenz amplifiziert werden konnte. Amplifikationsprodukte der erwarteten Fragmentgröße wurden bei 10 der 33 untersuchten Individuen aus dem Massengrab (14. Jh.) und bei den zwei Skeletten der frühmittelalterlichen Begräbnisstätte bei Aschheim (6. Jh.) erhalten.

Unter Anwendung von BLASTN ergab sich für die aus den Amplifikationsprodukten abgeleiteten Sequenzen eine 94 – 100 %ige Übereinstimmung mit der modernen *Y. pestis pla*-Sequenz in GenBank. Die beobachteten Abweichungen beruhen wahrscheinlich auf Basensubstitutionen, die durch Desaminierung von Cytosin in der *template* DNA bedingt sind (Hofreiter et al. 2001). Durch den prinzipiellen Nachweis *Y. pestis*-spezifischer DNA-Sequenzen war es daher möglich, die Hypothese zu bestätigen, dass das untersuchte Massengrab aus dem 14. Jh. die sterblichen Überreste von Pestopfern enthielt. Darüber hinaus lieferte der Nachweis von *Y. pestis pla*-DNA in zwei frühmittelalterlichen Skelettfunden aus Aschheim (6. Jh.) einen molekulargenetisch gestützten Hinweis auf das Vorliegen von *Y. pestis* in Süddeutschland während der ersten schriftlich belegten Pestpandemie.

Es wurde keine moderne *Y. pestis*-DNA in das Institut eingeführt. Alle begleitenden Extraktions- und PCR-Kontrollen blieben negativ.

Hinsichtlich zweier Skelettfunde wurde zusätzlich eine "Suizid-PCR" (Raoult et al. 2000) durchgeführt. Im Rahmen des Suizid-PCR-Protokolls dürfen Primerpaare im Labor nur einmal verwendet werden, so dass jegliche erhaltenen Amplifikationsprodukte aus dem gegenwärtigen Experiment resultieren müssen und nicht auf einer Kontamination mit Amplifikationsprodukten vorheriger Untersuchungen beruhen können. Die nach Anwendung der Suizid-PCR erhaltenen Amplifikationsprodukte bestätigten den unabhängigen Nachweis von *Y. pestis*-DNA in verschiedenen Skelettfunden.

Die vor kurzem veröffentlichte vollständige Genomsequenz von *Y. pestis* (Parkhill et al. 2001) liefert einen Einblick in die mit dem Auftreten dieser pathogenen Spezies assoziierten genetischen Ereignisse. Die Einbeziehung der aus menschlichen Skelettüberresten erhaltenen *Y. pestis* DNA-Sequenzen kann zum Verständnis der Evolution dieses Krankheitserregers beitragen.

Acknowledgements

We thank Dr. Dorit Reimann and Dr. Karl-Heinz Rieder, Bayerisches Landesamt für Denkmalpflege, for providing the samples.

Bibliography

Achtman M., Zurth K., Morelli G., Torrea G., Guiyoule A. & Carniel E., 1999.
  Yersinia pestis, the cause of plague, is a recently emerged clone of *Yersinia pseudotuberculosis*. *Proceedings of the National Academy of Sciences USA* **96**: 14043-14048.

Altschul S.F., Madden T.L., Schäffer A.A., Zhang J., Zhang Z., Miller W. & Lipman D.J., 1997.
  Gapped BLAST and PSI-BLAST: a new generation of protein database search programs. *Nucleic Acids Research* **25**: 3389-3402.

Biraben J.N. & Le Goff J., 1975.
  The Plague in the Early Middle Ages. In: Forster R. & Ranum O. (eds). *Biology of Man in History*: 48-80. Baltimore, Md.: The Johns Hopkins University Press.

Campbell J., Lowe J., Walz S. & Ezzell J., 1993.
  Rapid and specific identification of *Yersinia pestis* by using a nested polymerase chain reaction procedure. *Journal of Clinical Microbiology* **31**: 758-759.

Cowan C., Jones H.A., Kaya Y.H., Perry R.D. & Straley S.C., 2000.
  Invasion of epithelial cells by *Yersinia pestis*: Evidence for a *Y. pestis*-specific invasin. *Infection and Immunity* **68**: 4523-4530.

Haas C.J., Zink A., Molnar E., Szeimies U., Reischl U., Marcsik A., Ardagna Y., Dutour O., Pálfi G. & Nerlich A.G., 2000.
  Molecular evidence for different stages of tuberculosis in ancient bone samples from Hungary. *American Journal of Physical Anthropology* **113**: 293-304.

Hinnebusch J. & Schwan T.G., 1993.
  New method for plague surveillance using polymerase chain reaction to detect *Yersinia pestis* in fleas. *Journal of Clinical Microbiology* **31**: 1511-1514.

Hofreiter M., Jaenicke V., Serre D., von Haeseler A. & Pääbo S., 2001.
  DNA sequences from multiple amplifications reveal artifacts induced by cytosine deamination in ancient DNA. *Nucleic Acids Research* **29**: 4793-4799.

Hu P., Elliott J., McCready P., Skowronski E., Garnes J., Kobayashi A., Brubaker R.R. & Garcia E., 1998.
Structural organization of virulence-associated plasmids of *Yersinia pestis*. *Journal of Bacteriology* **180**: 5192-5202.

Kolman C.J., Centurion-Lara A., Lukehart S.A., Owsley D.W. & Tuross N., 1999.
Identification of *Treponema pallidum* subspecies *pallidum* in a 200-year-old skeletal specimen. *The Journal of Infectious Diseases* **180**: 2060-2063.

Lalueza Fox C., 1996.
Analysis of ancient mitochondrial DNA from extinct aborigines from Tierra del Fuego-Patagonia. *Ancient Biomolecules* **1**: 43-54.

Lindahl T., 1993.
Instability and decay of the primary structure of DNA. *Nature* **362**: 709-715.

Parkhill J., Wren B.W., Thomson N.R., Titball R.W., Holden M.T., Prentice M.B., Sebaihia M., James K.D., Churcher C., Mungall K.L., Baker S., Basham D., Bentley S.D., Brooks K., Cerdeno-Tarraga A.M., Chillingworth T., Cronin A., Davies R.M., Davis P., Dougan G., Feltwell T., Hamlin N., Holroyd S., Jagels K., Karlyshev A.V., Leather S., Moule S., Oyston P.C., Quail M., Rutherford K., Simmonds M., Skelton J., Stevens K., Whitehead S. & Barrell B.G., 2001.
Genome sequence of *Yersinia pestis*, the causative agent of plague. *Nature* **413**: 523-527.

Perry R.D. & Fetherston J.D., 1997.
Yersinia pestis – Etiologic agent of plague. *Clinical Microbiology Reviews* **10**: 35-66.

Raoult D., Aboudharam G., Crubezy E., Larrouy G., Ludes B. & Drancourt M., 2000.
Molecular identification by "suicide PCR" of *Yersinia pestis* as the agent of medieval black death. *Proceedings of the National Academy of Sciences USA* **97**: 12800-12803.

Reimann D., Düwel K. & Bartel A., 2000.
Vereint in den Tod – Doppelgrab 166/167 aus Aschheim. In: Bayerisches Landesamt für Denkmalpflege und Gesellschaft für Archäologie in Bayern, Hrsg. *Das archäologische Jahr in Bayern 1999*: 83-85. Stuttgart: Konrad Theiss.

Salo W.L., Aufderheide A.C., Buikstra J. & Holcomb T.A., 1994.
Identification of *Mycobacterium tuberculosis* DNA in a pre-Columbian Peruvian mummy. *Proceedings of the National Academy of Sciences USA*. **91**: 2091-2094.

Schumacher J., 1985.
Schnelle Silberfärbung von Nukleinsäuren. *LKB Sonderdruck RE-040*.

Sodeinde O.A. & Goguen J.D., 1989.
Nucleotide sequence of the plasminogen activator gene of *Yersinia pestis*: Relationship to *ompT* of *Escherichia coli* and gene *E* of *Salmonella typhimurium*. *Infection and Immunity* **57**: 1517-1523.

Sodeinde O.A., Subrahmanyam Y.V.B.K., Stark K., Quan T., Bao Y. & Goguen J.D., 1992.
A surface protease and the invasive character of plague. *Science* **258**: 1004-1007.

Winkle S., 1997.
Geißeln der Menschheit: Kulturgeschichte der Seuchen. 2. Aufl. Düsseldorf/Zürich: Artemis und Winkler.

Yang D.Y., Eng B., Waye J.S., Dudar J.C. & Saunders S.R., 1998.
Technical note: Improved DNA extraction from ancient bones using silica-based spin columns. *American Journal of Physical Anthropology* **105**: 539-543.

Yersin A., 1894.
La peste bubonique à Hong Kong. *Annales de l'Institut Pasteur* **8**: 662-667.

Zierdt H., Hummel S. & Herrmann B., 1996.
Amplification of human short tandem repeats from medieval teeth and bone samples. *Human Biology* **68**: 185-199.

Palaeoenvironmental interpretation of fish remains from the Wadi Howar region, northwest Sudan

N. Pöllath, SFB 389, Köln, and Staatssammlung für Anthropologie
und Paläoanatomie, München

J. Peters, Staatssammlung für Anthropologie und Paläoanatomie, München

Abstract / Zusammenfassung

Fish bone finds from Early to Middle Holocene sites in the Middle Wadi Howar region of northwest Sudan were studied and compared with contemporaneous assemblages from the Central Sudanese Nile Valley. Despite a rich ichthyofaunal spectrum, species diversity in the archaeological samples from the Wadi Howar is decidedly lower than those of samples found in sites located along the Nile. Based upon the ecological requirements of the species recognised in the samples, it is possible to describe the main freshwater biotopes that once characterized the study area.

Fischknochenfunde aus früh- bis mittelholozänzeitlichen Fundstellen in der Mittleren Wadi Howar-Region (Nordwestsudan) wurden analysiert und die Ergebnisse zeitgleichen Ichthyoarchaeofaunen aus dem zentralsudanesischen Niltal gegenübergestellt. Dabei stellte sich heraus, dass die Diversität in den erstgenannten Fundstellen deutlich niedriger ist, obwohl insgesamt betrachtet eine beachtliche Artenvielfalt vorliegt. Anhand der ökologischen Ansprüche der im Fundgut nachgewiesenen Arten ist es möglich, Aussagen über die wichtigsten Süßwasserbiotope zu machen, die das Landschaftsbild im Arbeitsgebiet einst charakterisierten.

Keywords: Eastern Sahara, Sudan, Wadi Howar, Holocene, Ichthyofauna, species diversity, palaeoecology
Ostsahara, Sudan, Wadi Howar, Holozän, Ichthyofauna, Artenvielfalt, Paläoökologie

Introduction

In 1995, a long-term scientific research program named "Arid Climate, Adaptation and Cultural Innovation" (SFB 389 – ACACIA) was established at the University of Cologne, Germany. Funded by the German Research Council (Deutsche Forschungsgemeinschaft), it focuses on the arid zones in northeast (Egypt, Sudan, Chad, Libya) and southwest Africa (Namibia). Its main objectives are (1) to reconstruct the Holocene climate, (2) to verify the impact of climatic changes on the terrestrial ecosystem and (3) to trace human adaptations relative to these climatic changes.

One region of focus is the Wadi Howar region in northwest Sudan. It is part of the Libyan Desert (Eastern Sahara) and is located at its southern fringe (http://www.uni-koeln.de/sfb389) (fig. 1). Geoscientific research has shown that this part of the Sahara went through several climatic changes, from hyperaridity at the end of the Pleistocene, to humid conditions during the Early and Middle Holocene, and back to a hyperarid state in the course of the Late Holocene (Kröpelin 1999; Pachur et al. 1990; Pachur & Hoelzmann 2000). A significant increase in precipitation that occurred around 9 ^{14}C kyr BP resulted in the development of vast drainage systems in the Eastern Sahara (Pachur & Kröpelin 1987; Kröpelin 1993, 1999; Pachur 1999; Pachur & Peters 2001).

The West Nubian Palaeolake and the Wadi Howar represent two major water bodies within the Holocene palaeodrainage system of northwestern Sudan, the latter being the Nile's most important tributary from the Sahara during the climatic optimum. It originates in an almost unexplored area in eastern Chad and reaches the river Nile after some 1,100 km, just north of Ed Debba (Pachur & Kröpelin 1987). Only during the peak of the climatic optimum may the Wadi Howar have been a perennial river. Later on, it developed into a chain of flood pools, swamps and lakes that were interconnected only seasonally (Kröpelin 1993). Today, while the lower part of the wadi is almost free of vegetation except in years with exceptional rainfalls, its middle section, though lacking surface water, still has contracted vegetation, characterized by acacia trees and shau bushes.

The West Nubian Palaeolake is located about 150 km north of the Middle Wadi Howar (fig. 1). It was the largest coherent lake in the Sudanese part of the Sahara with depths up to 15 m and a maximum extent of 5330 km² during the Holocene climatic optimum (Hoelzmann et al. 2001). To its east existed a number of smaller palaeolakes which may have been interconnect-

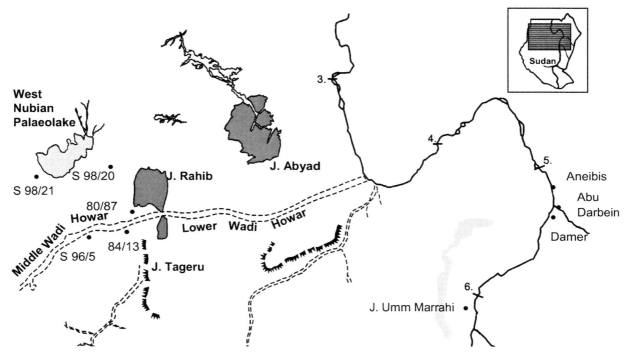

Fig. 1: Map of Northern Sudan showing the sites mentioned in the text.

ed seasonally in years with sufficient precipitation (Hoelzmann et al. 2001).

As mentioned, the Holocene climatic optimum re-established the connection between the Nile and the Lower Wadi Howar, enabling Nilotic animal and plant species associated with aquatic environments to disperse into the middle and upper sections of the wadi. Until now, however, the presence of a diverse ichthyofauna in the West Nubian Palaeolake cannot be satisfactorily explained because a hydrologic connection with the nearby Middle Wadi Howar that would have enabled the migration of fish is considered very unlikely due to geographic barriers.

During a series of field campaigns, more than 1800 archaeological sites were discovered in the study area (Keding 1997a,b, 1998; Jesse 1998, in press). Their relative chronologies were established using the ceramic sequence recorded at the "Conical Hill" site S 84/24 (Gabriel et al. 1985), and a series of ^{14}C dates from different sites provided absolute time frames (Hoelzmann et al. 2001). From this, it can be concluded that the human presence in the study area dates back to the beginning of the 5th millennium calBC. The human groups frequenting the Middle Wadi Howar region at this time appear to have been (semi-)sedentary foragers who subsisted primarily through hunting and gathering. Fishing also played an important role, as indicated by the numerous fish bone finds (Van Neer 1989a,b; Van Neer & Uerpmann 1989). During the later occupation phases, between 4000 and 2500 calBC, people continued to exploit the available aquatic resources, though subsistence was essentially pastoral. With increasing aridity and landscape deterioration, together with their domestic flocks (cattle, sheep, goats), they had to abandon ecologically unfavourable areas, such as the Palaeolake region, and restrict their activities to areas with sufficient pasture, e.g., along the Wadi Howar. Apart from cattle herding, subsistence now included small livestock husbandry, and to a minor extent, hunting (Berke 2001), whereas fishing became increasingly less important as aquatic habitats dried up. ^{14}C dates indicate that human settlement in the Wadi Howar region ceased in the first half of the 2nd millennium calBC (Hoelzmann et al. 2001, fig. 11).

Material and methods

The fish bone samples used and discussed in this study come from sites located either along the Middle Wadi Howar or in the West Nubian Palaeolake area (fig. 1). Site S 98/20 is situated on the shore of a smaller lake near El Atrun, which can be dated to the early stage of occupation on the basis of the ceramics. The same applies to site S 98/21, discovered on the southeastern shore of the West Nubian Palaeolake. As we are dealing with two sites of similar age and pertaining to the same ecosystem, we decided to group the material to enhance sample size (table 1).

According to its ceramic contents, site S 80/87 located at Jebel Rahib (fig. 1), also represents the early occupation

Palaeoenvironmental interpretation of fish remains 257

	Wadi Howar region				Central Nile Valley			
	Palaeolake S 98/20 & 21	Jebel Rahib 80/87	Middle Wadi Howar 84/13-9	Middle Wadi Howar S 96/5-1	Abu Darbein	El Damer	Aneibis	Jebel Umm Marrahi
Date	ca 6000-5300 BP	ca 6000 BP	ca 4500-4000 BP	ca 3300 BP		ca 9000-7000 BP		
Protopterus aethiopicus					4	46	50	1
Polypterus sp.	4		40		12	53	53	216
Heterotis niloticus			23				1	158
Hyperopisus bebe						5	2	
Mormyrops anguilloides					1		4	
Mormyridae indet.			14		41	140	38	14
Gymnarchus niloticus						37	6	8
Hydrocynus sp.					1	3		
Alestes sp.			8		3	54		
Characidae indet.			3			13		
Distichodus sp.					3	58	3	
Citharinus sp.						6		
Citharinidae indet.			2			5	4	
Barbus bynni			8					
Barbus sp.						20	6	
Labeo sp.			4			27	2	
Cyprinidae indet.			39		34	333	24	55
Bagrus bajad					4	2	1	
Bagrus docmak					2	2	1	
Bagrus sp.			10		50	247	48	36
Chrysichthys sp.			2	1				
Auchenoglanis sp.			18	1	1	1		6
Clarotes laticeps			1			4	1	
Schilbeidae indet.			4					
Clariidae indet.	5	385	1189	33	74	926	261	
Clarias sp.		41		2	11	71	9	540
Heterobranchus sp.						1		
Synodontis serratus							2	
Synodontis sorex					1	2		
Synodontis batensoda					1			
Synodontis membranaceus					3			
Synodontis schall					5	6	13	
Synodontis sp.	1	6	172		301	261	455	66
Malapterus electricus						1	1	
Lates niloticus	22	5	45	2	32	217	176	239
Tilapiini indet.	15	109	1896	58	3	42	22	89
Hemichromis fasciatus			13					
Parachanna								2
Tetraodon fahaka			8		1	1	3	1

Table 1: Taxonomic composition of fish bone samples from the Wadi Howar region and the Central Sudanese Nile Valley

phase (Jesse 1998). Excavated in 1980 by members of the University of Cologne (B.O.S.-Project; Kuper 1981), the site yielded 554 identifiable fish remains (table 1).

Another important site at the southern edge of the Middle Wadi Howar, Djabarona 84/13, was discovered and excavated by the B.O.S. team in 1984 (fig. 1). Ceramics and faunal remains indicate that the site dates to the pastoral phase (Keding 1997a). In terms of sample size (number of identified specimens/NISP = 3491) and number of taxa (n = 17), this ichthyofaunal assemblage is the richest studied so far, and hence of particular importance. The fish bones from Djabarona 84/13 were analysed and published by Van Neer (1989a,b; Van Neer & Uerpmann 1989) (table 1).

Site S 96/5, located alongside the Middle Wadi Howar to the west of Djabarona 84/13 (fig. 1), was occupied toward the end of the Holocene wet phase when subsistence had already shifted to primarily small livestock husbandry. Among the 97 bone fragments identified, the presence of at least five taxa could be evidenced (table 1).

Identification of the fish remains was carried out using morphological criteria and comparative specimens housed at the State Collection for Anthropology and Palaeoanatomy in Munich (Staatssammlung für Anthropologie und Paläoanatomie München). The size of the fossil fishes was estimated by comparing them with the corresponding bones of fishes of known size from the modern collection.

The taxonomic compositions of the samples analysed in this study have been compared with those of ichthyofaunal assemblages from sites in the Central Nile Valley (fig. 1), i.e., from Abu Darbein, Aneibis and El Damer on the right bank of the Nile near its confluence with the Atbara river (Peters 1995), and from Jebel Umm Marrahi on the left bank of the Nile, ca. 50 km north of the Sudanese capital of Khartoum (Gautier et al. 2002). Each of these assemblages yielded more than 1,000 identified specimens, and the number of fish taxa recognised averages 20 (table 1). All sites date to the "Early Khartoum" period, a cultural phase that is clearly older (7th millennium calBC; see Haaland & Magid 1995) than the occupation stage represented by the oldest archaeological inventories considered in this study (S 98/20, S 98/21, 80/87).

To describe species diversity within biological systems, ecologists frequently use a measure combining data on numbers of categories (taxa) and abundance within each category. Diversity by this definition reflects the degree of uncertainty in predicting the identity of an individual picked at random from a community, in other words, the heterogeneity of the samples (Reitz & Wing 1999, 105-6).

The diversity of the samples analysed in this study has been calculated using the Shannon-Wiener function:

$$H' = -\sum_{i=1}^{s} (p_i)(\log p_i)$$

In this formula H' stands for the *information content of the sample*, p_i for *the relative abundance of the ith taxon within the sample* and s for *the number of taxonomic categories*. With this measure of diversity, more taxonomic categories will lead to greater diversity values when samples show the same degree of equitability in abundance. Moreover, samples with an even distribution of abundance between taxa have a higher diversity than samples with the same number of taxa but completely distorted distributions. Based upon our present-day knowledge of the ecological demands of the different fish taxa recognized in the samples, a reconstruction of the aquatic habitats in the study area can be attempted.

Results and discussion

A list of taxa identified from the sites mentioned above is given in table 1, together with the number of identified specimens (NISP) of each taxon. Most remains could be identified beyond the family level, in many cases to the genus, and sometimes even to the species level. Fish genera/taxa recognized include *Polypterus*, *Chrysichthys*, *Auchenoglanis*, *Clarias*, *Synodontis*, *Lates*, Tilapiini and *Tetraodon*. Table 1 also contains the data published from other relevant sites (Van Neer 1989a,b; Peters 1995; Gautier et al. 2002). It also shows that the assemblages from the Central Nile Valley are richer in taxa. Moreover, the fish bone samples from the Wadi Howar region essentially yielded evidence of the presence of two catfish genera (*Clarias*, *Synodontis*), one cichlid tribe (Tilapiini) and the Nile perch (*Lates niloticus*). If not an artefact of sample size, direct comparison of the assemblages thus reveals that diversity, on the average, was decidedly lower in the northwestern Sudanese Holocene palaeodrainage system than in the contemporaneous main Nile. The ichthyofaunal record of Djabarona 84/13, where at least 17 taxa could be identified, seems to contradict this assumption, but as stated, taxon frequency is equally important when evaluating diversity. Using the Shannon-Wiener-function, it can be illustrated that the ichthyofaunas from the Wadi Howar region indeed have significantly lower diversity indices than the values calculated for the fish bone assemblages from the Central Nile Valley (fig. 2).

A number of considerations, for example, fishing techniques, fishing gear as well as the size of the Nile perch (fishes that are over 1.5m long that seldom leave the main Nile channel and hence had to be caught using rafts) lead us to believe that ecological rather than cultural differ-

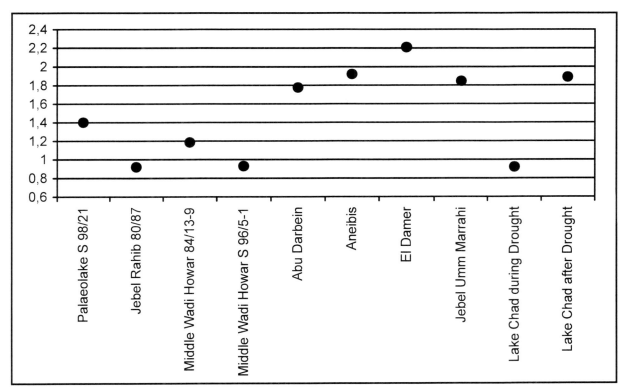

Fig. 2: Diversity (H') within ichthyofaunas from modern Lake Chad and from early to middle Holocene sites in the Wadi Howar region and in the Central Sudanese Nile Valley.

ences account for this phenomenon. To verify this assumption, it is necessary to take a closer look at the habitat requirements of the species present, since these may provide information about the ecology of the water body.

In arid ecosystems water bodies are characterised by (daily and seasonal) fluctuations in water temperature, high evaporation rates, considerable turbidity and other natural phenomena, all of which affect water chemistry. This certainly applies to shallow waters in tropical, (semi)arid zones. Such aquatic habitats are far from stable, as is illustrated for instance, by the occasional mass mortalities of fish in Lake Chad (Bénech et al. 1976). However, studies on Nilo-Sudanic (and other Afro-tropical) fishes demonstrate that some species possess adaptations for coping with extreme changes in water chemistry. In many freshwater fishes, temperature influences, among other things, spawning behaviour, egg maturation and individual growth. Few species, however, can withstand extreme water temperatures (> 35°) and/or a wide temperature range. Only eurytherm species, e.g., the widespread catfish (*Clarias gariepinus*) and the tilapiine species (*Oreochromis niloticus* and *Tilapia zillii*) are able to do so. They can tolerate water temperatures up to as high as 40°C and to as low as about 8°C (Lévêque 1997). To illustrate the importance of temperature for spawning, two members of the catfish family Clariidae can be compared. While in *Heterobranchus longifillis* spawning will only take place when the water temperature is between 25 and 29°C, the eurytherm *Clarias gariepinus* can reproduce within a much wider temperature range of 19 to 31°C (Lévêque 1997). In sum, adverse conditions prevailing in shallow, still, or slowly moving tropical inland waters, which can warm up very fast during day and cool down considerably during night, will cause less problems to populations of tilapias and *Clarias* than to other species.

Salinity and oxygen contents of the water largely correlate with water temperature. Species with high oxygen demands are, for example, the Nile perch (*Lates niloticus*), Bagrid catfishes and Mormyrids (e.g., Fish 1956). In the case of *Lates*, mass mortalities due to heavy winds or strong storms which stir up the organic bottom sediment and cause deoxygenation of the water have been recorded (Worthington 1929). However, a number of Nilo-Sudanic species possess morphological and/or physiological adaptations for surviving in low concentrations of dissolved oxygen. (*Brachy*)*Synodontis*, for example, are able to enhance their haemoglobin production in order to ensure sufficient oxygen supply (Green 1977), whereas the haemoglobin of tilapiines show significantly higher oxygen affinity compared to other species (Fish 1956). Clariid catfishes and the African lungfish *Protopterus aethiopicus* developed additional respiratory organs for aerial breathing (Lévêque 1997); hence, adult *Clarias gariepinus* can obtain up to 50% of their total oxygen requirements from the air.

In the arid tropics, the high evaporation rate can cause an increase in salinity in shallow, seasonal water bodies. Species able to survive high salinity levels are *Clarias gariepinus*, *(Brachy)Synodontis* and tilapias. On the contrary, members of the family Mormyridae (*Mormyrus*, *Mormyrops* etc.,) will avoid water bodies even with low level of salinity (Bénech et al. 1983).

Apart from these abiotic factors, other species-specific characteristics, such as feeding and reproductive behaviour, will result in a separation, in space and time, of species frequenting the same water body. Reproductive cycles, for example, are linked to the seasons and to the existence of favourable conditions for the survival of eggs and larvae. In African freshwater ecosystems, therefore, species can be separated according to their spawning strategies. Whereas spawning in most species occurs at the beginning of the rainy season, when inundated floodplains provide large areas with abundant food and cover for newly hatched juveniles, the adults of certain taxa (e.g., *Lates*, *Bagrus*) will leave the main river later and only for a very short period. Again, other taxa, like Clariids, tilapias, *Polypterus* and *Protopterus*, etc., because of their tolerance of adverse conditions, will stay on the inundated plains until the waters start to recede, even at the risk of being cut off from the main water body when the temporary pools dry out. According to their feeding and/or spawning behaviour, Nilo-Sudanic fish taxa can therefore be roughly classified into two categories, i.e., as "floodplain dwellers" or "open water forms" (Van Neer 1989b). Representative of the first category are *Clarias* and tilapiini, and mature *Lates niloticus*, *Synodontis* and Bagrids (fig. 3) of the second.

From table 1 it can be seen that remains from Clariids and tilapias form the bulk of the fish bone samples collected in sites along the Middle Wadi Howar, and this suggests that the ecological requirements of the two taxa were largely met within the respective site catchments. To a certain extent, the abundance of these taxa mirror the constraints imposed by the early to mid-Holocene environment. However, isolated bone remains of *Lates niloticus*, a species with particular ecological demands (high O_2-concentration, low salinity), were also identified in each assemblage. This testifies to more advantageous conditions–at least temporarily–in certain parts of the aquatic ecosystem. Considering the relative abundance of the taxa identified and their ecological requirements (fig. 3), it can be concluded that the mid-Holocene Middle Wadi Howar palaeodrainage system essentially consisted of shallow, temporary water bodies. However, there is also evidence for the presence of some deeper, larger and better oxygenated freshwater habitats in the Wadi Howar catchment. Given the low frequencies of fishes associated with these environments, these water bodies may not have been necessarily located in the neighbourhood of the sites. During the rainy seasons, fish presumably migrated from a larger water body upstream (Upper Wadi Howar?) to the inundated areas downstream.

On the basis of taxonomic composition and relative frequencies (table 1; Peters et al. 2002), much more stable environmental conditions can be postulated for the West Nubian Palaeolake compared to the Middle Wadi Howar drainage system.

Living conditions, however, were far from optimal for most taxa, as can be deduced from their size distributions (fig. 4). Indeed, *Clarias*, *Lates* and tilapias from the Wadi Howar region are, on the average, smaller than their relatives that frequented the contemporaneous Holocene Nile. Comparatively poor living conditions in the Wadi Howar drainage system could have been the cause, since growth rates correlate with environmental factors (Lévêque 1997), but these seem to have differed between species. Whereas *Clarias* from sites in the study area reached almost the same size as their Nile relatives, the tilapias and the Nile perch remained significantly smaller, implying that the habitat was not optimal. With respect to the Nile perch, the largest individuals are recorded from sites along the main Nile, whereas the smallest ones were from the Middle Wadi Howar sites, with *Lates* from the West Nubian Palaeolake occupying an intermediate position. From this we can deduce that the latter biotope represented a much more stable aquatic environment compared to the Middle Wadi Howar, but as such, did not offer optimal living conditions for the species, hence limiting its size.

One final remark: an example of the extent to which climatic changes can affect fish species diversity in inland waters of tropical Africa comes from Lake Chad. In figure 2 we added two diversity values obtained from modern Lake Chad (Bénech et al. 1983), the first being calculated on the basis of experimental catches at the end of the drought period in 1975, the second at the onset of a new equilibrium since 1977. Of interest is the fact that the first value corresponds well to those obtained for the Wadi Howar sites, while the second fits the values calculated for the Central Nile Valley sites better. From a methodological point of view, a comparison of archaeological samples with modern zoological data is problematic because of fundamental differences between fossil and modern samples as well as sampling methods. For instance, while ichthyologists can sample active present-day freshwater ecosystems, archaeozoologists can only analyse the leftovers of prehistoric human groups. Moreover, fossil fish bone assemblages have been modified by pre- and postdepositional processes that cause differential preservation, e.g., a bias in favour of skeletal remains of species with more robust skeletons, such as *Lates niloticus*. Nevertheless, the phenomena observed in

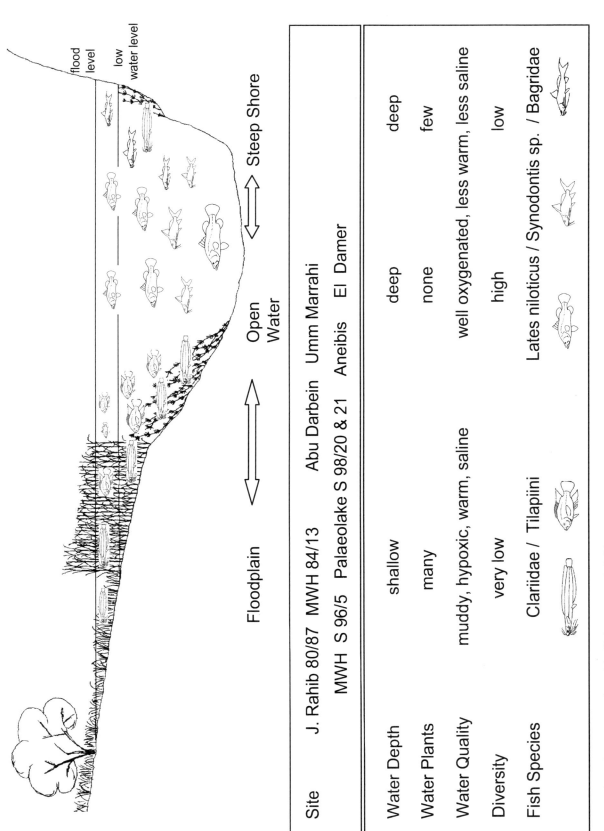

Fig. 3: Schematic reconstruction of the Holocene aquatic habitats.

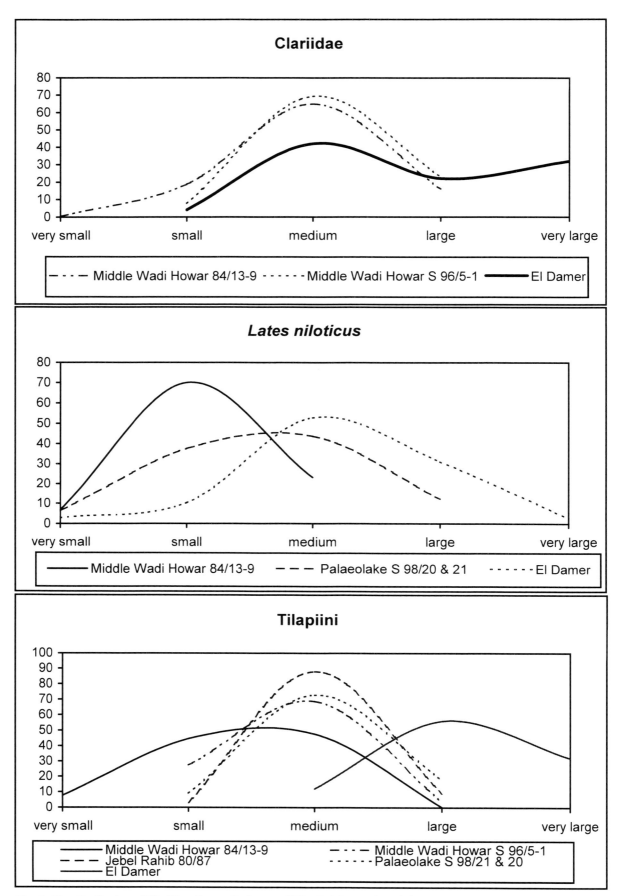

Fig. 4: Size distribution of Clariidae, *Lates niloticus,* and Tilapiini from selected sites.

the Lake Chad ecosystem provide an interesting model for interpreting the Holocene ichthyofaunal record in the Middle Wadi Howar region. Looking at the taxonomic composition of the fish fauna from Lake Chad at the end of the long drought period (Bénech et al. 1983), it may be observed that the habitat features of the mid-Holocene water bodies in the Middle Wadi Howar region may not have been very different at all.

Conclusions / Schlussfolgerungen

Based upon species composition, taxonomic diversity and size distribution of fish remains collected from archaeological sites in the Middle Wadi Howar region, it can be concluded that during the Middle Holocene, the Middle Wadi Howar landscape had a number of shallow, well-vegetated smaller lakes and pools, exploited for fish by prehistoric man on a seasonal rather than on a permanent basis. During the rainy season, the different water bodies along the wadi course may have been interconnected, enabling the (re)colonization by the fish fauna, conceivably from a larger, ecologically more favourable aquatic habitat, probably located upstream. Compared to the Middle Wadi Howar, the lake landscape to the north clearly offered more stable hydrologic conditions, enabling fishing on a year round basis.

Although we know that the Wadi Howar served as a migration path for Nilo-Sudanic fish during the Holocene climatic optimum, the precise way in which aquatic vertebrates dispersed into the West Nubian palaeodrainage system, in particular into the Palaeolake region, is still not well understood. To clarify this issue, bone samples from a series of sites located along the upper and lower sections of the wadi and dating to the consecutive occupation stages need to be investigated. The results of the ichthyofaunal analysis would also enable us to trace in more detail environmental change on a (sub)regional scale as climatic deterioration progressed, and would contribute to our understanding as to why prehistoric man's subsistence strategies changed throughout the Holocene.

Aufgrund der tierartlichen Zusammensetzung, taxonomischen Diversität und Größe der aus archäologischen Fundplätzen in der Mittleren Wadi Howar-Region geborgenen Fischreste wird ersichtlich, dass die Landschaft des mittleren Wadi Howar während des mittleren Holozäns durch untiefe Gewässer mit üppigem Uferbewuchs charakterisiert war, die es dem prähistorischen Menschen ermöglichten, zu bestimmten Jahreszeiten zu fischen. Während der Regenzeit bestanden Verbindungen zwischen den ansonst isolierten Gewässern, die eine Verbreitung der Fischfauna, wohl aus einem im oberen Wadi-Abschnitt befindlichen See mit günstigeren ökologischen Bedingungen, ermöglichten. Im Vergleich zum Mittleren Wadi Howar weist die Seenlandschaft nördlich davon deutliche stabilere hydrologische Verhältnisse auf, sodass man dort das ganze Jahr hindurch fischen konnte.

Obwohl wir wissen, dass die Nilo-Sudanesische Fischfauna während des holozänen Klimaoptimums über das Wadi Howar in das Paläodrainage-System Westnubiens einwanderte, konnte bislang nur wenig Konkretes über die Verbreitung der aquatischen Wirbeltierfauna, beispielsweise in das Paläoseengebiet, in Erfahrung gebracht werden. Zur Klärung dieser Fragestellung sind auch Analysen von Knochenansammlungen aus Fundstellen entlang des Ober- bzw. Unterlaufes des Wadi Howar, die die verschiedenen Bewohnungsphasen repräsentieren, notwendig. Die Untersuchung der Ichthyofaunen würde es darüber hinaus ermöglichen, Veränderungen der Landschaft infolge fortschreitender Klimaverschlechterung detaillierter und auf (sub)regionaler Ebene zu erfassen, was dazu beitragen würde, die sich im Laufe des Holozäns ändernden Subsistenzstrategien des prähistorischen Menschen eingehender als bisher zu verstehen.

Acknowledgements

The financial support of the Deutsche Forschungsgemeinschaft (DFG) for this project (SFB 389/Teilprojekt A2) is gratefully acknowledged.

References

Bénech V., Lemoalle J. & Quensière J., 1976.
Mortalité de poissons et conditions de milieu dans le lac Tchad au cours d'une période de sécheresse. *Cahiers ORSTOM (Office de la Recherche Scientifique et Technique Outre Mer), Série Hydrobiologie* 10 : 119-130.

Bénech V., Durand J.-R. & Quensière J., 1983.
Fish communities of Lake Chad and associated rivers and floodplains. In: Carmouze J.-P., Durand J.-R. & Lévêque C. (eds). *Lake Chad. Ecology and productivity of a shallow tropical ecosystem. Monographiae biologicae* 53: 293-356. The Hague, Boston, Lancaster: Junk Publishers.

Berke H., 2001.
Gunsträume und Grenzbereiche. Archäozoologische Beobachtungen in der Libyschen Wüste, Sudan und Ägypten. In: Gehlen B., Heinen M. & Tillmann A. (eds). *Zeit-Räume. Gedenkschrift für Wolfgang Taute. Archäologische Berichte* 14: 237-256. Bonn: Habelt.

Fish G.R., 1956.
Some aspects of the respiration of six species of fish from Uganda. *Journal of Experimental Biology* 33: 186-195.

Gabriel B., Kröpelin S., Richter J. & Cziesla E., 1985.
Parabeldünen am Wadi Howar. *Geowissenschaften in unserer Zeit* **3**: 105-122.

Gautier A., Linseele V. & Van Neer W., 2002.
The fauna of the early Khartoum occupation on Jebel Umm Marrahi (Khartoum Province, Sudan). In: Jennerstraße 8 (Hrsg.). *Tides of the Desert. Contributions to the Archaeology and Environmental History of Africa in Honour of Rudolph Kuper. Africa Praehistorica* 14: 337-344. Köln: Heinrich-Barth-Institut.

Green J., 1977.
Haematology and habitats in catfish of the genus *Synodontis*. *Journal of Zoology* **182**: 39-50.

Haaland R. & Magid A.A., 1995.
Radiocarbon dates. In: Haaland, R. & Magid A.A. (eds). *Aqualithic sites along the rivers Nile and Atbara, Sudan*: 47-51. Bergen: Alma Mater.

Hoelzmann P., Keding B., Berke H., Kröpelin S. & Kruse H.-J., 2001.
Environmental change and archaeology. Lake evolution and human occupation in the eastern Sahara during the Holocene. *Palaeogeography, Palaeoclimatology, Palaeoecology* **169**: 193-217.

Jesse F., 1998.
Zur Wavy Line-Keramik in Nordafrika unter besonderer Berücksichtigung des Wadi Howar (Sudan) und dort des Fundplatzes Rahib 80/87. Diss. phil., Universität zu Köln.

Jesse F., in press.
Early Khartoum Ceramics in the Wadi Howar, Northwest-Sudan. In: Krzyzaniak L. & Kobusiewicz M. (eds). *Recent Research into the Stone Age of north-eastern Africa*. Poznan: Poznan Archaeological Museum.

Keding B., 1997a.
Djabarona 84/13. *Untersuchungen zur Besiedlungsgeschichte des Wadi Howar anhand der Keramik des 3. und 2. Jahrtausends v.Chr. Africa Praehistorica* 9. Köln: Heinrich-Barth-Institut.

Keding B., 1997b.
Prehistoric investigations in the Wadi Howar region: a preliminary report on the 1995-1996 season. *Kush* **1997**: 33-46.

Keding B., 1998.
The Yellow Nile. New Data on Settlement and the Environment in the Sudanese Eastern Sahara. *Sudan and Nubia* **2**: 2-12.

Kröpelin S., 1993.
Zur Rekonstruktion der spätquartären Umwelt am Unteren Wadi Howar (Südöstliche Sahara/NW Sudan). Berliner Geographische Abhandlungen 54.

Kröpelin S., 1999.
Terrestrische Paläoklimatologie heute arider Gebiete. Resultate aus dem Unteren Wadi Howar (Südöstliche Sahara/Nordwest-Sudan). In: Klitzsch E. & Thorweihe U. (eds). *Nordost-Afrika. Strukturen und Ressourcen. Ergebnisse aus dem Sonderforschungsbereich "Geowissenschaftliche Probleme in ariden und semiariden Gebieten"*: 226-506. Weinheim: Wiley-VCH.

Kuper R., 1981.
Untersuchungen zur Besiedlungsgeschichte der östlichen Sahara. Vorbericht über die Expedition 1980. *Beiträge zur Allgemeinen und Vergleichenden Archäologie* **3**: 215-275.

Lévêque C., 1997.
Biodiversity dynamics and conservation. The freshwater fish of tropical Africa. Cambridge: Cambridge Univ. Press.

Pachur H.-J., 1999.
Paläo-Environment und Draninagesysteme der Ostsahara im Spätpleistozän und Holozän. In: Klitzsch E. & Thorweihe U. (Hrsg.). *Nordost-Afrika: Strukturen und Ressourcen. Ergebnisse aus dem Sonderforschungsbereich "Geowissenschaftliche Probleme in ariden und semiariden Gebieten"*: 366-445. Weinheim, New York, Chichester, Brisbane, Singapore, Toronto: Wiley-VCH.

Pachur H.-J. & Hoelzmann P., 2000.
Late Quaternary palaeoecology and palaeoclimates of the Eastern Sahara. *Journal of African Earth Sciences* **30**: 929-939.

Pachur H.-J. & Kröpelin S., 1987.
Wadi Howar: Paleoclimatic evidence from an extinct river system in the southeastern Sahara. *Science* **237**: 298-300.

Pachur H.-J., Kröpelin S., Hoelzmann P., Goschin M. & Altmann N., 1990.
Late Quaternary fluvio-lacustrine environments of Western Nubia. *Berliner Geowissenschaftliche Abhandlungen (A)* **120.1**: 203-260.

Pachur H.-J. & Peters J., 2001.
The position of the Murzuq Sand Sea in the palaeodrainage system of the Eastern Sahara. *Palaeoecology of Africa and the Surrounding Islands* **27**: 259-290.

Peters J., 1995.
Mesolithic subsistence between the 5th and the 6th Nile Cataract. The archaeofaunas from Abu Darbein, El Damer and Anabeis (Sudan). In: Haaland R. & Magid A.A. (eds). *Aqualithic sites along the rivers Nile and Atbara, Sudan*: 178-244. Bergen: Alma Mater.

Peters J., Pöllath N. & Driesch A. von den, 2002.
Ichthyological Diversity in the Holocene Palaeodrainage Systems of Western Nubia. In: Jennerstraße 8 (Hrsg.). *Tides of the Desert. Contributions to the Archaeology and Environmental History of Africa in Honour of Rudolph Kuper. Africa Praehistorica* 14: 325-335. Köln: Heinrich-Barth-Institut.

Reitz E.J. & Wing E.S., 1999.
Zooarchaeology. Cambridge Manuals in Archaeology. Cambridge.

Van Neer W., 1989a.
Holocene fish remains from the Sahara. *Sahara* **2**: 61-68.

Van Neer W., 1989b.
Recent and fossil fish from the Sahara and their palaeohydrological meaning. *Palaeoecology of Africa and the Surrounding Islands* **20**: 1-18.

Van Neer W. & Uerpmann H.-P., 1989.
Palaeoecological significance of the Holocene faunal remains of the B.O.S.-missions. In: Kuper R. (ed). *Forschungen zur Umweltgeschichte der Ostsahara. Africa Praehistorica* 2: 309-341. Köln: Heinrich-Barth-Institut.

Worthington E.B., 1929.
A Report on the Fishing Survey of Lakes Albert and Kioga. London: The Crown Agents for the Colonies.

Holocene faunas from the Eastern Sahara:
Past and future zoogeographical implications

Joris Peters, Angela von den Driesch,
Staatssammlung für Anthropologie und Paläoanatomie, München

Abstract / Zusammenfassung

After a period of hyperaridity at the end of the Pleistocene, extensive palaeodrainage systems became active in the Eastern Sahara during the Holocene climatic optimum. A rise in water table lead to the formation of rivers and lakes in different regions of the study area. This large-scale climatic change is reflected in the faunal record and illustrated by analyses of vertebrate remains from early to mid-Holocene fluviatile and lacustrine deposits and from contemporaneous archaeological sites. The faunal data have been evaluated in terms of taxonomic diversity and composition, palaeozoogeography, and palaeoecological conditions during the first half of the Holocene. The results of this study primarily contribute to our knowledge of the past, but as the Wadi Howar region was proclaimed a National Park in 2001, archaeozoological data may also be relevant for future decisions regarding nature conservation and animal species reintroduction.

Der hyperariden Phase in der Ostsahara am Ende vom Pleistozän folgte ein Klimaoptimum, wodurch sich im Frühholozän weitläufige Paläodrainagesysteme entwickeln konnten. Infolge der Anhebung des Grundwasserspiegels bildeten sich in den verschiedenen Regionen des Forschungsgebietes Flüsse und Seen. Diese weiträumige klimatische Veränderung spiegelt sich auch in der Fauna wider, wie es aufgrund der Analyse von Resten von Wirbeltieren aus früh- bis mittelholozänen Fluss- und Seesedimenten sowie aus zeitgleichen archäologischen Fundstellen ersichtlich wird. Ihre Auswertung zielte darauf, Aussagen über das Tierartenspektrum bzw. die Artenvielfalt sowie die Verbreitung der Tiere einschließlich ihrer Lebensräume zu treffen. Die in dieser Studie vorgestellten Ergebnisse mögen zunächst nur aus faunenhistorischer Sicht relevant erscheinen, sie werden jedoch durch die 2001 erfolgte Gründung des Wadi Howar Nationalparks zweifelsohne auch in Zukunft zur Klärung von Fragen bezüglich der Erhaltung der Natur einschließlich der Wiedereinbürgerung von Arten herangezogen werden können.

Keywords: Eastern Sahara, Holocene, climatic optimum, vertebrate fauna, palaeozoogeography, palaeoenvironment

Ostsahara, Holozän, Klimaoptimum, Wirbeltierfauna, Paläozoogeographie, Paläoenvironment

Introduction

More than 50 % of the earth's surface between 10°E and the Nile Valley receive an annual precipitation of less than 5 mm and is therefore considered hyperarid. During the last 12,000 years, the Sahara has undergone drastic climatic changes, from hyperarid conditions at the end of the Pleistocene to semi-humid conditions during the first half of the Holocene, and back to hyperaridity in the course of the last 3,000 years. At the beginning of the 8th millennium calBC, the tilt of the earth's axis was stronger than it is today, and the perihelion was at the end of July (Berger 1978). This led to stronger insulation in the northern Hemisphere during summers and to an increase in the amplitudes of the seasonal cycles, thus enhancing the land-ocean temperature contrast, which in turn amplified the African and Indian monsoons (e.g., Kutzbach & Otto-Bliesner 1982; Kröpelin & Petit-Maire 2000).

From aerial photographs and satellite images as well as field work data, it can be shown that the endorheic state in large parts of the Eastern Sahara that resulted from the hyperaridity during terminal Pleistocene times ended between 9500 and 9000 BP. Higher precipitation rates and increased run off in the Saharan mountain ranges (e.g., Tibesti, Ennedi, Gilf Kebir, Messak Settafet, Acacus), in particular, resulted in the formation of extensive palaeodrainage systems, with some water courses that ran more than 800 km, e.g., Wadi Howar (Fig. 1) (Pachur & Kröpelin 1987; Pachur & Wünneman 1996). Apart from these drainage systems implying an (at least seasonal) increase in mean annual precipitation, evidence for large-scale climatic change in the Eastern Sahara also comes from a series of palaeolakes (Pachur 1999; Hoelzmann et al. 2001). They indicate a comparatively high groundwater table and imply relatively stable hydrological conditions during the first half of the Holocene. The largest of these former lakes in the East-

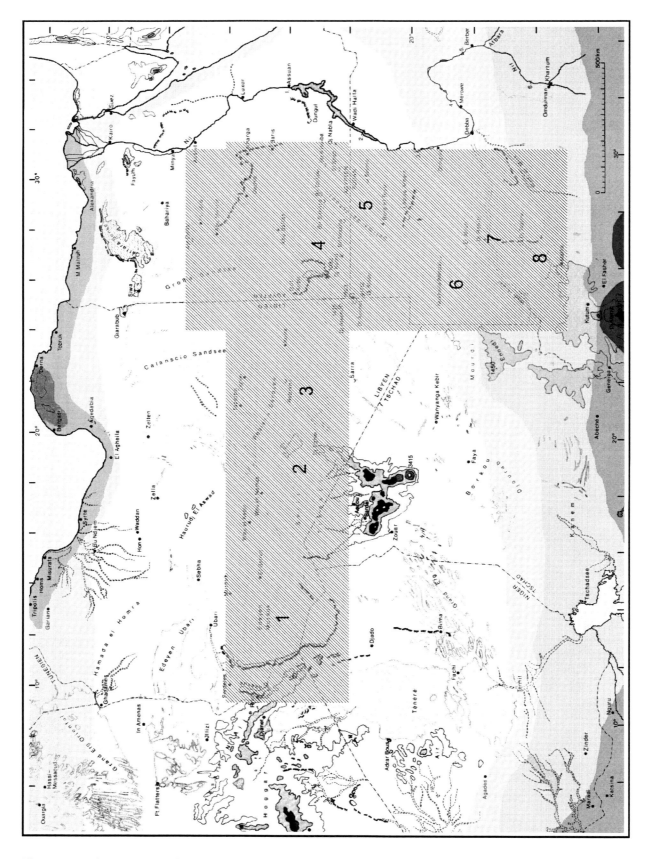

Fig. 1: Map of the Eastern Sahara indicating the present-day positions of the sub-Saharan biomes and the two transects from which the study materials have been collected: (1) eastern foreland of the Acasus mountains and the Erg of Murzuq; (2) northern foreland of the Tibesi mountains; (3) Kufra Basin; (4) Western Desert; (5) Laqyia area and Selima Sandsheet; (6) West Nubian Palaeolake area; (7) Wadi Howar; (8) Jebel Tageru area and Meidob Hills.

ern Sahara was located in West Nubia. It had an estimated maximum surface of some 5,330 km² during the Holocene climatic optimum (Hoelzmann et al. 2001).

The Holocene hydrological conditions favoured the dispersion of species, associated with biomes located to the north (Mediterranean vegetation) and to the south (Sahel vegetation, Sudan Savanna) of the present-day desert zone, into the Eastern Sahara. This is reflected, for example, by the animals depicted in rock art (Fig. 2, 3) and by the faunal and botanical records from archaeological sites in southwestern Egypt and northwestern Sudan (e.g., Gautier 1982; Gautier & Van Neer 1982, 1989; Peters 1987; Neumann 1989; Van Neer & Uerpmann 1989; Berke 2001; Peters et al. 2002).

In this contribution, the results of the analyses of vertebrate remains from early to mid-Holocene semi-lacustrine, fluviatile and limnetic deposits, collected within the framework of the interdisciplinary project, "Geoscientific Research in Arid and Semi-Arid Areas" (*Sonderforschungsbereich* SFB 69), are presented. Combined with (un)published information on vertebrate archaeofaunas excavated in early to mid-Holocene contexts in the Eastern Sahara and adjacent areas (Nile Valley, eastern Sudan) by other projects (e.g., B.O.S.; SFB 389; both University of Cologne), the faunal data (cf. table 1) can be evaluated in terms of species composition and diversity, palaeozoogeography, vegetation and climate in the first half of the Holocene.

Material and methods

Sampling of vertebrate remains in natural deposits *grosso modo* took place along two major transects. The first transect runs about 1500 km in an west-east direction along the Tropic of Capricorn, the second for some 1200 km in a north-south direction along latitude 26-28°E (Pachur 1999). The West-East Transect provided vertebrate remains from different palaeodrainage systems that were known to have existed in the first half of the Holocene in (1) the eastern foreland of the Acacus mountains and the Erg of Murzuq, (2) the northern foreland of the Tibesti mountains, (3) the Kufra Basin and (4) the (Egyptian) Western Desert (Fig. 1).

From an ecogeographical viewpoint, the first transect runs throughout a single ecozone that is characterised by desert conditions, whereas the second one crosses different ecozones, i.e., from the present-day hyperarid Western Desert in the north (< 5 mm precipitation p.a.) to the Sahelian landscape near the Meidob hills in the south (ca. 100 mm precipitation p.a.). The southern edge of the Western Desert encompasses Selima Oasis and Dry Selima (Fig. 1). Further south lies western Nubia, bordered on the east by the Sudanese Nile Valley and to the west

Fig. 2: African buffaloes, Wadi Alamasse. (Lutz & Lutz 1995, fig. 107).

by the eastern foreland of the Ennedi mountain range. West Nubia was once characterized by an extensive network of water courses, including the most important tributary of the Saharan Nile, the Wadi Howar (Fig. 1; see also Pöllath & Peters, this vol.), in the (late) Quaternary. Faunal remains have been collected along the middle section of the Wadi Howar as well as in areas located to its north (West Nubian Palaeolake system) and to its south (Wadi Magrur, Jebel Tageru, Meidob hills).

As said, a good proportion of the faunal remains discussed here come from fluviatile, limnetic, and semi-lacustrine sediments. They were usually collected by handpicking, though in some cases bulk samples were taken. Given the nature and origin of these deposits, the samples essentially yielded remains from species associated with aquatic and riparian habitats. Although part of the vertebrate remains collected along the seasonal water courses, or wadis, represent secondary depositions, e.g., animal bones transported downstream by occasional high floods, some lake shore sediments also indicate primary contexts since vertebrate skeletons that were still partly intact were found.

Fig. 3: Giraffe herd, Wadi Aramas. (Lutz & Lutz 1995, fig. 108).

The animals encountered in the lake and wadi deposits died naturally, most likely as a result of changes in (micro)environmental conditions, such as short-term (daily, seasonal) or long-term fluctuations in temperature; water chemistry (e.g., salinity, O_2-concentration); or water levels (e.g., Van Neer 1989a, 1989b; Pachur & Peters 2001; Pöllath & Peters, this vol.). Taphonomic processes resulted in the preservation of remains of organisms with mineralised exo- or endoskeletons, i.e., ostracods, molluscs, fish, reptiles, birds and mammals. Except for the shells of invertebrates and certain compact bones of vertebrates, such as dermal plates of crocodiles and carpals and tarsals of mammals, the bulk of the bone material studied was heavily fragmented and degraded by post-depositional weathering, e.g., as a result of wind and thermoclastic erosion.

Since the shell and bone samples were obtained from sediments with a calibrated ^{14}C age ranging between the 9th and the 3rd millennium BC, an early to mid-Holocene age can also be postulated for the associated faunal remains. This is confirmed by a series of ^{14}C-dates on shells and bone apatite (Pachur 1999; Kröpelin 1999, Pachur & Peters 2001).

The animal remains submitted for study have been identified with the aid of (1) the comparative collections housed at the State Collection of Anthropology and Palaeoanatomy and (2) osteological features described in literature (e.g., Boessneck et al. 1964; Peters 1988; Peters et al. 1997). During analysis, it was noted that the samples yielded comparably few remains of smaller vertebrates, i.e., of animals with an average body weight below 500 g. The fact that the animal remains were primarily collected by handpicking might explain this discrepancy, though fossilisation processes and differential preservation certainly favoured larger species. Hence, evaluation of taxonomic richness between extant and fossil faunal communities of the Eastern Sahara was limited to taxa with an average body weight above 500 g, even if a taxon possesses animals both above and below that weight.

Holocene vertebrate spectrum of the Eastern Sahara

Up to now, nearly 100 vertebrate taxa have been recognised in faunal samples from the Early to Middle-Holocene Eastern Sahara, i.e., 23 fish, 8 reptile, 17 bird and 49 mammalian taxa, as well as four domestic mammals, i.e., Cattle, Sheep, Goat and Dog (Table 1). The absence of Dromedary, Horse and other domestic species is not surprising, since their appearance in northeast Africa postdates the Middle Holocene (e.g., Peters & von den Driesch 1997; von den Driesch & Peters, in press). Apart from the Dorcas and Dama gazelle, Addax, Scimitar oryx and Fennec, which are well-adapted to survive (semi)desert living conditions, the majority of the taxa identified in the samples are at present associated with biomes located to the south of the Eastern Sahara (see below). Interestingly, the species spectrum observed in the natural deposits overlaps well with the one recorded from early to mid-Holocene archaeological sites in the Eastern Sahara (e.g., Gautier 1982; Gautier & Van Neer 1982, 1989; Peters 1987; Van Neer & Uerpmann 1989; Berke 2001) (Table 1).

A comparison of the faunal data from the various subregions reveals interesting differences, not only in a north-south direction, as may be expected because of latitude-related changes in precipitation and vegetation cover, but also in a west-east direction along the Tropic of Capricorn. The number of Artiodactyl taxa once present in the Erg of Murzuq (n = 15) and in the Western Desert (n = 7) (Table 1) is a good illustration of the latter phenomenon, which is not related to sample size, since the number of bone fragments analysed from the Western Desert is more than ten times higher (> 100,000). Qualitative differences between the western and eastern part of the Eastern Sahara are also reflected by the presence of species with higher ecological demands, in particular those associated with aquatic habitats, e.g., Bohor reedbuck, Hippopotamus, Nile crocodile etc. (Table 2). Thus, the complete absence of any of these taxa in the Western Desert on the one hand and their relative abundance in the Erg of Murzuq (n = 8) on the other hand – the Kufra Basin occupying an intermediate position (n = 4) – is indicative of a west-east gradient in palaeoclimatic conditions along the Tropic of Capricorn as well. The presence of vast mesic habitats in southwest Libya in Early to Middle Holocene times is also illustrated by rock engravings of Nile crocodile (Fig. 4) and Hippopotamus (Fig. 5) in the Messak Settafet mountains.

As already mentioned, latitude-related changes in precipitation and vegetation cover in the past can be expected as well, considering the present-day situation in sub-Saharan Africa. Ungulate taxonomic composition and diversity within the different faunal assemblages is but one example that illustrates a north-south precipitation gradient during the Holocene climatic optimum (Fig. 6).

Palaeoenvironment

Based on the ecological requirements of the taxa identified, it is possible to describe the landscapes that once characterised the Eastern Sahara. An illustration of this can be found in the contribution on the early to mid-Holocene ichthyofauna from West Nubia (Pöllath & Peters, this vol.). The fish taxa present are indicative of a variety of habitats, from seasonal, shallow, well-vegetated ponds and lakes up to large, permanent, stable, and hence ecologically more favourable, aquatic ecosystems. Nevertheless, species richness and diversity in the West Nubian palaeodrainage system remained significantly lower compared to the contemporaneous Central Sudanese Nile (Peters 1995; Gautier et al. 2002), implying that heterogeneity of the habitats, both spatially and temporally, was decidedly higher in the main river.

A detailed palaeoenvironmental interpretation of all faunal data lies beyond the scope of this paper; therefore, it was decided to focus only the archaeozoologically least known part of the study area, i.e., the Erg of Murzuq. Apart from the vertebrate taxa recognized in the fossil assemblages (Table 1), animals depicted in rock art in the surrounding mountain ranges (Messak Settafet; Tadrart Acacus) can also be used to reconstruct past biotopes. Taxonomic richness in rock art is remarkable (e.g. Frobenius 1937; Lutz & Lutz 1995; Muzzolini 1995; Le Quellec 1998), with most wild species depicted being confined today to the Aethiopic realm, e.g., African elephant (*Loxodonta africana*), White rhinoceros (*Ceratotherium simum*), African wild ass (*Equus africanus*), Hippopotamus (*Hippopotamus amphibius*), Giraffe (*Giraffa camelopardalis*), (Giant) African buffalo (*Syncerus caffer / S. (c.) antiquus*), Hartebeest (*Alcelaphus buselaphus*), Barbary sheep (*Ammotragus lervia*), Leopard (*Panthera pardus*), Ostrich (*Struthio camelus*) and the Nile crocodile (*Crocodylus niloticus*). Palaearctic species include Wild cattle (*Bos primigenius*) and probably Wild boar (*Sus scrofa*). Engravings of domestic cattle predominate in rock art (Fig. 7), but Sheep and Dog (Fig. 8) are also depicted. It should be mentioned, however, that opinions differ about the age of the rock engravings in the mountain ranges of southwest Libya. A Late Pleistocene age for the initial stage of rock art (so-called *Bubalus antiquus* phase), postulated by Mori (1965), seems very unlikely since the climate was hyperarid and living conditions even drier than today (Kröpelin 1999). Whether the earliest naturalistic depictions of game species were made by early Holocene hunter-gatherer groups (e.g., Lutz & Lutz 1995), or whether the petroglyphs postdate the 5th millennium calBC arid spell and should therefore be attributed to cattle pastoralist societies (Muzzolini 1995; Le Quellec 1998, 232 ff.), is still debated. Whichever the case, it can be safely assumed that the fauna depicted in rock art and the species recognised by osseous remains from natural deposits and archaeological sites (Gautier & Van Neer 1982; Corridi 1998) are broadly contemporaneous. All these data can therefore be used for palaeoenvironmental reconstructions.

Considering the ichthyofauna, taxonomic diversity in the Erg of Murzuq and in the Acacus mountains is poor compared with the spectrum observed in Western Nubia (Table 1). Migration out of the main drainages certainly influenced fish species composition, with selection being against the species less well-adapted to adverse conditions. Though we do not know the exact number

TRANSECT	WEST – EAST				NORTH – SOUTH			
	Erg of Murzuq – Acacus mountains	Tibesti forelnad	Kufra Basin	Western Desert	Selima Oasis - Dry Selima – Laqiya area	West Nubian Palaeolake region	Middle Wadi Howar – Jebel Rahib	Jebel Tageru – Wadi Magrur – Meidob hills
FISH								
Polypterus sp.	-	-	-	-	-	+	+	+
Heterotis, *Heterotis niloticus*	-	-	-	-	-	-	+	-
Mormyridae indet.	-	-	-	-	-	-	+	-
Alestes sp.	-	-	-	-	-	-	+	-
Characidae indet.	-	-	-	-	-	-	+	-
Citharinidae indet.	-	-	-	-	-	-	+	-
Bynni, *Barbus bynni*	-	-	-	-	-	-	+	-
Labeo sp.	-	-	-	-	-	+	+	+
Cyprinidae indet.	-	-	-	-	-	-	+	+
Bagrus sp.	-	-	-	-	-	+	+	+
Chrysichthys sp.	-	-	-	-	-	-	+	-
Auchenoglanis sp.	-	-	-	-	-	-	+	+
Widehead catfish, *Clarotes laticeps*	-	-	-	-	-	-	+	-
Schilbeidae indet.	-	-	-	-	-	-	+	-
Clarias sp.	+	-	+	-	-	+	+	+
Heterobranchus sp.	-	-	-	-	-	+	-	+
Clarias/Heterobranchus	+	-	+	-	-	+	+	+
Wahrindi, *Synodontis schall*	-	-	-	-	-	+	-	-
Synodontis sp.	+	-	+	-	-	+	+	+
Nile perch, *Lates niloticus*	-	-	-	-	-	+	+	+
Banded jewelfish, *Hemichromis fasciatus*	-	-	-	-	-	-	+	-
Tilapiini indet.	+	-	+	-	-	+	+	+
Globefish, *Tetraodon lineatus*	-	-	-	-	-	+	-	+
REPTILES								
African soft-shelled turtle, *Trionyx triunguis*	-	-	-	-	-	+	+	+
Pelusios sp.	-	-	-	-	-	-	+	-
Cyclanorbis sp.	-	-	-	-	-	-	+	-
Testudo/Geochelone	+	-	-	+	+	+	-	+
Nile monitor, *Varanus niloticus*	-	-	-	-	-	-	+	-

Tab. 1a: Subregions of the Eastern Sahara and their early to middle Holocene faunal spectra

TRANSECT	WEST – EAST				NORTH – SOUTH			
	Erg of Murzuq – Acacus mountains	Tibesti forelnad	Kufra Basin	Western Desert	Selima Oasis – Dry Selima – Laqiya area	West Nubian Palaeolake region	Middle Wadi Howar – Jebel Rahib	Jebel Tageru – Wadi Magrur – Meidob hills
Varanus sp.	-	-	-	+	-	-	-	-
Nile crocodile, *Crocodylus niloticus*	+	-	-	-	-	+	+	+
African rock python, *Python sebae*	-	-	-	-	-	+	+	+
BIRDS								
Ostrich, *Struthio camelus*	+	-	-	+	+	-	-	+
Great crested grebe, *Podiceps cristata*	+	-	-	-	-	-	-	-
Little grebe, *Podiceps ruficollis*	+	-	-	-	-	-	-	-
Podiceps sp.	+	-	-	-	-	-	-	-
Common moorhen, *Gallinula chloropus*	+	-	-	-	-	-	-	-
Great egret, *Casmerodius albus*	-	-	-	-	-	-	-	+
Yellow-billed stork, *Mycteria ibis*	+	-	-	-	-	-	-	-
Spur-winged goose, *Plectopterus gambensis*	-	-	-	-	-	-	-	+
Mallard, *Anas platyrhynchos*	+	-	-	-	-	-	-	-
Anas sp.	+	-	-	-	-	+	-	-
Ferruginous duck, *Aythya nyroca*	+	-	-	-	-	-	-	-
Vanellus sp.	-	-	-	-	-	-	-	+
African fish eagle, *Haliaeetus vocifer*	+	-	-	-	-	-	-	-
Falconiformis indet.	+	-	-	-	-	-	-	-
White stork, *Ciconia ciconia*	+	-	-	-	-	-	-	-
Helmeted guinea fowl, *Numida meleagris*	-	-	-	-	-	-	-	+
Common quail, *Coturnix coturnix*	-	-	+	-	-	-	-	-
WILD MAMMALS								
Brown hare, *Lepus capensis*	+	-	-	+	-	-	-	-
Lepus sp.	-	-	+	+	-	+	+	+
North African crested porcupine, *Hystrix cristata*	-	-	-	-	-	-	-	+
Greater cane rat, *Thryonomys swinderianus*	-	-	-	-	-	+	+	+
Golden jackal, *Canis aureus*	+	-	-	+	-	-	-	+
Rueppell's fox, *Vulpes rüppelli*	-	-	-	+	-	-	-	-
White-tailed mongoose, *Ichneumia albicauda*	--	-	-	-	-	-	+	+
Striped hyena, *Hyaena hyaena*	-	-	-	+	-	-	-	+
Spotted hyena, *Crocuta crocuta*	-	-	-	-	-	-	-	+

Tab. 1b: Subregions of the Eastern Sahara and their early to middle Holocene faunal spectra

TRANSECT	WEST – EAST				NORTH – SOUTH			
	Erg of Murzuq – Acacus mountains	Tibesti forelnad	Kufra Basin	Western Desert	Selima Oasis - Dry Selima – Laqiya area	West Nubian Palaeolake region	Middle Wadi Howar – Jebel Rahib	Jebel Tageru – Wadi Magrur – Meidob hills
Wild cat, *Felis silvestris*	-	-	-	+?	+	-	-	+
Serval, *Leptailurus serval*	+	-	-	-	-	-	-	-
Caracal, *Felis caracal*	-	-	-	+	-	-	-	-
Leopard, *Panthera pardus*	+	-	-	-	-	-	-	-
Aardvark, *Orycteropus afer*	-	-	-	-	-	+	-	+
African Elephant, *Loxodonta africana*	+	-	-	+	-	+	+	+
Cape hyrax, *Procavia capensis*	-	-	-	+	-	-	-	-
African wild ass, *Equus africanus*	+	-	-	-	+	-	-	-
Burchell's zebra, *Equus burchellii*	-	-	-	-	-	-	-	+
Equidae indet.	-	-	-	-	-	+	-	+
White rhinoceros, *Ceratotherium simum*	-	-	-	-	-	-	-	+
Black rhinoceros, *Diceros bicornis*	-	-	-	-	-	-	-	+
Ceratotherium/Diceros	+	-	-	+	-	-	-	+
Warthog, *Phacochoerus aethiopicus*	+	-	-	-	-	+	+	-
Hippopotamus, *Hippopotamus amphibius*	+	-	-	-	-	+	+	+
Giraffe, *Giraffa camelopardalis*	-	-	-	+	+	+	+	+
Small bovids:								
Sylvicapra/Ourebia	+	-	-	-	-	-	-	-
Dorcas gazelle, *Gazella dorcas*	+	+	+	+	+	+	-	-
Unidentified *Small Bovids*	+	+	+	+	+	+	+	+
Medium bovids:								
Bohor reedbuck, *Redunca redunca*	+	-	+	-	-	+	-	+
Kob, *Kobus kob*	-	-	-	-	-	-	-	+
Kobus sp.	-	-	-	-	-	+	-	+
Slender-horned gazelle, *Gazella leptoceros*	-	-	-	+	-	-	-	-
Red-fronted gazelle, *Gazella rufifrons*	+	-	-	-	-	-	+	+
Barbary sheep, *Ammotragus lervia*	+	-	-	+	+	+	-	-
Unidentified *Medium Bovids*	+	-	+	+	+	+	-	+
Large Bovids:								
Addax, *Addax nasomaculatus*	+	-	+	+	+	-	-	-
Scimitar-horned oryx, *Oryx dammah*	+	-	-	+	+	+	-	+
Hartebeest, *Alcelaphus buselaphus*	+	-	-	+	-	-	+	+

Tab. 1c: Subregions of the Eastern Sahara and their early to middle Holocene faunal spectra

TRANSECT	WEST – EAST				NORTH – SOUTH			
	Erg of Murzuq – Acacus mountains	Tibesti forelnad	Kufra Basin	Western Desert	Selima Oasis - Dry Selima – Laqiya area	West Nubian Palaeolake region	Middle Wadi Howar – Jebel Rahib	Jebel Tageru – Wadi Magrur – Meidob hills
Alcelaphus/Damaliscus	+	-	-	-	-	+	-	-
Waterbuck, *Kobus ellipsiprymnus*	+	-	-	-	-	-	-	+
Roan antilope, *Hippotragus equinus*	+	-	-	-	-	-	-	+
Dama gazelle, *Gazella dama*	-	-	+	+	+	-	-	-
Unidentified *Large Bovids*	+	-	+	+	+	+	+	+
Very large Bovids :								
Giant eland, *Taurotragus derbianus*	-	-	-	-	-	+	-	+
African buffalo, *Syncerus caffer*	+	-	-	-	-	+	+	+
Giant buffalo, *Syncerus (caffer) antiquus*	+	-	-	-	-	+	+	+
Syncerus sp.	-	-	-	-	+	-	-	+
Unidentified *Very large Bovids*	+	-	+	-	-	+	-	+
DOMESTIC MAMMALS								
Dog	-	-	-	-	+	-	+	-
Cattle	+	-	+	+	+	+	+	+
Sheep	+	-	-	+	+	-	+	-
Sheep/Goat	+	-	-	+	+	+	+	-
Goat	-	-	+	+	+	+	-	-

Tab. 1d: Subregions of the Eastern Sahara and their early to middle Holocene faunal spectra

Taxon	Erg of Murzuq	Kufra Basin
Clariid, *Clarias/Heterobranchus* sp(p).	+	+
Mochokid, *(Brachy)synodontis* sp(p).	+	+
Cichlid, *Tilapia/Sarotherodon* sp(p).	+	+
Nile crocodile, *Crocodylus niloticus*	+	-
African River Eagle, *Haliaeetus vocifer*	+	-
Hippopotamus, *Hippopotamus amphibius*	+	-
Bohor Reedbuck, *Redunca redunca*	+	+
Waterbuck, *Kobus ellipsiprymnus*	+	-

Tab. 2: Taxa associated with Holocene aquatic environments along the West-East Transect

Fig. 4: Crocodile, Wadi Imrawen. (Lutz & Lutz 1995, fig. 119).

Fig. 5: Hippopotamus, Upper Wadi Aramas. (Lutz & Lutz 1995, fig. 118).

of fish taxa that reached the Erg of Murzuq during the climatic optimum, only three taxa (*Clarias, (Brachy) Synodontis, Tilapiini*) have been recorded in the lacustrine sediments until now, and which can be best explained by the ecological bottlenecks the migrating ichthyofauna went through. The small mean size of these individuals is indicative of unstable aquatic conditions, such as poorly oxygenated water bodies with

A: Laqiya Area – Selima Sandsheet
   Equus, Giraffa

B: West Nubian Palaeolake
   *Loxodonta, Hippopotamus,
   Giraffa, Phacochoerus, Redunca, Kobus,
   Alcelaphini, Taurotragus, Syncerus*

C: Middle Wadi Howar
   *Loxodonta, Hippopotamus, Giraffa,
   Phacochoerus, Redunca, Kobus,
   Alcelaphus, Taurotragus, Syncerus*

D: Jebel Tageru – Meidob Hills
   *Loxodonta, Equus, Giraffa, Diceros,
   Ceratotherium, Phacochoerus, Hippopotamus,
   Redunca, Kobus, Alcelaphus, Hippotragus,
   Taurotragus, Syncerus*

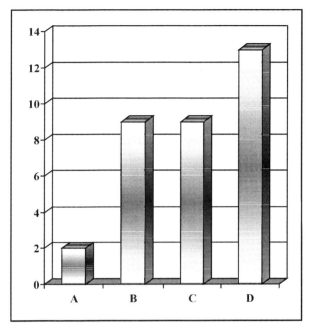

Fig. 6: Early to Middle Holocene Ungulate taxonomic composition in the Eastern Sahara in a North-South direction along 26-28°E.

Fig. 7: Cattle herd, Wadi Takabar. (Lutz & Lutz 1995, fig. 159).

comparatively high salinity levels. Though a good tolerance of hypoxic conditions and salinity has been noted in *(Brachy)Synodontis*, members of this taxon are less well-adapted to cope with high salinity levels compared to *Clarias* and *Tilapiini* (Lévèque 1997). Interestingly, all *(Brachy)Synodontis* remains collected so far in the Erg of Murzuq come from sediments of the somewhat larger, more permanent palaeolakes that stretched over 3 km and had estimated maximum depths of ca. 10 m, whereas remains of the ecologically more tolerant catfish and tilapias have also been found on ancient shores of much smaller water bodies.

Fig. 8: Dog chasing an ostrich, Wadi In Elobu. (Lutz & Lutz 1995, fig. 124).

Palaeoenvironmental information also comes from the other vertebrate groups. The crocodile remains found, for example, are mostly of individuals with lengths below 2.5 m. Smaller crocodiles mainly take up residence in swamps and backwaters, feeding essentially on (cat)fish (Branch 1988). Following the ecological requirements of the different bird taxa (table 1), larger (African fish eagle, Great crested grebe) as well as smaller (Moorhen, Pintail) aquatic habitats were present, mainly with shallow, slow-moving (Yellow-billed stork, Ferruginous duck), fresh to brackish waters (Ferruginous duck, Pintail), bordered by trees (African fish eagle), reeds and other aquatic vegetation (Great crested grebe, Pintail, Ferruginous duck, Moorhen), and surrounded by grasslands and marshes (White stork, Yellow-billed stork, Moorhen, Quail). Most birds present are known to rely on a broad spectrum of plants and/or animals (molluscs, amphibians etc.), with the possible exception of the African fish eagle, which normally prefers surface-feeding fish up to 1 kg, like catfish, tilapias or lungfish (Brown et al. 1982, 313).

As to the terrestrial mammals, their habitat requirements indicate that the dominant vegetation type in the Erg of Murzuq were grasslands and sparse, open woodlands with an abundance of shrubs and some stands of smaller trees (Hare, Serval, Warthog, Hartebeest, Roan Antelope, Korrigum, Oribi, Crowned duiker, Red-fronted gazelle), whereas near permanent water bodies large trees and shrubs were found, forming quite dense shady habitats (Leopard, Buffalo, Waterbuck). The vegetation associated with the lakes, marshes and swamps consisted of moist grasslands (Hippopotamus, Buffalo, Waterbuck, Korrigum), water plants, tall grasses and reeds (Bohor reedbuck, Hippopotamus). Based on this, it can be concluded that the palaeodrainage system and the vegetated dunes of the Erg of Murzuq, as well as the surrounding mountainous areas and wadis, formed a complex, heterogeneous ecosystem with a multitude of habitats and niches. The formation of this landscape during the Holocene wet phase enabled animal species, at that time confined to biomes located to the south and the north of the present-day desert, to (re)colonise the northwestern part of the Eastern Sahara.

Palaeozoogeographical aspects

Different palaeozoogeographical issues can be addressed using the data collected in this project. In this paper, however, only aspects of the geographic origin of the aquatic and herbivorous vertebrate communities will be discussed.

In the Sahara, relict populations of Nilo-Sudanic freshwater fishes still occur today, although the number of species in these residual permanent water bodies of the Eastern Sahara is very low. The fishes present belong to the few taxa (*Clarias, Tilapia*) with low ecological demands and particular physiological adaptations (Fish 1956; Green 1977; Van Neer 1989b; Lévèque 1990, 1997; see also Pöllath & Peters, this vol.). In view of the actual distribution patterns of Nilo-Sudanic fish taxa, it can be assumed for the western part of the Eastern Sahara (Erg of Murzuq, Acacus mountains, Kufra Basin) that the presence of these taxa reflects a northern extension of a large, present-day basin located to the south. The palaeodrainage system of the Erg of Murzuq was conceivably interconnected with the Chad Basin, which extended much further to the north during the Holocene wet phase. Thus, the Nilo-Sudanic fish taxa present in the Erg of Murzuq most likely originated from the Mega-Chad palaeodrainage system (Fig. 9).

In the southeastern part of the Eastern Sahara, colonisation of the early Holocene drainage system by fish and other aquatic vertebrates (Nile crocodile, turtles, Nile monitor) most likely did not take place from the south, but rather from the east. It was the largest and northernmost tributary of the Nile, the Wadi Howar, that served as a migration path for Nilo-Sudanic fish (Van Neer 1989a, b). The presence of a diverse Nilo-Sudanic ichthyofauna in the palaeolakes to the north of the Middle Wadi Howar, however, still cannot be satisfactorily explained, since geographic barriers prevented a colonisation out of this part of the palaeodrainage system.

With respect to the terrestrial flora and associated fauna, classification of extant biomes grade from areas of high annual rainfall (forests) to those that are extremely dry (deserts). Between these two extremes, vegetation structure decreases, in density of trees and other woody species, as rainfall decreases. Landscape features, fires, herbivores as well as humans have an impact on plant cover, and consequently, on animal life in extant communities. During the Holocene climatic optimum, effects of human impact may have been less severe in the Eastern Sahara than in the late Holocene because human population density was lower. Moreover, during the Holocene wet phase, human groups probably avoided particular regions for health reasons (malaria, sleeping sickness, etc.), meaning areas where the palaeodrainage systems caused the formation of vast swamplands. Such a scenario could be invoked to explain why, in the West Nubian Palaeolake region, traces of Holocene habitation in form of hunter-fisher-gatherer camp sites only appear toward the end of the 6th millennium calBC. This would be when the large coherent lake was already segmented into smaller water bodies as a result of desiccation (Hoelzmann et al. 2001), whereas along the Central Sudanese Nile, human groups with a similar lifestyle can be traced back archaeologically as early as the 8th millennium calBC.

As said, vegetation structure becomes less dense in tandem with decreases in rainfall, and this in turn influences species composition in faunal communities. For palaeozoogeographical studies, strict herbivorous species are of particular interest since they will only occur in biomes where their favoured food plants occur. Table 3 gives an overview of the different herbivore species found in the Early to Middle Holocene assemblages of the Eastern Sahara and the vegetation zones they generally frequent.

The herbivore records of the Erg of Murzuq and the adjacent mountain ranges thus indicate environmental conditions comparable to those prevailing today in the southern Sahel (250-500 mm precipitation/year) and/or in the northern Sudan Savanna zone (500-600 mm precipitation/year). Further to the east, species composition in the Kufra Basin and the Tibesti foreland suggests that the vegetation was a mixture of a northern (100-250 mm precipitation/year) and a southern Sahel, whilst the contemporaneous faunal samples from the Western Desert (southwest Egypt) are indicative of a northern Sahel type of vegetation. As to the North-South Transect, the fauna of the West Nubian Palaeolake and the Middle Wadi Howar region imply a southern Sahel/northern Sudan Savanna faunal community, whereas the southernmost part of the study area (Wadi Magrur, Jebel Tageru, Meidob Hills) most likely witnessed a Sudan Savanna vegetation during the Early Holocene. On the basis of this information, the extent of the northward shift of the sub-Saharan biomes during the Holocene optimum can be assessed. Today, the vegetation zones run almost parallel to the line of latitude in sub-Saharan Africa. Provided this was also the case in the past, the following assumptions can be made: to maintain a southern Sahel vegetation and its associated faunal community in the Erg of Murzuq, it can be inferred that the area must have received at least some 250-400 mm of precipitation per year. If so, a northward shift of the present-day sub-Saharan biomes of some 900 to 1200 km can be assumed. It is noteworthy that at present, a mammalian fauna similar to the one described for the early/middle Holocene Erg of Murzuq is recorded some 12-13° to the south, namely from northeast Nigeria (11-12°N; 12-15°E) (Happold 1987, 355 ff.). No doubt, the northward shift of the sub-Saharan biomes during the Holocene wet phase correlates with the northerly extension of the Chad Basin due to the formation of Mega Chad, which at its maximum, had a surface somewhat similar to the present-day Caspian Sea (Pachur 1999). To maintain a mixture of a northern and southern Sahel vegetation in the Kufra Basin and in the Tibesti foreland, some 150 to 250 mm of precipitation must be postulated, whereas in the even drier Western Desert, annual precipitation may not have exceeded 100-200 mm (e.g., Peters 1987; Kröpelin 1989; Neumann 1989; Van Neer & Uerpmann 1989). A northward shift of 600 to 800 km of the present-day biomes would be necessary to explain the species recognised in the

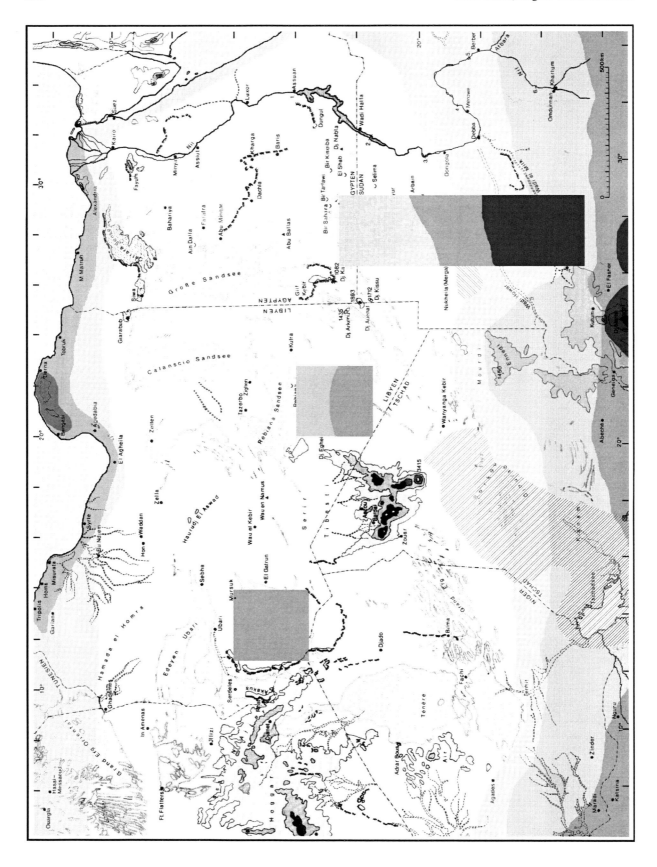

Fig. 9: Estimated position of the biomes during the Holocene climatic optimum on the basis of faunal remains from natural deposits and archaeological sites in the Eastern Sahara.

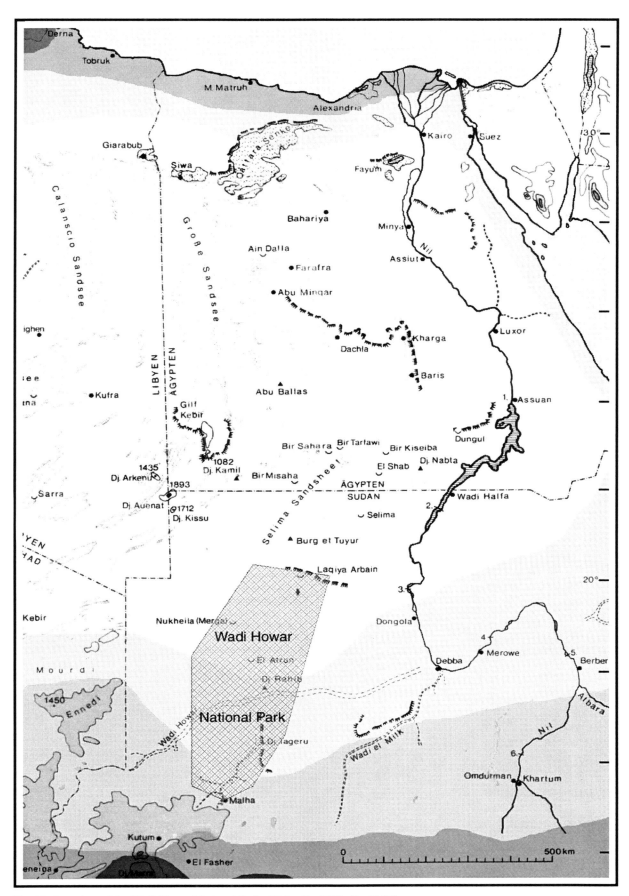

Fig. 10: Location of the Wadi Howar National Park. (Kröpelin 1999, fig. 11.24).

faunal samples from the last two regions (Kröpelin 1989, 1999; Neumann 1989).

To the south of the Western Desert, environmental conditions improved with latitude. The early to mid-Holocene herbivore community associated with the West Nubian Palaeolake and the Middle Wadi Howar, for example, implies an annual precipitation of some 350-500 mm, which would correspond to a northward shift of the present-day biomes of some 400-550 km. The Sudan Savanna herbivore community found in the southernmost part of the study area is indicative of some 500 to 600 mm of precipitation and would imply a shift of the actual biomes of some 300 to 400 km to the north. These estimates fit well with the results obtained in the frame of (palaeo)botanical research (Wickens 1982; Neumann 1989) and with the climatic reconstructions made using faunal remains from a series of early to mid-Holocene archaeological sites to the east of the Nile (Peters 1989).

Archaeozoological research and the future

As mentioned, long-term geo-scientific and archaeological research programmes were established in the 1980s, with the aim to better understand the climatic and cultural past of the Eastern Sahara. Within the frame of this project, cooperation between scientists from different countries as well as with government officials, in particular from the host countries (Sudan, Egypt), were established. As seen, one area of focus lies in western Nubia. In due course, scientists became aware of the immense natural (e.g., Pachur & Kröpelin 1987; Pachur et al. 1990; Kröpelin 1993, 1999; Pachur 1999) and cultural (e.g. Kuper 1981; Keding 1997; Hoelzmann et al. 2001) values of this hyperarid, (semi)desert region and much effort was put into convincing the Sudanese government of its uniqueness. This has resulted in their recent decision to designate the Wadi Howar area a National Park. With a surface of some 100,000 km^2, the Wadi Howar NP is one of the largest National Parks in Africa (fig. 10).

Future nature conservation measures will aid the protection of the scarce vegetation against destruction by domestic herbivores (sheep, goats) or by bush fires. Since precipitation seems to be on the increase in the area (S. Kröpelin, pers. comm. 2002), most likely as a result of the rise in global temperatures, plant cover and living conditions might even improve. If so, the not-too-distant future might witness the reintroduction of the herbivore species that once roamed West Sudan in large numbers. If the latter scenario becomes reality, archaeozoological research in the Wadi Howar region would not have contributed only to our understanding of its past, but can also be used to plan its future.

Summary / Zusammenfassung

During the last 12,000 years, the Eastern Sahara has passed from a hyperarid state to semiaridity and again to a hyperarid state. From aerial photographs, satellite images, and systematic field surveys within the frame of a long-term geo-scientific project, it could be shown that at the end of the Pleistocene, the endorheic state in large parts of the Eastern Sahara came to an end because of the formation of extensive palaeodrainage systems, that sometimes covered distances of more than 800 km. This was the case in western Nubia, where a temporary connection existed between the Central Sudanese Nile and Jebel Marra via the palaeodrainage system of Wadi Tageru and Wadi Howar (Pachur & Wünneman 1996). The existence of such extensive drainages systems and their associated limnetic accumulations indicate an increase in precipitation and a rise in water table.

Climatic change during Holocene times would also be evidenced if the rise in water table would be accompanied by changes in the Early and Middle Holocene biota of the Eastern Sahara. The faunal remains analysed in the course of this study have been collected during field work between 1980 and 2000 in fluviatile, limnetic, and semi-lacustrine sediments along two major transects. The first transect runs in a west-east direction along the Tropic of Capricorn, the second in a north-south direction along latitude 26-28°E. Identification of the bone remains was carried out mainly with the aid of the modern skeletal collection housed at the State Collection for Anthropology and Palaeoanatomy in Munich. Their interpretation focuses upon faunal spectrum and diversity, past landscapes, zoogeography, climate and vegetation in Early and Middle Holocene times. Where relevant, the results of the analyses of animal remains from well-dated archaeological contexts in the study area have been considered too.

Nearly 100 vertebrate taxa could be recognized in the samples (Table 1). Species spectrum and habitat requirements of the vertebrates identified clearly show that the Early and Middle Holocene witnessed much more favourable environmental conditions than today. From the faunal record it can moreover be deduced that species and habitat diversity increased in an east-west as well as in a north-south direction.

Based on palaeobotanical and geo-scientific research, it could be shown that in the eastern part of the Eastern Sahara, tropical monsoon rains may have reached 800 km further to the north during the Holocene climatic optimum (Kröpelin 1989, 1999; Neumann 1989), enabling the development of plant cover in the form of grass, thorn trees and shrub vegetation in the present-day hyperarid desert regions of southwest Libya, south Egypt and northwest Sudan. Moreover, vegetation cover would become progressively more dense southwards (Neumann 1989). These assumptions fit the faunal record (Peters 1987, Van Neer & Uerpmann 1989; Berke 2001; this study). As such, annual precipitation in the Western Desert during the climatic optimum may have been 200 mm at best, whereas West Nubia may even have received up to 500 mm p.a.. Based on geomorphological (Kröpelin 1993, 1999), (palaeo)botanical (e.g. Wickens 1982; Neumann 1989) and (archaeo)zoological (Peters 1987; Van Neer & Uerpmann 1989; Berke 2001; this study) research, a northward shift of the biomes parallel to their actual positions of 600 to 800 km during the early Holocene and of 400-550 km during the middle Holocene can be postulated for the North-South-Transect.

The analysis of the vertebrate remains from the Erg of Murzuq and the Tibesti foreland enable a reconstruction of the landscape in the western part of the Eastern Sahara. Many vertebrates recognized from bone remains are also depicted in contemporaneous rock art. During the Holocene optimum, animal species with comparably high ecological demands, in particular from sub-Saharan Africa, could have migrated along the south(west)-north(east) oriented water courses into the former desert to reach the Tropic of Capricorn. This would not have been possible without the northern extension of the Chad Basin (Mega Chad). Based on the biotope requirements of the species identified, it can be inferred that a southern Sahel vegetation (350-500 mm annual precipitation) or even a northern Sudan Savanna (500-600 mm annual precipitation) characterized the landscape at that time. Considering the present-day positions of these biomes, this would mean a northward shift of some 900 to 1200 km. Compared to the eastern part of the region, early to mid-Holocene ecogeographical conditions in the western part of the Eastern Sahara thus provided much more favourable conditions for the dispersion of Aethiopian mammals.

One of the most important results of the research cooperation between German and Sudanese scientist and officials is designation of the Wadi Howar region as a National Park by the Sudanese Government in 2001. If the increase in precipitation observed here during the past years (S. Kröpelin, pers. comm. 2002) continue, the not-too-distant future might witness the reintroduction of herbivore game species. If the latter scenario becomes reality, archaeozoological data from this region will be valuable in planning its future.

In den letzten 12,000 Jahren unterlag die östliche Sahara einem Klimawandel von einem hyperariden über einen semi-ariden bis wiederum zu einem hyperariden Stadium. Anhand von Luft- und Satellitenaufnahmen sowie durch systematische Felduntersuchungen im Rahmen des Berliner geowissenschaftlichen Sonderforschungsbereiches (SFB 69) konnte gezeigt werden, dass dieser endorheische Zustand in verschiedenen Teilgebieten dieser Region durch Drainagesysteme mit Lauflängen von über 800 km aufgehoben wurde. In Westnubien, beispielsweise, bildete das östlich vom Djebel Marra ausgehende Drainagesystem des Wadi Tageru ein Tributär des Wadi Howar und somit einen Anschluss an den Nil (Pachur & Wünneman 1996). Die Existenz solcher ausgedehnten Drainagesysteme und der in sie eingeschalteten, in das Frühholozän datierende limnische Akkumulationen sprechen für eine Anstieg des Niederschlags- und Grundwasserniveaus in dieser Zeit.

Der Nachweis einer Klimaänderung im Holozän würde ebenfalls erbracht sein, wenn sich die genannten Phänomene auch in den früh- bis mittelholozänen Taphozönosen der östlichen Sahara widerspiegeln. Die im Rahmen dieser Studie analysierten Wirbeltierreste wurden während Feldforschungen im Zeitraum von 1980 bis 2000 aus holozänzeitlichen fluviatilen, limnischen und semilakustrischen Sedimenten entlang zweier Transekte geborgen: Das erste Transekt verläuft von West nach Ost in Höhe des Wendekreises, das zweite von Nord nach Süd entlang 26-28°E. Die Bestimmung der Knochenfunde erfolgte im wesentlichen mit Hilfe der Rezentskelettsammlung der Staatssammlung für Anthropologie und Paläoanatomie, München. Ihre Auswertung betrifft das Tierartenspektrum bzw. die Artenvielfalt, die damalige Landschaft, und die Verbreitung der Tiere sowie Klima und Vegetation im Früh- und Mittelholozän. Zur Klärung dieser Fragestellungen fanden gegebenenfalls die für die einzelnen Regionen relevanten Forschungsergebnisse an Tierknochen aus datierten archäologischen Kontexten Berücksichtigung.

Bislang wurden knapp 100 Wirbeltiertaxa nachgewiesen (Table 1). Artenspektrum und Biotopansprüche der nachgewiesenen Tierarten belegen einwandfrei, dass im Früh- und Mittelholozän weitaus günstigeren Umweltbedingungen herrschten als heute. Parallel zu der in ost-westliche bzw. in nord-südliche Richtung verlaufende Zunahme der Artenvielfalt ist ersichtlich, dass auch die Ansprüche an das Biotop gehobener werden.

Paläobotanischen und geowissenschaftlichen Untersuchungen zufolge, sollen die tropischen Monsunregen während der holozänzeitlichen Feuchtphasen in dem östlichen Teil der Ostsahara bis zu 800 km weiter nach Norden vorgedrungen sein (Kröpelin 1989, 1999; Neumann 1989), was ausgereicht haben sollte, um die heute hyperariden Wüsten Südwestlibyens, Südägyptens und des Nordwestsudans mit einer Gras- bzw. Strauchvegetation und dornenbewehrten Bäumen zu bedecken. Richtung Süden wird außerdem die Vegetationsdecke zunehmend geschlossener (Neumann 1989). Diese Rekonstruktion findet sich aufgrund von Faunenanalysen bestätigt (Peters 1987, Van Neer & Uerpmann 1989; Berke 2001; dieser Beitrag). Entsprechend ist während des Klimaoptimums für die Westwüste mit einer jährlichen Niederschlagsmenge von 100 bis 200 mm, in Westnubien sogar mit bis zu 500 mm zu rechnen. Aufgrund geomorphologischer (Kröpelin 1993, 1999), (paläo)botanischer (z.B. Wickens 1982; Neumann 1989) und faunenhistorischer (Peters 1987; Van Neer & Uerpmann 1989; Berke 2001; dieser Beitrag) Untersuchungen kann entlang des Nord-Süd-Transektes eine breitenkreisparallele, nördliche Verschiebung der sich heute südlich der Sahara befindenden Biome von bis zu 600-800 km im Frühholozän und bis zu 400-600 km im Mittelholozän angenommen werden.

Die Analyse der Faunenreste aus dem Erg von Murzuq und dem Tibesti Vorland ermöglicht es, auch für dieses Gebiet konkrete Aussagen über Tierwelt und Landschaft in der westlichen Ostsahara zu treffen. Viele der anhand von Knochenfunden nachgewiesenen Tierarten lassen sich auch in der zeitgleichen Felskunst nachweisen. Längs der süd(west)-nord(ost) orientierten Gerinnebahnen konnten während der holozänen Feuchtphase ökologisch anspruchsvollere Tierarten, insbesondere aus Afrika südlich der Sahara, in die ehemalige Wüste bis hin zum nördlichen Wendekreis vordringen. Ein Zusammenhang mit der nördlichen Ausdehnung des Tchadsees (Mega-Tchad) im Holozän liegt indes auf der Hand. Aufgrund der Biotopansprüche der nachgewiesenen Arten dürfte zur Zeit des holozänen Klimaoptimums die Vegetation weitgehend einer südlichen Sahel (350-500 mm Niederschlag/Jahr) bzw. einer nördlichen Sudansavanne (500-600 mm Niederschlag/Jahr) entsprochen haben. In Anbetracht ihrer heutigen Positionen muss deshalb zum Zeitpunkt des frühholozänen Klimaoptimums von einer Verschiebung der Biome in Richtung Norden von 900 bis 1200 km ausgegangen werden. Somit wird ersichtlich, dass die früh- und mittelholozänen ökogeographischen Bedingungen im westlichen Teil der Ostsahara das Vordringen der Äthiopisfauna erheblich mehr begünstigten als dies im östlichen Teil der Region der Fall war.

Als eines der herausragendsten Ergebnisse der Forschungskooperation zwischen deutschen und sudanesischen Wissenschaftlern und Behörden ist der 2001 von Seiten der Sudanesischen Regierung proklamierten Wadi Howar Nationalpark zu erwähnen. Falls die seit einigen Jahren beobachtete, höhere Niederschlagsmenge in der hyperariden Wadi Howar Region (S. Kröpelin, pers. Mitt. 2002) weiterhin anhält, und es dadurch zu einer Regeneration der Vegetation kommen würde, könnte die Wiedereinführung von Wildtieren langfristig Realität werden. Trifft dies zu, werden die dort erhobenen, archäozoologischen Daten sogar für die Zukunft des Wadi Howar von Bedeutung sein.

Acknowledgements

Vertebrate remains from natural deposits and archaeological sites have been collected by numerous scientists, to whom we are indebted, in particular to H.-J. Pachur, S. Kröpelin, R. Kuper, B. Keding, and F. Jesse. W. Van Neer kindly provided unpublished data on fish remains from the B.O.S. missions. Funds provided by the Deutsche Forschungsgemeinschaft to carry out this analysis (Pe 424-4; SFB 389/TP A2) are gratefully acknowledged.

References

Berger A., 1978.
　Long-term variation of caloric insulation resulting from earth's orbital elements. *Quaternary Research* **9**: 136-167.

Berke H., 2001.
　Gunsträume und Grenzbereiche. Archäozoologische Beobachtungen in der Libyschen Wüste, Sudan und Ägypten. In: Gehlen B., Heinen M. & Tillmann A. (Hrsg.). *Zeit-Räume. Gedenkschrift für Wolfgang Taute. Archäologische Berichte 14*: 237-256. Bonn: Selbstverlag der Deutschen Gesellschaft für Ur- und Frühgeschichte e.V. / Dr. R. Habelt.

Boessneck J., Müller H.-H. & Teichert M., 1964.
　Osteologische Unterscheidungsmerkmale zwischen Schaf (*Ovis aries* Linné) und Ziege (*Capra hircus* Linné). *Kühn-Archiv* **78**: 1-129.

Branch B., 1988.
　Bill Branch's Field Guide to the Snakes and other Reptiles of Southern Africa. Cape Town: Struik Publishers.

Brown L.H., Urban E.K. & Newman K., 1982.
　The Birds of Africa. Vol. 1. London: Academic Press.

Corridi C., 1998.
　Some new archaeozoological data from the Tadrart Acacus, Libya (9th to 5th millennium B.P.). In: Gautier A. (ed). *Animals and people in the Holocene of North Africa. ArchaeoZoologia* 9: 41-47. Grenoble: La Pensée Sauvage

Driesch A. von den & Peters J., in press.
　Frühe Pferde- und Maultierskelette aus Auaris (Tell el-Dab'a, östliches Nildelta).

Fish G.R., 1956.
Some aspects of the respiration of six species of fish from Uganda. *Journal of Experimental Biology* **33**: 186-195.

Frobenius L., 1937.
Ekade Ektab, die Felsbilder Fezzans. Leipzig: Harrassowitz.

Gautier A., 1982.
Neolithic faunal remains in the Gilf Kebir and the Abu Hussein Dunefield, Western Desert, Egypt. In: El-Baz. F. & Maxwell T.A. (eds). *Desert landforms of southwest Egypt: a basis for comparison with Mars*: 335-339. Washington: NASA Scientific and Technical Information Branch.

Gautier A., Linseele V. & Van Neer W., 2002.
The fauna of the early Khartoum occupation on Jebel Umm Marrahi (Khartoum Province, Sudan). In: Jennerstraße 8 (Hrsg.). *Tides of the Desert. Contributions to the Archaeology and Environmental History of Africa in Honour of Rudolph Kuper. Africa Praehistorica* 14: 337-344. Köln: Heinrich-Barth-Institut.

Gautier A. & Van Neer W., 1982.
Prehistoric fauna from Ti-n-Torha (Tadrart Acacus, Libya). *Origini* **12**: 87-127.

Gautier A. & Van Neer W., 1989.
Animal Remains from the Late Paleolithic Sequence at Wadi Kubbaniya. In: Wendorf F., Schild R. (assembl.) & Close A.E. (ed). *The prehistory of Wadi Kubbaniya. Vol. 2. Stratigraphy, paleoeconomy, and environment*: 119-161. Dallas: Southern Methodist University Press.

Green J., 1977.
Haematology and habitats in catfish of the genus *Synodontis*. *Journal of Zoology* **182**: 39-50.

Happold D.C.D., 1987.
The Mammals of Nigeria. Oxford: Clarendon Press.

Hoelzmann P., Keding B., Berke H., Kröpelin S. & Kruse H.-J., 2001.
Environmental change and archaeology: lake evolution and human occupation in the Eastern Sahara during the Holocene. *Palaeogeography, Palaeoclimatology, Palaeoecology* **169**: 193-217.

Keding B., 1997.
Djabarona 84/13. *Untersuchungen zur Besiedlungsgeschichte des Wadi Howar anhand der Keramik des 3. und 2. Jahrtausends v.Chr. Africa Praehistorica* 9. Köln: Heinrich-Barth-Institut.

Kröpelin S., 1989.
Untersuchungen zum Sedimentationsmilieu von Playas im Gilf Kebir (Südwest-Ägypten). In: Kuper R. (Hrsg.). *Forschungen zur Umweltgeschichte der Ostsahara. Africa Praehistorica* 2: 183-305. Köln: Heinrich-Barth-Institut.

Kröpelin S., 1993.
Zur Rekonstruktion der spätquartären Umwelt am Unteren Wadi Howar (Südöstliche Sahara/NW Sudan). Berliner Geographische Abhandlungen **54**.

Kröpelin S., 1999.
Terrestrische Paläoklimatologie heute arider Gebiete : Resultate aus dem Unteren Wadi Howar (Südöstliche Sahara/Nordwest-Sudan). In: Klitzsch E. & Thorweihe U. (Hrsg.). *Nordost-Afrika: Strukturen und Ressourcen. Ergebnisse aus dem Sonderforschungsbereich "Geowissenschaftliche Probleme in ariden und semiariden Gebieten"*: 446-506. Weinheim: Wiley-VCH.

Kröpelin S. & Petit-Maire N., 2000.
Paleomonsoon variations and environmental change during the late Quaternary. *Global Planetary Change* **26**: vii-viii.

Kuper R., 1981.
Untersuchungen zur Besiedlungsgeschichte der östlichen Sahara. Vorbericht über die Expedition 1980. *Beiträge zur Allgemeinen und Vergleichenden Archäologie* **3**: 215-275.

Kutzbach J.E. & Otto-Bliesner B.L., 1982.
The sensitivity of the African-Asian monsoonal climate to orbital parameter changes for 9000 years BP in a low resolution General Circulation Model. *Journal of Atmosphere Science* **39**: 1177-1188.

Lévêque C., 1990.
Relict tropical fish fauna in Central Sahara. *Ichthyological Exploration of Freshwaters* **1**: 39-48.

Lévêque C., 1997.
Biodiversity dynamics and conservation. The freshwater fish of tropical Africa. Cambridge: Cambridge Univ. Press.

Le Quellec J.-L., 1998.
Art rupestre et préhistoire du Sahara. Le Messak libyen. Paris : Payot.

Lutz R. & Lutz G., 1995.
Das Geheimnis der Wüste. Die Felskunst des Messak Sattafet und Messak Mellet – Libyen. Innsbruck: Golf Verlag.

Mori F., 1965.
Tadrart Acacus, Arte rupestre e culture del Sahara preistorico. Turin : Einaudi.

Muzzolini A., 1995.
Les images rupestres du Sahara. Toulouse, printed by the author.

Neumann K., 1989.
Vegetationsgeschichte der Ostsahara im Holozän. Holzkohlen aus prähistorischen Fundstellen. In: Kuper R. (Hrsg.). *Forschungen zur Umweltgeschichte der Ostsahara. Africa Praehistorica* 2: 13-181. Köln: Heinrich-Barth-Institut.

Pachur H.-J., 1999.
Paläo-Environment und Drainagesysteme der Ostsahara im Spätpleistozän und Holozän. In: Klitzsch E. & Thorweihe U. (Hrsg.). *Nordost-Afrika: Strukturen und Ressourcen. Ergebnisse aus dem Sonderforschungsbereich "Geowissenschaftliche Probleme in ariden und semiariden Gebieten"*: 366-445. Weinheim, New York, Chichester, Brisbane, Singapore, Toronto: Wiley-VCH.

Pachur H.-J. & Kröpelin S., 1987.
Wadi Howar: paleoclimatic evidence from an extinct river system in the south-eastern Sahara. *Science* **237**: 298-300.

Pachur H.-J., Kröpelin S., Hoelzmann P., Goschin M. & Altmann N., 1990.
Late Quaternary fluvio-lacustrine environments of Western Nubia. *Berliner Geowissenschaftliche Abhandlungen (A)* **120.1**: 203-260.

Pachur H.-J. & Peters J., 2001.
The position of Murzuq Sand Sea in the palaeodrainage system of the Eastern Sahara. *Palaeoecology of Africa and the Surrounding Islands* **27**: 259-290.

Pachur H.-J. & Wünnemann B., 1996.
Reconstruction of the Palaeoclimate along 30°E in the Eastern Sahara during the Pleistocene/Holocene transition. *Palaeoecology of Africa and the surrounding Islands* **24**: 1-32.

Peters J., 1987.
The Faunal Remains collected by the Bagnold-Mond Expedition in the Gilf Kebir and Jebel Uweinat in 1938. *Archéologie du Nil Moyen* **2**: 251-264.

Peters J., 1988.
Osteomorphological features of the appendicular skeleton of African buffalo, *Syncerus caffer* (Sparrman, 1779) and of domestic cattle, *Bos primigenius* f. taurus Bojanus, 1827. *Zeitschrift für Säugetierkunde* **53**: 108-123.

Peters J., 1989.
Faunal Remains and Environmental Change in Central and Eastern Sudan from Terminal Pleistocene to Middle Holocene Times. *Academiae Analecta, Mededelingen van de Koninklijke Academie voor Wetenschappen, Letteren en Schone Kunsten van België, Klasse der Wetenschappen* **51**(4): 121-148.

Peters J., 1995.
Mesolithic subsistence between the 5th and the 6th Nile Cataract. The archaeofaunas from Abu Darbein, El Damer and Aneibis (Sudan). In: Haaland R. & Magid A.A. (eds). *Aqualithic sites along the rivers Nile and Atbara, Sudan*: 178-244. Bergen: Alma Mater.

Peters J. & Driesch A. von den, 1997.
The two-humped camel (*Camelus bactrianus*): new light on its distribution, management and medical treatment in the past. *Journal of Zoology London* **242**: 651-679.

Peters J., Pöllath N. & Driesch A. von den, 2002.
Ichthyological Diversity in the Holocene Palaeodrainage Systems of Western Nubia. In: Jennerstraße 8 (Hrsg.). *Tides of the Desert. Contributions to the Archaeology and Environmental History of Africa in Honour of Rudolph Kuper. Africa Praehistorica 14:* 325-335. Köln: Heinrich-Barth-Institut.

Peters J., Van Neer W. & Plug I., 1997.
Comparative postcranial osteology of Hartebeest (*Alcelaphus buselaphus*), Scimitar oryx (*Oryx dammah*) and Addax (*Addax nasomaculatus*), with notes on the osteometry of Gemsbok (*Oryx gazella*) and Arabian oryx (*Oryx leucoryx*). *Annales du Musée royal de l'Afrique Centrale, Sciences Zoologiques* **280**.

Van Neer W., 1989a.
Holocene fish remains from the Sahara. *Sahara* **2**: 61-68.

Van Neer W., 1989b.
Recent and fossil fish from the Sahara and their palaeohydrological meaning. *Palaeoecology of Africa and the Surrounding Islands* **20**: 1-18.

Van Neer W. & Uerpmann H.-P., 1989.
Palaeoecological significance of the Holocene faunal remains of the B.O.S. missions. In: Kuper R. (Hrsg.). *Forschungen zur Umweltgeschichte der Ostsahara. Africa Praehistorica* **2**: 309-341. Köln: Heinrich-Barth-Institut.

Wickens G.E., 1982.
Palaeobotanical Speculations and Quaternary Environments in the Sudan. In: Williamson M.A.J. & Adamson D.A. (eds). *A Land between Two Niles:* 23-50. Rotterdam: Balkema.

Documenta Archaeobiologiae: Instructions for authors

1. ***Documenta Archaeobiologiae*** aims to publish original papers, focus articles, and short notes in German and English covering the interaction between the natural sciences and archaeology, with particular emphasis upon research carried out with bio-archaeological collections.

2. ***Submission of manuscripts.*** All material for publication should be sent to The Editors, *Documenta Archaeobiologiae,* Staatssammlung für Anthropologie und Paläoanatomie, Karolinenplatz 2a, D-80333 München. The submission of a manuscript will be taken to imply that the material is original and that no similar paper is being, or will be, submitted elsewhere. When accepted, the copyright of a paper becomes the property of the Staatssammlung whose permission must be obtained to reproduce material therefrom. A Copyright Transfer Agreement will be sent to the author(s) when the manuscript is accepted for publication. Serialized studies can be submitted but should be discussed with the editors before submission.

3. ***Presentation.***

   (i) Three copies of the typescript must be provided, double-spaced throughout on one side of paper, with a wide margin all round. Some important features for preparing electronic documents for disk submittal are given below. All papers must be serially numbered and securely fasted together.

   The typescript should follow the conventional form (Introduction – Material and Methods – Results – Discussion – Conclusions – Summary – Acknowledgements – Bibliography) but must include:
   (a) Title page giving a concise specific title with the name(s) of the author(s) and the institution(s) where the work was carried out. A short title for page headings must be provided.
   (b) Abstract and keywords. A short abstract should be submitted with the paper, together with 4-7 keywords suitable for indexing. The abstract should not contain more than 300 words and should be intelligible without reference to the main text. Abstract and keywords will be published in English and German.
   (c) Extensive summary (up to 5 pages for large contributions) is obligatory and should give a succinct account of the subject with an in-depth presentation of the results and the conclusions. This summary will be published in English and German to ensure maximum distribution within the scientific community.

   (ii) *Conventions.* The metric system must be used and SI units where appropriate. In the use of abbreviations and symbols British Standards 1991 Part 1: 1954 should be followed. Whole numbers one to nine should be spelled out (except in the Methods section) and numbers 10 onwards given in figures but this is not essential if many numbers appear together. If specific collections and/or comparative specimens are used, the institution in which the material is housed must be given, together with details of the registration assigned to it.

   (iii) *Tables and captions for illustrations* should be typed separately at the end of the manuscript and ***their required positions indicated in the margins of the text.***

4. ***Illustrations.*** All illustrations will be reduced to a maximum size of A4. Outsize artwork (i.e. needing more than 50% reduction) must not be submitted. A metric scale should be included on each illustration. The name(s) of the author(s) and the number of the figure must be marked on the back of all illustrations and the orientation given if necessary. Captions should be listed on a separate sheet.

   (i) *Line drawings.* These must be of a high enough standard for direct reproduction. They should be prepared in black (Indian) ink on white card or tracing paper, preferably A4 size. Any necessary lettering may be inserted in pencil on the original, or written on an accompanying photocopy.

   (ii) *Photographs.* High quality glossy prints of maximum contrast must be submitted, preferably of the final size required. Any labels and scale-lines should be accurately indicated on a duplicate set of photographs.

   Illustrations, whether photographs or line drawings, should be numbered in one series in their order of mention. They are referred to as Fig. 1, Fig. 2, etc., and any sub-sections as (a), (b), etc. Only high-quality photocopies should be submitted for reviewing purposes. **Original illustrations will be required for printing but should only be sent if the paper has been accepted.** In the case of acceptance, authors must also provide usable electronic files for all figures (see below).

5. **Tables.** These must be presented to fit the page size without undue reduction. Oversize tables will not be accepted. For specifications concerning electronic files see below.

6. **References** in the text should be by name(s) and date(s), thus: Brown 1987, 1990; Brown & Reed 1987; Brown et al. 1987a, b. In the bibliography references should be arranged first alphabetically under author(s) name(s) and then in chronological order if several papers by the same author(s) are cited. The full title of the paper must be given together with the first and last pages. **Journal titles should be given in full.** Book titles should be followed by the place of publication and the publisher. If different from the author cited, the name of the editor(s) of the book should be given. The reference list should be checked against the text to ensure (a) that the spelling of the authors' names and the dates given are consistent and (b) that all authors quoted in the text (in date order if more than one) are given in the reference list and vice versa.

Examples

Books:
Zeuner F., 1963.
A History of Domesticated Animals. London: Hutchinson.

Contributions to books:
Wickens G., 1982.
Palaeobotanical Speculations and Quaternary Environments in the Sudan. In: Williams M.A.J. & Adamson D.A. (eds). *A land between Two Niles: Quaternary Geology and Biology of the Central Sudan*: 23-51. Rotterdam: Balkema.

Journals:
Van Neer W., 1989.
Recent and fossil fish from the Sahara and their palaeohydrological meaning. *Palaeoecology of Africa* **20**: 1-18.

Notes. Footnotes and/or endnotes should be avoided, but where necessary should be numbered in the order in which they appear in the text and listed at the end of the typescript (endnotes).

Proofs. Since typescripts are submitted as electronic files, proofs will only be sent if considered necessary by the editors.

Offprints. 15 offprints will be supplied to the author(s).

Guide for authors preparing electronic documents for disk submittal

DO

- Enter text in the style and order of the Journal.
- Use MS Word for Windows 2000 for PC.
- Use Times New Roman for the text font (11 pt) and the symbols.
- Indicate on the hard copy by hand any special characters or accents that still need to be incorporated in the text.
- Use the word processing formatting features to indicate **Bold**, *Italic*, Maths, ^{Superscript} and $_{Subscript}$ characters.
- Type headings in the style of the Journal.
- Type references in the correct order and style of the Journal.
- Use the TAB key **once** for paragraph indents.
- Insert figure captions and tables at the end of the file.
- Save any tables, diagrams, figures, graphs or illustrations generated electronically as separate files and not embedded into the text file.
- Use MS Word for Windows or MS Excel to generate tables.
- Submit Graphs as MS Excel files.
- Submit all other illustrations as electronic files (300x300 dpi for black-and-white photographs, 800x800 dpi for graphics) in .tif format.
- Check the final copy of your paper carefully. In the case of a mismatch between disk and hard copy, the hard copy will be taken as the definitive version.

DO NOT

- Enter carriage returns to obtain spacing between lines, paragraphs, references etc. The space required is generated automatically by the typesetters.
- Use double spaces after each sentence within a paragraph.
- Use the automatic page numbering, running titles and footnote features of your word processing programme. Only the hard copies should be numbered by hand at the bottom of the page.

Disk submittal

The disk should not be sent until the paper has been accepted and should contain the final revised version of the manuscript. After the manuscript has gone through the review and editing stages, copy final version onto a clean DOS-formatted disk. To avoid confusion do not copy any irrelevant and/or back-up files onto the disk. Apple Mac users should ensure the disk wastebasket is empty before submitting the disk. Use the first-named author's name for the disk label and file name.